Disaster Risk Reduction in Mexico

Diana Sánchez-Partida ·
José-Luis Martínez-Flores ·
Santiago-Omar Caballero-Morales ·
Patricia Cano-Olivos

Disaster Risk Reduction in Mexico

Methodologies, Case Studies, and Prospective Views

Springer

Diana Sánchez-Partida
Universidad Popular Autónoma del Estado de Puebla (UPAEP)
Puebla, Mexico

José-Luis Martínez-Flores
Universidad Popular Autónoma del Estado de Puebla (UPAEP)
Puebla, Mexico

Santiago-Omar Caballero-Morales
Universidad Popular Autónoma del Estado de Puebla (UPAEP)
Puebla, Mexico

Patricia Cano-Olivos
Universidad Popular Autónoma del Estado de Puebla (UPAEP)
Puebla, Mexico

ISBN 978-3-030-67294-2 ISBN 978-3-030-67295-9 (eBook)
https://doi.org/10.1007/978-3-030-67295-9

This Springer imprint is published by the registered company Springer Nature Switzerland AG
The registered company address is: Gewerbestrasse 11, 6330 Cham, Switzerland

Synopsis

This book recognizes and describes the circumstances and challenges of Mexico in case of natural disaster offering empirical risk-reduction methods in critical cases in Mexico.

Mexico for its geological and geographical situation is tackled annually by different natural hazards, including severe rains, floods, volcanic eruptions, earthquakes, among others. These events are increasing over time in both frequency and impact (mainly socially and economically).

The proposals considered here include real and detailed analysis, a set of models, frameworks, strategies, and findings in the three stages of the disaster (before–during–after). During the stage "pre-disaster," this book proposes recommendations for specific communities to be resilient by using risk-reduction methods. For example, this book suggests how to find a convenient and safe location of the Regional Humanitarian Response Depot (RHRD). This position will be critical in order to keep safe and scatter all necessary goods among the victims after an emergency. This book proposes other recommendations such as location and creation of an export logistics cluster, analyses of inventory levels on humanitarian relief for vulnerable municipalities, forecasting medicines, or planning of evacuation routes, and so forth.

In the "disaster" stage, the documentation of what happened in the case of September 19 (19S) earthquake in Mexico is presented for knowing the help process from each one of the participants or actors that intervened in this stage. In this research, it was detected that the incorporation of information and communications technologies (ICTs) was a crucial factor for the flow of information that helps the volunteers in the decision making of humanitarian logistics. Also, in this stage, a framework of key performance indicators (KPIs) that can measure this process is proposed because it opens up possibilities to provide the necessary information for improvement. With both proposals were detected challenges and opportunities for better planning to support the victims of this disaster in the best way.

Finally, the "post-disaster' stage includes a case study that describes how the intervention of multidisciplinary volunteers led by a non-governmental organization implemented several recovery and reconstruction strategies, obtaining the solution by diverse scientific methods with the objective of recovering the livelihoods of some particular communities of Mexico affected by the September 7 earthquake. Also, in this stage is proposed a theoretical mathematical model that would help allocation procedure of resources for restoring the affected community that takes into account the preferences of it.

Additionally, ways to create resilience in the main economic sectors of the country such as agriculture and industry are proposed. These proposals offer an evaluation framework, which will allow detecting disruption activities and their negative impacts. Then, the results will guide to elaborate a continuity business plan letting the farmers and companies be more resilient. Moreover, for last, the new trends for Mexico as a result of the climate change are described and made suggestions for mitigating the possible disasters.

We believe that this book can give support in the decision making of the society, national government, non-governmental organization, enterprises, and so, and protect the human life, the communities, and the livelihoods. Although the methodology of some of these proposals has been made at a regional level, they are highly replicable to other highly populated societies with similar socioeconomic structures.

Besides, this book can be the basis for generating more innovative proposals by researchers, graduate students, academics, professionals, and practitioner, since the creation of this antecedent is vital to obtain a better planning and a better collaboration between all the actors of the humanitarian chain, and in this way, minimize the risks and maximize the resilience of Mexico.

Contents

Part II During the Disaster

Part III After the Disaster

Part IV Ways to Create Resilience in the Economic Activities

About the Authors

Diana Sánchez-Partida is Professor–Researcher and Academic Director of the Postgraduate in Logistics and Supply Chain Management at Universidad Popular Autonoma del Estado de Puebla (UPAEP) in Mexico. She is the leader of the Humanitarian Logistics Group in the same institution. She received a Ph.D. in Logistics and Supply Chain Management. She has been granted a doctorate and post-doctorate scholarship by CONACyT. Since 2018, she has been a member of the National Council of Researchers(SNI level 1) in Mexico. Her research areas of interest are Disaster Risk Reduction, Humanitarian Logistics, Resilience in Economic Activities and Logistics Operations, among others.

José-Luis Martínez-Flores is a researcher of the Postgraduate in Logistics and Supply Chain Management at Universidad Popular Autonoma del Estado de Puebla (UPAEP) in Mexico. He received a Ph.D. in Mathematics. His objectives in research are the optimization and implementation of models for problems in the field of logistics through the use of information technologies. He is a member of the National Council of Researchers (SNI level 1) in Mexico.

Santiago-Omar Caballero-Morales is Professor-Researcher in the Department of Logistics and Supply Chain Management at Universidad Popular Autonoma del Estado de Puebla (UPAEP) in Mexico. In 2009, he received a Ph.D. in Computer Science from the University of East Anglia in the UK. Since 2011, he has been a member of the National Council of Researchers (SNI level 1) in Mexico. His research interests are quality control, operations research, combinatorial optimization, pattern recognition, analysis and simulation of manufacturing processes, and human–robot interaction.

Patricia Cano-Olivos is Professor–Researcher in the Department of Logistics and Supply Chain Management at Universidad Popular Autonoma del Estado de Puebla (UPAEP) in Mexico. She received a Ph.D. in Logistics and Supply Chain

Management. Her research interests are risks in the supply chain and logistics management models. She is a member of the National Council of Researchers (SNI level 1) in Mexico.

Part I
Before the Disaster

Chapter 1
The Most Frequent Natural Disasters and Their Tendency in Mexico from a Perspective Based on Humanitarian Logistics

Hugo Romero-Montoya, Diana Sánchez-Partida, José-Luis Martínez-Flores, and Patricia Cano-Olivos

Abstract This paper presents a detailed analysis of humanitarian logistics and its components; the specification is to be able to identify and can be characterized once the disasters caused by natural phenomena are interconnected. From a qualitative methodology with an exploratory and non-experimental descriptive approach, analyze in detail three information bases referring to natural phenomena data and their implications. The first of these was the Center of research in epidemiology and disasters (EM-DAT), considered as an international database, second as a national level, the National Center for Disaster Prevention (CENAPRED), and the third also considered a national level the drought monitor in Mexico (CONAGUA). Besides, several articles related to the subject matter were reviewed. Among the identified components are the following; the humanitarian logistics cycle according to the reviewed literature, the phases of disaster management in Mexico, the areas of logistics decision, the types of natural phenomena, and future trends of natural phenomena that could affect Mexico in a period very close. Finally, it carries out a statistic of the behavior of the components analyzed, as well as a series of conclusions in this regard. This analysis will help to be able to characterize the components involved in disaster risk management as well as the response of humanitarian logistics to different conditions and unforeseen situations of risk of which is the next type of disaster with the most significant possibility of occurrence and repercussions in Mexico.

H. Romero-Montoya (✉)
Department of Strategic Planning and Technology Management, Universidad Popular Autónoma del Estado de Puebla, (UPAEP University), 17 Sur 901, Barrio de Santiago, CP 72410 Puebla, Puebla, Mexico
e-mail: hugo.romero@upaep.edu.mx

D. Sánchez-Partida · J.-L. Martínez-Flores · P. Cano-Olivos
Department of Logistics and Supply Chain Management, Universidad Popular Autónoma del Estado de Puebla (UPAEP University), 17 Sur 901, Barrio de Santiago, CP 72410 Puebla, Puebla, Mexico

D. Sánchez-Partida et al. (eds.), *Disaster Risk Reduction in Mexico*,
https://doi.org/10.1007/978-3-030-67295-9_1

Keywords Humanitarian logistics · Trends in natural phenomena · Mexico · Humanitarian aid

1.1 Introduction

Currently, natural disasters are part of the global problems, and these cause suffering and havoc among those who come into direct contact with them (Zary et al., 2014). So it is necessary to be clear that the natural events they cause. These disasters differ from each other, acquiring relevance for specific groups of humans that are affected (Capacci & Mangano, 2015). Besides, that disasters of any type are generally events of great magnitude and that have a tremendous negative impact on society and the environment (Viera et al., 2012), thus impeding the development of countries, mainly from the poorest (CENAPRED, 2019), hence the importance of its analysis and monitoring.

Humanitarian logistics is a discipline that helps to mitigate the ravages of natural disasters. It refers to the process of planning, implementation, and control of the efficient and profitable flow and storage of goods and materials, including information related from the point of consumption with in order to meet the requirements of the final beneficiary (Ye & Liu, 2013). Although for others like (Costa et al., 2012) humanitarian logistics is also what happens:

> Immediately after the occurrence of disasters, humanitarian operations begin with the intention of providing rapid assistance to victims in different ways, such as rescuing the wounded and/or stranded, collecting and undoing corpses, resource allocation, provision of assistance food, shelter, medical care, and restore access to remote locations.

Recently, research on humanitarian logistics is becoming a critical factor in devising new ways of managing aid operations among multiple stakeholders (Ransikarbum & Mason, 2016). Its importance lies in the ability to mitigate the direct and indirect effects caused by the appearance of catastrophes within an unpredictable context.

At the same time, the factors that are directly related to natural disasters such as the growth of the world population, the increase in the number and size of cities, the use of land, the stress of ecosystems, the global trend in urban planning and the gradual deterioration of the environment. They are currently considered as elements that make the world population more susceptible, exposing them to suffer the action of disasters in greater magnitude and aggressiveness (Huang et al., 2015; Zary et al., 2014) facilitating risk situations and the effects that cause the suffering of the population.

On the other hand, all this suggests, even more, the consequences of anthropic action as a determining factor for many catastrophic phenomena, especially those related to climate (Capacci & Mangano, 2015).

It should be mentioned that since the 1950s, the number of people affected due to their effect has increased in proportion to around 235 million people per year on average since the 1990s (Boonmee et al., 2017). In parallel, the total insured and uninsured losses caused by natural disasters multiplied by seven between 1980 and 2015 (Jais et al., 2017).

On the other hand, those disasters that are related to the climate have also increased in number and magnitude, reversing development achievements. That is why for the last decade, the UN reported that more than 700,000 women, men, and children lost their lives, more than 1.4 million were injured, and around 23 million were left homeless as a result of disasters (Unidas, 2017) of this nature.

It must also be said that according to the data presented by the research center on disaster epidemiology (EM-DAT) of the Catholic University of Louvain, the trend that most impacts worldwide is led by hydrometeorological issues and their derivations (floods, landslides, and storms). Figure 1.1 specifies the composition in the distribution of the distress that occurred in the last 19 years.

Also, if the typology of natural phenomena is considered, the most significant impacts per continent are distributed according to their nature. For example, concerning hurricanes and tropical cyclones, America is the most affected; in Asia due to earthquakes and tsunamis; in Europe due to floods or other atmospheric phenomena, while in African countries, mainly in sub-Saharan Africa, deaths are due to droughts (Capacci & Mangano, 2015). It should be mentioned that in addition to these places, some specialists identify the Middle East as an area that could be highly vulnerable to natural hazards (especially drought) (Majewski & Heigh, 2010).

Meanwhile, at a national level, the tendency of natural disasters is very similar to that registered at a global level, the frequency of disasters in Mexico has increased in recent decades (mainly those of hydrometeorological origin), as well as their economic cost to from the eighties (Abeldaño Zúñiga & González Villoria, 2018).

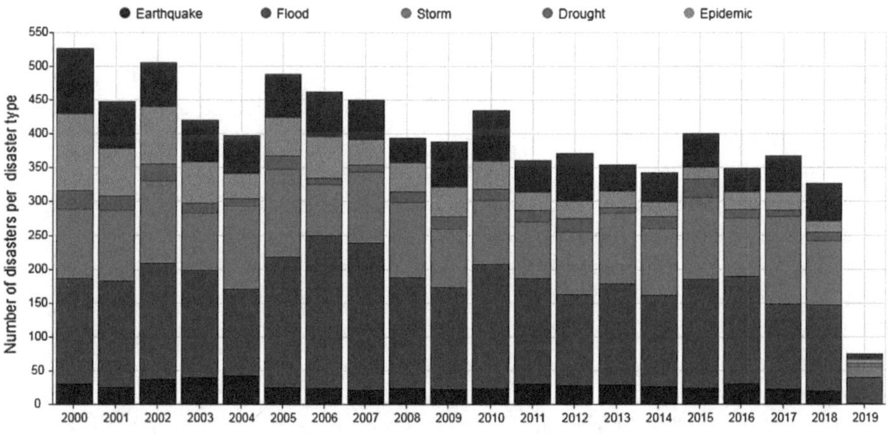

Fig. 1.1 Distribution of the types of natural disasters that have occurred in the world in the last 19 years extracted from the database of emergency and EM-DAT events Catholic University of Louvain (UCL) www.Emdat.be (Brussels Belgic)

All this is because Mexico is a country located in a place where it is frequently impacted by the formation of tropical cyclones in the Pacific coast and the Gulf of Mexico. Also, it is impacted by other types of disasters that are associated with instability of the soils (Abeldaño Zúñiga & González Villoria, 2018).

According to the National Institute of Statistics and Geography (INEGI), Mexico tends to be exposed to an intense rainy season, and about 90% of disasters are of the hydrometeorological type. These affect the southeast region of the country, so 75% National funds for disasters have been spent to alleviate this region (Cruz Benítez et al., 2015), all this shows that in Mexico catastrophic phenomena occur frequently and with results that They directly affect vulnerable populations (Abeldaño Zúñiga & González Villoria, 2018).

Although the evidence that exists about natural disasters in Mexico is closely related to hydrometeorological phenomena such as floods, landslides, heavy rains, and hurricanes, some claim the meteorological drought. That is, the climatic condition in which the accumulated rains in a season or years they are significantly below average, it constitutes one of the most significant climatic hazards for Mexico (Magaña-Rueda et al., 2018) due to its social, economic and food implications that it represents.

It should also be said that since the 1970s–1980s, desertification has been a topic of extensive media coverage and considered one of the most severe problems affecting the dry, semi-arid, and dry sub-humid regions of the planet due to its environmental and environmental implications. Socioeconomic (Diéguez et al., 2014).

At present, drought is one of the problems that is causing the most damage in Mexican society, mainly in the north of the country, where historically, this hydrometeorological phenomenon occurs (Esparza, 2014). While it is true that drought is not a spectacular phenomenon like other natural events (a hurricane or a tornado, for instance), its long-term impacts are often more substantial and more devastating (Ortega-Gaucin, 2013).

The effects of the drought, in economic and social terms, are a function of the impact on the various economic and productive sectors, associated with the supply and demand of goods and services they offer, and the imbalance due to natural and low conditions—humidity, which occurs when demand exceeds supply (Ortega-Gaucin, 2013).

In particular, Mexico is especially vulnerable to natural disasters due to the existence of areas that will be impacted by droughts (Northwest); by floods (Southeast) due to extreme weather events on both coastlines. Whose intensity could increase depending on the economic and social structure, especially the states that are most affected (Ochoa Lupián & Ayvar Campos, 2015) or also because of climatic variability that consists of those changes on the climate that depend on atmospheric conditions extremes that far exceed the standard averages (Serrano Vincenti et al., 2017). If climate change is creating different problems that go beyond the immediacy of a natural disaster, for example, the most significant impact it is producing is and will be on the population. One of its consequences will be involuntary and compulsive mobility, which, in turn, influences and will influence territorial planning (Rua, 2013).

Mexico is a developing country where irrigation and seasonal agriculture is an essential source of employment and economic income. The problem of drought, according to Ortega-Gaucin (2013), must also be considered as a significant disaster and, therefore, evaluated from the perspective of humanitarian logistics, since this seems to be a problem that has not been analyzed with sufficient depth and importance.

It is vital to consider that humanitarian logistics operations should be planned accordingly to the nature of natural disasters to obtain the maximum response and a minimum loss of resources (Safeer et al., 2014). They were carefully distinguishing each one of the factors that characterize them and, thus, be able to establish concrete and practical conditions before their appearance.

This research aims to define and identify: What are the natural disasters with the most significant impact at the national level? What factors are involved in this type of natural disaster? What is the trend regarding the type of natural disasters that will impact Mexico in the coming years? Moreover, what types of impact do disasters have in the country?

The answers to these questions will help to be able to characterize the main factors involved in natural disasters in Mexico, which would help strengthen risk management not only based on known natural phenomena but also concerning those that very likely occur in the future.

1.2 Literature Review

1.2.1 Natural Disasters

The need to explain the origin of disasters as well as the effects of climate on society and ecosystems have often led to the naturalist paradigm suggesting that disasters are caused by "atypical" weather and climate conditions, without repairing in the high vulnerability of society and its economy to variations in the climate (Rueda & Neri, 2012).

The World Health Organization (WHO) defines a "disaster" as any event that causes damage, destruction, ecological disturbance, loss of human lives, human suffering, deterioration of health, and health services on a sufficient scale to ensure a response extraordinary from outside the affected community or area. For example, earthquakes, hurricanes, tornadoes, volcanic eruptions, fires, floods, blizzards, drought, terrorism, chemical spills, nuclear accidents are included among the causes of disasters, and all have devastating effects in terms of personal injuries and property damage (Safeer et al., 2014) which in turn have different characteristics in the management of risks and problem solving during the performance of humanitarian aid.

On the other hand, (Scarpin & De Oliveira Silva, 2014) argue that disasters can be classified into four types:

1. Sudden natural appearance: such as earthquakes, tornadoes, hurricanes;
2. Naturally slow onset, such as hunger, drought, and poverty;
3. The result of a sudden start caused by men, such as the terrorist attack, the coup d'etat and the escape of chemical substances;
4. The slow result of human origin, such as political crises and the refugee crisis.

The perception of gravity is often associated with the rapidity of events that, as in the case of earthquakes, landslides or hurricanes, concentrate their effects in a limited period. And they are characterized by manifesting themselves unexpectedly and with consequences that persist for a long time. The phenomena due to in large measure, to human activities, such as air pollution, desertification processes, among others, do not have tangible territorial, economic, or social in the short-term and are perceived as less dangerous (Capacci & Mangano, 2015), but this does not mean that they are not harmful.

The greenhouse effect and climate change have been present for a few years at the center of scientific discussions and are phenomena that develop very slowly. For what is in the background, concerning those who, as we have seen, have tangible economic and social consequences immediately (Capacci & Mangano, 2015), but in recent times its effects are becoming devastating problems for society, which makes them red foci for disaster analysis.

1.2.2 Humanitarian Logistics and Natural Disasters

Large-scale disasters can lead to a massive loss of lives and the means for it, as already mentioned above. In such disasters, humans perish due to multiple causes such as drowning, fire, and lack of food and medicine (Chakravarty, 2014). Most of these occur unexpectedly concerning time, place, and intensity (Safeer et al., 2014).

Due to this enormous escalation of disasters, more attention has been paid to the need for management (Bozorgi-Amiri & Khorsi, 2016). That is capable of reducing complications due to the increase in the number of scenarios and the complexity of the development of activities and humanitarian logistics processes.

That is why, in order to organize humanitarian logistics effectively, it has become a significant economic challenge (Chandes & Paché, 2009), which to date involves the search for new forms of action and not only from Conventional means, it is a new situation. For a long time, the priority concern was to collect donations to have sufficient financial resources and thus face the effects of a natural earthquake or famine catastrophe or a human-made catastrophe, war or coup d'etat (Chandes & Paché, 2009). According to Beamon and Kotleba (2006) the picture is currently different, and the logistical needs of humanitarian organizations are often exceeding the capabilities of current emergency.

Due to the nature of disaster-related activities, humanitarian logistics is carried out in a complex environment, characterized by the pressure of time and filled with various sources of uncertainty (Gómez et al., 2017). It is not possible to predict natural disasters, but measures can be taken to deal with such complex crises and reduce the impact of natural disasters on people and society (Chiappetta Jabbour et al., 2017).

1.2.3 The Cycle of Humanitarian Logistics and the Phases for the Disaster in Mexico

Today the world faces increasingly complex and interrelated challenges, which require sustained responses (United, 2017), the management of humanitarian logistics implies a systematic approach to deal with natural disasters and human-made disasters (Ahmadi et al., 2015). The disaster management cycle has been used to define response and recovery logistics (David Swanson & Smith, 2013). The life cycle of humanitarian operations is linked to the timing of such events. There are pre-disaster decisions/operations made before the occurrence of the disaster and post-disaster decisions/operations made after the onset of the disaster (Moreno et al., 2017; Boonmee et al., 2017).

The disaster management cycle consists of four main phases: mitigation, preparedness, response, and recovery. The mitigation phase involves long-term efforts to prevent the occurrence of disasters or reduce their effects (Ahmadi et al., 2015) while the focus of preparedness is the response before a disaster occurs (Ransikarbum & Mason, 2016). These operations carried out before a disaster occurs called pre-disaster operations and play an instrumental role in strategic planning (location of facilities and pre-stock positioning) or disaster mitigation (evacuation) (Caunhye et al., 2012).

In the response phase, immediate relief operations are carried out after a disaster. These include the location of alternative care facilities, distribution of relief items, mass evacuation and transportation, and treatment of wounded (Caunhye et al., 2016). Finally, the recovery phase refers to the restoration of disaster-affected systems through activities such as infrastructure repair and reconstruction and debris management (Caunhye et al., 2016).

Most authors agree on these four phases to identify the disaster cycle, although some change their terms a bit. For example, (Awan & Shafiq, 2015) refer to the phases as mitigation, preparation, response, rehabilitation, and development, while (David Swanson & Smith, 2013) determine that the immediate response operation is divided into demand management, supply management, and compliance. However, in the end, all the authors focus on the four elements of Fig. 1.2.

Fig. 1.2 The cycle of humanitarian logistics based on the phases of the own elaboration catastrophe

The phases of humanitarian logistics constitute a spiral that guides the use of resources from the supply chain resulting in a kind of task execution cycle that entails not only the comprehensive approach to the incident but also a way to provide for the use of elements keys. As well as the most suitable strategy, locating the incident in different stages, times, and circumstances of progress, which deliberately becomes a guide facilitating the choice of strategies in the face of the different problems that could happen again.

In Mexico, the cycle of humanitarian logistics is configured from a series of phases very similar to those mentioned above. However, in which case its main objective is to strengthen the integral management of risks in the country through a series of actions and efforts that the government carries out from institutional means such as the law of the national civil protection system, which stipulates within its content the phases of the disaster such as (CENAPRED, 2019):

> A set of actions aimed at the identification, analysis, evaluation, control, and reduction of risks, considering them for their multifactorial origin and in a permanent construction process, which involves the three levels of government, as well as the sectors of society. These facilitate the realization of actions aimed at the creation and implementation of public policies, strategies, and procedures integrated to the achievement of sustainable development guidelines. It fights the structural causes of disasters and strengthens the resilience or resilience capacities of society.

This set of actions can be summarized in three moments; the first moment Before, the second moment During and the third moment After each moment, in turn, are subdivided into phases, this classification is summarized in the information presented in Tables 1.1, 1.2 and 1.3.

The details presented in Table 1.1 establish a clear picture of all the joint actions that the government develops to establish actions preceding the overwhelming impact of a disaster on the population, which causes significant effects and alterations between it. The first moment is the phase with the most stages; there are five approaches oriented to the prevention of risks and the possibilities that these entails.

Table 1.1 The definition and the actors established in the first moment (BEFORE), according to data contained in the material of the Mexico X platform comprehensive risk management course (CENAPRED, 2019)

Moment	Phase	Definition	Actors
First moment: BEFORE	Identification	It is the first step to establish comprehensive risk management, and it is defined as a set of actions and procedures that are carried out in a specific town or geographic area to obtain information about natural or technological hazards and conditions of vulnerability	Three government levels
	Foresight	The foresight is to become aware of the risks that may be caused and the need to face them through the stages of risk identification, prevention, mitigation, preparedness, emergency care, recovery, and reconstruction	The foresight is to become aware of the risks that may be caused and the need to face them through the stages of risk identification, prevention, mitigation, preparedness, emergency care, recovery, and reconstruction
	Prevention	Set of actions and mechanisms implemented in advance of the occurrence of the disturbing agents, in order to know the dangers or risks, identify them, eliminate them or reduce them; avoid or mitigate its destructive impact on people, goods, infrastructure, as well as anticipate the social processes of their construction	The actors that have an impact on prevention are many, including the transport, agricultural, construction, and other sectors. It constitutes a much more extensive range than the forecast since it involves the authorities and the general population
	Mitigation	Mitigation is any action aimed at reducing the impact or damage in the presence of a disturbing agent on an affectible agent. Mitigating means are taking actions to reduce the effects of disasters before they occur	The actors involved in the mitigation stage are diverse, ranging from experts and scholars on the different aspects of risk to the authorities involved in the construction of infrastructure, health services, education
	Preparation	Preparation is the last preventive stage in comprehensive risk management. It constitutes those activities and measures taken in advance to ensure an effective response to the impact of a disturbing phenomenon in the short, medium, and long term. The preparation, unlike prevention, is made up of measures taken when the impact of the disturbing agent is imminent	Disaster preparedness is a permanent multisectoral activity involving institutional actors, both from the private sector and the public and social sector. The affected population and voluntary groups also participate

Table 1.2 The definition and the actors that the second moment establishes (DURING), according to data contained in the material of the Mexico X platform comprehensive risk management course (CENAPRED, 2019)

Moment	Phase	Definition	Actors
Second moment: DURING THE EVENT	Assistance	The aid is the response to help people at risk or victims of an accident, emergency or disaster by public or private specialized groups, or by internal civil protection units, as well as actions to safeguard the other agents Affectable	Municipal bodies, government instances, the participation of society, the Secretary of the Navy, and the Secretariat of National Defense (Sedena)

1.3 Discussion and Findings

By consulting the disaster records at the base of EM-DAT Center, contingency reports issued by part of the CENAPRED, the CONAGUA records collected by the drought monitor in Mexico and various publications related to the subject, it was possible to carry out a descriptive exploratory analysis of the conditions that currently prevail concerning natural disasters and their implications in Mexico.

The first data analyzed were those obtained from the international base of the Santé Publique de l'Université Catholique de Louvain, Belgium (EM-DAT), the data extracted from this base covered from 1985 to date, the result of this was a projection of 34 years of disasters in Mexico as well as its trends.

On the other hand, verified records also include factors such as; the frequency of disasters per year, the costs of the disaster, the number of deaths, the number of people affected, the total number of people affected (which is equivalent to the sum of the number of deaths plus that of the people affected), homeless and total injured.

Figure 1.3 shows the frequency bars assigned to each type of disaster observed according to the records obtained from EM-DAT in it. It can be seen in detail that the hydrometeorological phenomena lead to natural disasters in Mexico, which corroborates in some way what already was thought about the type of disasters that most impact the country.

Now, concerning the information observed in Fig. 1.3, the storms are the first to position themselves in the graph followed by the floods, later the earthquakes, and finally, the extreme temperature. It is necessary to consider that this group of natural disasters frames 80% of the observations collected in the 34 years analyzed, which already is considered as regards the volume of data analyzed.

Table 1.3 The definition and the actors established in the third moment (AFTER), according to data contained in the material of the comprehensive risk management course of the Mexico X platform (CENAPRED, 2019)

Moment	Phase	Definition	Actors
Third moment: AFTER	Retrieval	Retrieval is defined as the process that begins during the emergency, consisting of actions aimed at returning to the normality of the affected community It is a transition period between the time of onset and the end of the emergency. At this stage of comprehensive risk management, the essential public services, such as the supply of electricity, access to drinking water, urban drainage and communication and transport roads, must be restored. Likewise, the repair of damages in housing must be initiated with temporary solutions or relocations	When natural phenomena impact the country, local and state governments coordinate to support recovery efforts. If the damages exceed national capacities, international support can be requested. In this case, the World Bank is one of the international organizations that support countries through assessments of subsequent needs and planning recovery and reconstruction tasks
	Reconstruction	Finally, reconstruction is the transitory action aimed at reaching the environment of social and economic normality that prevailed among the population before suffering the effects produced by a disturbing agent in a given space or jurisdiction. This process should seek, as far as possible, the reduction of existing risks, ensuring the non-generation of new risks, and improving pre-existing conditions	The reconstruction phase can have a very variable duration according to the type of disturbing phenomenon presented, and the magnitude of its impact. Coordination between political actors and society is essential successfully to carry out reconstruction work so that there is a fair use of human and material resources. Similarly, professionals, scientists, and planners from the urban, economic, environmental, and civil protection development sectors must participate

After obtaining a perspective of the main disasters from an international database, the same evaluation process was carried out with the information, but now in the records that CENAPRED reports at the national level from two databases of records of disasters. The first so-called socioeconomic impact and the second

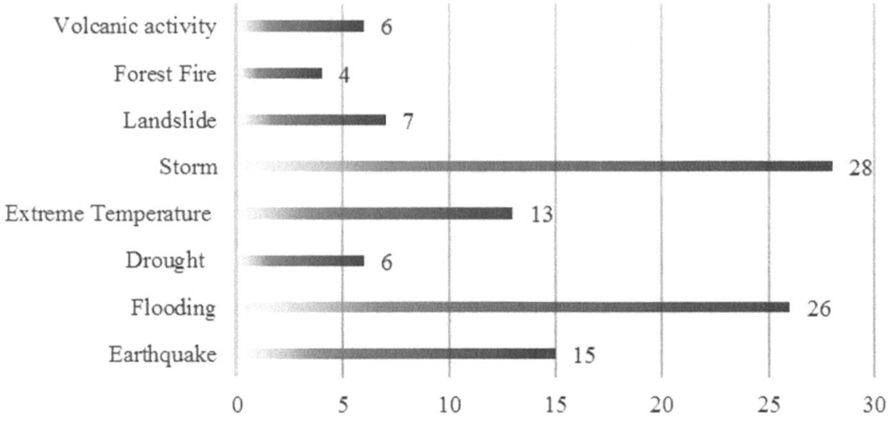

Fig. 1.3 Frequency of occurrence of natural distress in Mexico from 1985 to date. Own elaboration based on the data obtained from the EM-DAT base Centre de Recherche sur l'Epidémiologie des Desastres (CRED), L'École de Santé Publique de l'Université Catholique de Louvain, Bélgica

declarations of disasters, unlike previous data, CENAPRED records only cover an analysis horizon of 18 years (2000–2018).

In the same way as in EM-DAT, the CENAPRED base weights hydrometeorological disasters as those with the highest frequency above all others, this is evident if Fig. 1.4 is observed where the bar with the highest height corresponds to this type of event.

Although Fig. 1.4 shows a general classification of the types of disasters that are registered at the base of CENAPRED, within each item, there is another detailed division concerning the events that arise from this context. For example, for hydrometeorological contingencies, there is a division that distributes the different types of disasters; this classification is contained in Table 1.4.

Besides, it was also possible to identify which of the different types of disasters have the most significant impact according to the information obtained, and Table 1.5 shows the different classifications and their frequencies.

As previously stated, extreme rains continue to be the disasters that most impact the national territory, as well as those derived from them. Although on the other hand, it is evident to highlight that the frequencies observed by droughts and extreme temperatures also register a significant Weighting in Table 1.5, which establishes a condition for this type of hydrometeorological conditions.

There is no doubt that hydrometeorological disasters are the ones that cause the most considerable amount of affectations within the national territory, their effects are distributed nationally, and they establish a specific type of parameters regarding the type of hydrometeorological disturbances and the geographical location of the

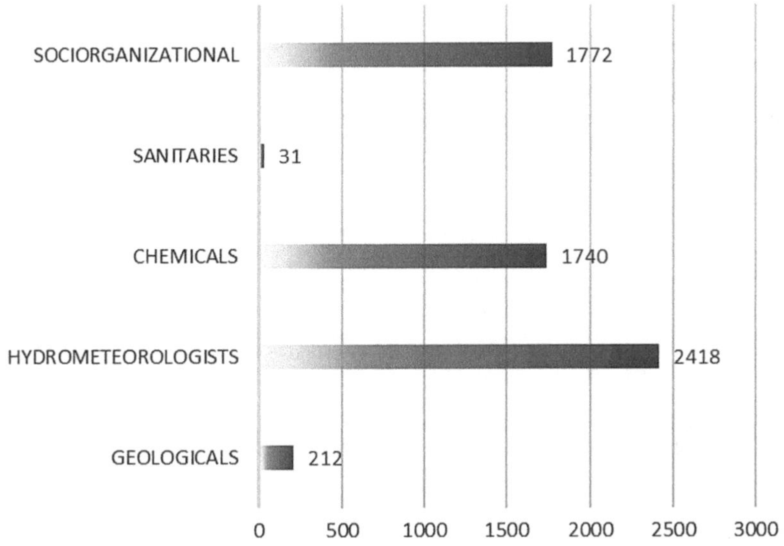

Fig. 1.4 Frequencies observed according to the types of distress that are recorded in the contingency base of the national destination center. Own elaboration based on data from the CENAPRED 2000–2015 base

Table 1.4 Classification of disasters, according to CENAPRED

HYDROMETEOROLOGISTS	HIDRO (ID)	Low temperatures
		Tropical cyclone
		Strong winds
		Severe storm or hailstorm, twister
		Flood
		Showers (EXTREME RAIN)
		Storm surge
		Drought
		Frost
		Extreme (high) temperature

affected state. So, in general, there are states of the republic that record a more considerable amount of hydrometeorological incidents, and these initiations were also determined to be able to establish a distribution of frequencies. These go from the state with the lowest record of them to the highest. Figure 1.5 shows this distribution of frequencies in ascending form for a bar graph to be able to appreciate in more detail this condition.

On the other hand, the records from the extreme rains and their derivations were finalized individually. It helped to identify which states of the republic have a more significant impact on this type of phenomenon. Figure 1.6 contains an ascending

Table 1.5 Frequencies recorded according to the type of hydrometeorological disaster with the data obtained from the socio-economic impact bases and declarations of emergency CENAPRED 2000–2018

Classification of disasters HIDRO	Frequency
Showers (EXTREME RAIN)	1444
Low temperatures	228
Severe or hailstorm, tornado	214
Tropical cyclone	174
Drought	145
Flood	141
Strong winds	93
Extreme temperature (high)	79
Frost	37
Storm surge	5

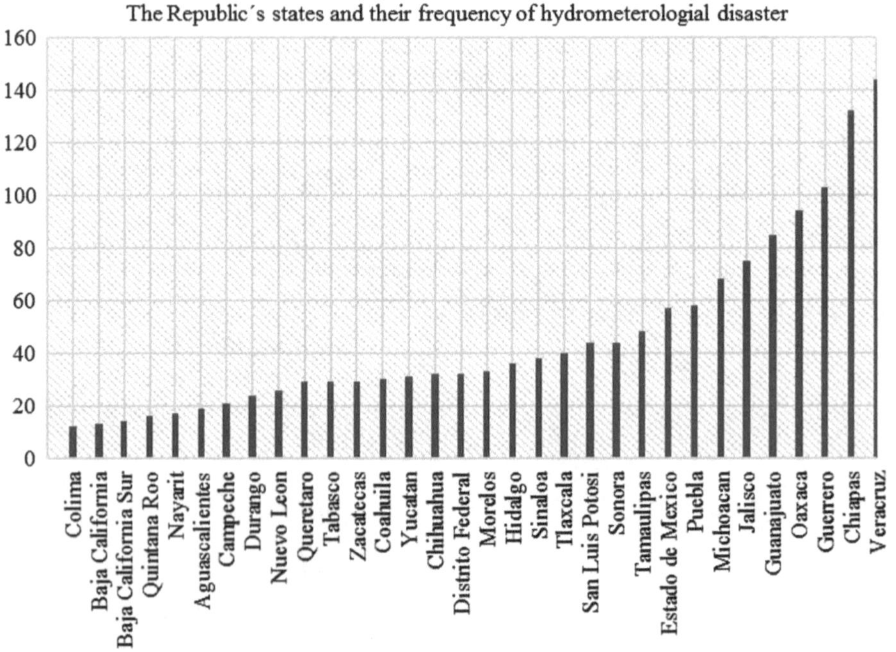

Fig. 1.5 Frequencies prioritized for the number of hydrometeorological disasters produced in the states of the Republic between 2000 and 2018 according to the records obtained from CENAPRED

graph that identifies the states of the Mexican republic that have a higher incidence of rains, winds, floods, and storms.

Generally, the southern states of the country like; Veracruz, Guerrero, Chiapas, and Oaxaca concentrate the highest number of incidents. However, as discussed

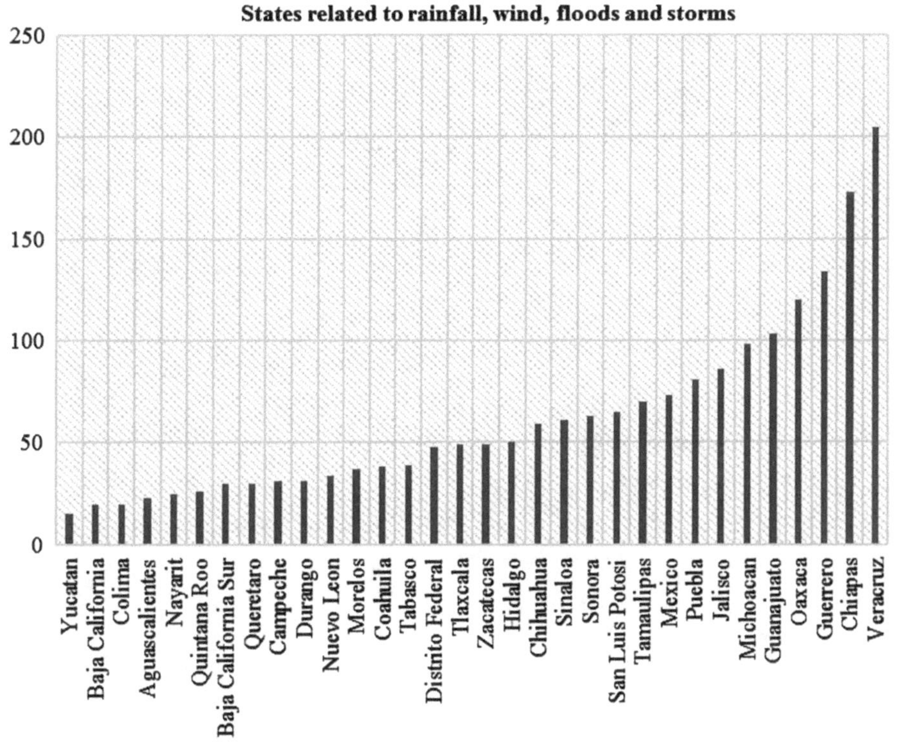

Fig. 1.6 States of the republic with the highest incidence of extreme rains and their derivations ordered of lesser number of incidents

below, these states not only participate in this type of hydrometeorological phenomena.

At the same time, droughts and extreme temperatures tend to occur in the north-central states of the republic as Chihuahua, San Luis Potosi, Michoacan, Durango, and Sonora. This behavior is very natural if we take into account that Mexico has a large part of its territory in the high-pressure strip of North latitude, so it has arid and semi-arid areas that coincide in Latitude with the regions of the great African, Asian and Australian deserts. Also, due to its orographic characteristics, these types of areas are located in the central highlands of the Mexican Republic (Ortega-Gaucin, 2013).

Nevertheless, although the behavior of the data obtained in Fig. 1.7 is similar to that established in the previous paragraph, the presence of droughts in states such as Puebla, Veracruz, Campeche Chiapas, and Tabasco is striking, since these republic states are characterized by be places cataloged with the presence of phenomena related to extreme rains, floods, and storms.

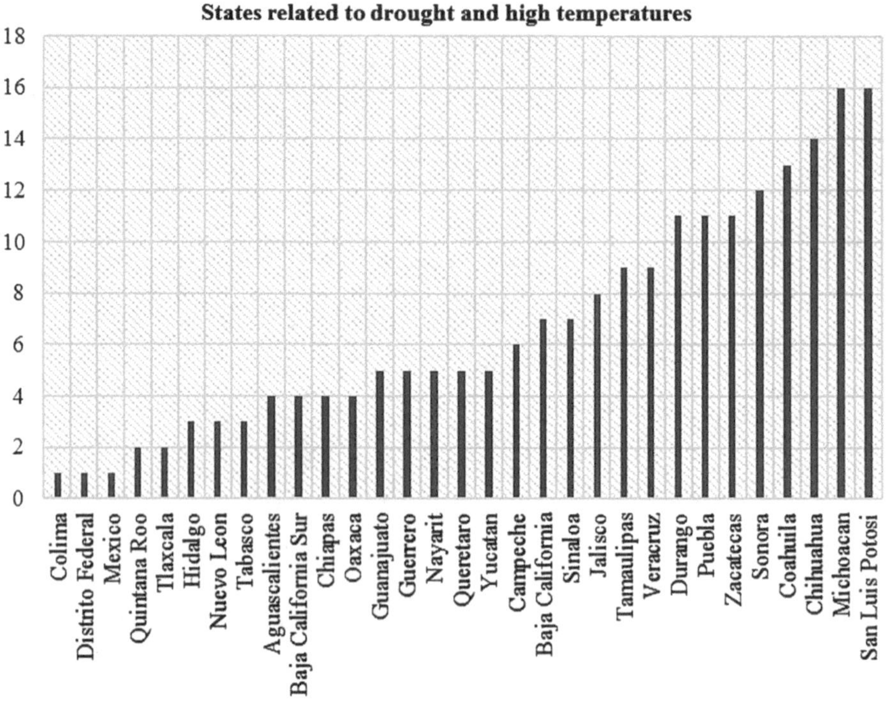

Fig. 1.7 States of the republic with more significant initiation of droughts and high temperatures accommodated from lowest to highest

The evaluation of the data obtained in the information base of CENAPRED concerning droughts and high temperatures seems to give a new pattern that could become a trend within the occurrence of hydrometeorological phenomena related to drought.

According to the last report issued on August 15th of the current year by CONAGUA through the drought monitor in Mexico, the state of the republic such as Veracruz, Yucatan, Chiapas, and Hidalgo among others are indicated as entities that present more than 40% from its affected territory (municipalities) due to some drought condition that ranges from abnormally dry—to exceptional drought. Figure 1.8 is a map of the conditions related to the percentage of areas with droughts in the federal entities of the Mexican Republic. It was published on August 21st, 2019, and shows the hydrometeorological conditions that currently prevail in the national territory for the drought phenomenon.

On the other hand, CONAGUA, through its drought monitor in Mexico, establishes an index of water vulnerability formed from:

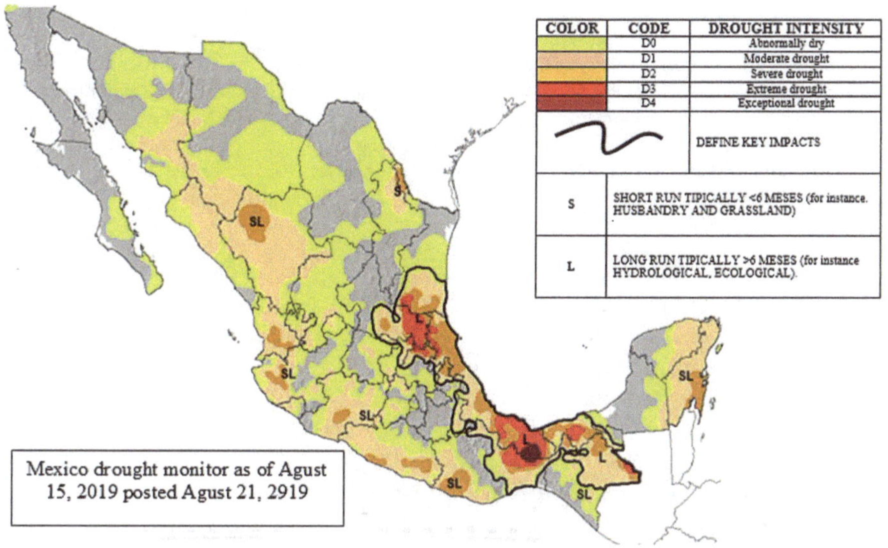

Fig. 1.8 Map issued from the data of the drought monitor in Mexico until August 15 of the year, in turn, the possible results range from abnormally dry to exceptional drought, where the yellow color is the lowest valuation and the crimson red higher (CONAGUA, 2019)

Table 1.6 Classification of drought indices according to their intensity and periods of appearance, according to CONAGUA, from the sub-management of climatology and climatological services (CONAGUA, 2019)

COLOR	CODE	INTENSITY OF DROUGHT
	D0	Abnormally dry
	D1	Moderate drought
	D2	Severe drought
	D3	Extreme drought
	D4	Exceptional drought
		DEFINE KEY IMPACTS
S	SHORT-RUN TYPICALLY <6 MESES (for instance. HUSBANDRY AND GRASSLAND)	
L	LONG RUN TYPICALLY >6 MESES (for instance, HYDROLOGICAL, ECOLOGICAL)).	

> Obtaining and interpreting various indices or indicators of drought such as the Standardized Precipitation Index (SPI) that quantifies the conditions of deficit or excess rainfall (30, 90, 180, 365 days), Rain Abnormality in Percentage of Normal (30, 90, 180, 365 days), Satellite Vegetation. Health Index (VHI) that measures the degree of vegetation stress through the observed radiance, the Leaky Bucket CPC-NOAA Soil Moisture Model that estimates soil moisture through a single layer hydrological model, the Normalized Index of Vegetation Difference (NDVI), the Average Temperature Anomaly, the Percentage of Water Availability in the dams of the country and the contribution of local experts (CONAGUA, 2019).

This index stipulates the percentages of affectations according to abnormally dry (D0), moderate drought (D1), severe drought (D2), extreme drought (D3) until exceptional drought (D4). Table 1.6 shows in more detail the conditions mentioned above, which are also considered in the map of Fig. 1.8.

In the same way, as with the EM-DAT and CENAPERD records, an exhaustive review was carried out on the data obtained from the CONAGUA drought monitor that covers from January 2003 to August 2019, except for August 2003 and February 2004 dates in which it is reported that there was no drought monitoring due to external factors, so the record was not prepared.

These data address a more realistic perspective on the behavior of droughts in the different states of the republic. Figure 1.9 prioritizes from minor to major the states of the Mexican Republic about the number of municipalities affected by drought. It means that the frequency observed corresponds to the number of municipalities affected in that state. Thus in the case of Oaxaca that It has the highest frequency, equal to 570 equivalents to the 570 municipalities that have been registered with drought during the specified time, and so on for each of the frequencies observed.

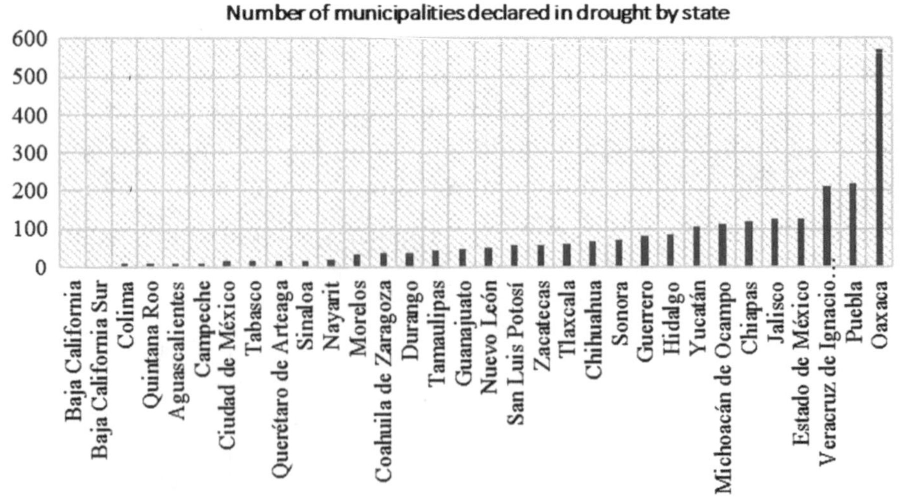

Fig. 1.9 Ascending distribution of the number of municipalities by states considered to have drought problems

Table 1.7 Records tabulated concerning the intensity of the drought counted for each state of the republic with affectations in measure of their different municipalities during the period from 2003 to 2019, the codes evaluated obey the nomenclature of Fig. 1.9 where the intensity of the drought

State	Drought intensity (2003–2019)				
	D0	D1	D2	D3	D4
Aguascalientes	362	204	95	51	5
Baja California	246	254	260	242	3
Baja California Sur	239	169	60	9	0
Campeche	807	478	152	49	1
Coahuila de Zaragoza	1974	990	491	474	187
Colima	748	354	73	14	0
Chiapas	8029	4181	1504	355	0
Chihuahua	4571	1940	780	648	244
Ciudad de México	1103	309	157	5	0
Durango	2254	1114	478	416	55
Guanajuato	2750	1406	714	228	10
Guerrero	6345	3015	319	14	0
Hidalgo	4965	1934	505	66	0
Jalisco	8285	4286	2134	247	49
Estado de México	7705	2938	768	109	0
Michoacán de Ocampo	8629	5100	1885	372	0
Morelos	1715	688	80	0	0
Nayarit	1533	739	399	82	0
Nuevo León	2844	946	410	486	13
Oaxaca	33,872	11,111	1903	663	5
Puebla	12,126	4063	906	90	0
Querétaro de Arteaga	1020	506	215	37	4
Quintana Roo	639	383	127	1	0
San Luis Potosí	3503	1844	732	123	0
Sinaloa	1350	849	387	106	58
Sonora	5382	3069	1874	986	168
Tabasco	899	883	456	246	5
Tamaulipas	2812	1227	396	270	41
Tlaxcala	2997	1440	58	19	0
Veracruz de Ignacio de la Llave	14,439	6105	2004	439	8
Yucatán	8041	3675	1446	599	56
Zacatecas	2683	1217	542	280	48

Now, Fig. 1.9 only determines the number of municipalities by the state of the republic that has been cataloged with any drought from D0 to D4, but Table 1.7 defines the type of droughts that have been specified for each state according to the intensity of the affectation to them.

Since the number of incidents has been recorded according to the intensity of the drought in each of the states. It is observed that Oaxaca is the state where the highest frequency prevails concerning abnormally dry conditions and moderate drought, which confirms by which is one of the most various entities in the graph of Fig. 1.9. On the other hand, Jalisco maintains the highest frequency regarding a severe drought. Finally, Sonora and Chihuahua maintain the first places regarding the extreme and exceptional drought, respectively. Although only a few states remain at the forefront of the different types of intensity in the drought, it does not mean that some cities in the southeast such as Veracruz, Yucatan, and downtown Michoacan do not appear with initiations not very far from the determined among which they mainly lead each type of drought intensity.

Besides, it should also be noted that some of the variables that can be consulted within the CENAPRED records. Include the following fields like Deaths, Affected population, Damaged Homes, Schools, Hospitals, Damaged crop area/grasslands, Paths affected (Km), Total damages (Millions of pesos), Total damages (Millions of dollars), Type of declaration, Substance involved, Source, Observations, Documented and Cycle time. Finally, Fig. 1.10 indicates the trend of hydrometeorological disasters during the period analyzed, taking into account that the data obtained from 2018 only cover the first months of the year clearly distinguishes a somewhat flared distribution with high frequencies in 2011, 2008 and 2004.

In addition to the data evaluated in the records of CENAPRED, EM-DAT, and CONAGUA, a perspective was added to complement the identified characteristics. It was carried out from a review of 39 related articles with humanitarian logistics; this search revealed that in most of the articles analyzed, the focus is on hydrometeorological disasters related to extreme rains, floods, landslides, and earthquakes. Only a few of them consider droughts according to their implications in agricultural activities, like the quality of human life in communities vulnerable to drought (Fernando et al., 2014) and the effects of drought on migration and mobility of

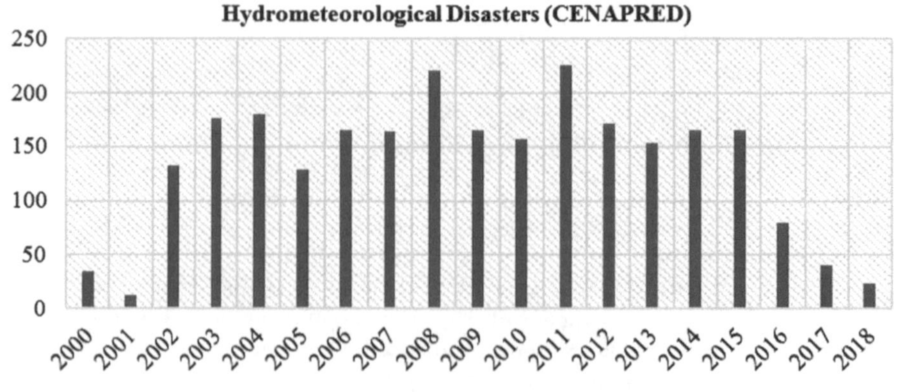

Fig. 1.10 Trend of hydrometeorological disasters per year according to CENAPRED records

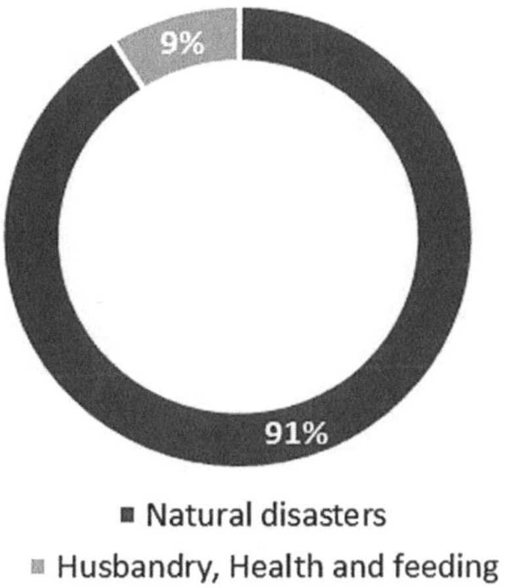

Fig. 1.11 Percentage distribution of the categories in which the different articles consulted was evaluated

people in African countries (Dufour et al., 2018; Rancourt et al., 2015) and regions of the world impacted by this type of hydrometeorological conditions.

It should be noted that in recent years more than 90% of displacements are related to weather disasters such as floods, storms, and extreme cold (Ochoa Lupián & Ayvar Campos, 2015), which is predictable if we consider all the previous results obtained from the analysis of hydrometeorological disasters. However, in the case of Mexico, studies show that it is especially vulnerable since there are areas that will be impacted by droughts (Ochoa Lupián & Ayvar Campos, 2015), those that have always been historically susceptible and new ones in the south of the country.

However, and according to the analysis carried out on the 39 articles related to humanistic logistics, it was determined that 91% of them focus directly on the effects caused by natural disasters, while the remaining 9% deals with problems related to agriculture, health and food, but within that 91% none of them is based on the hydrometeorological conditions of the drought (Fig. 1.11).

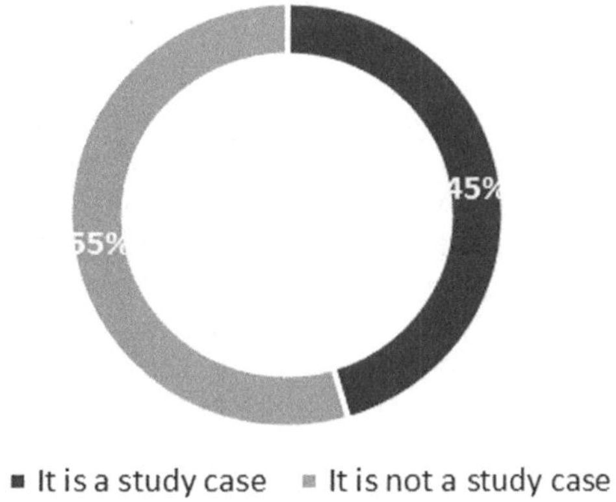

Fig. 1.12 Percentage distribution of the articles that present and do not present a case study

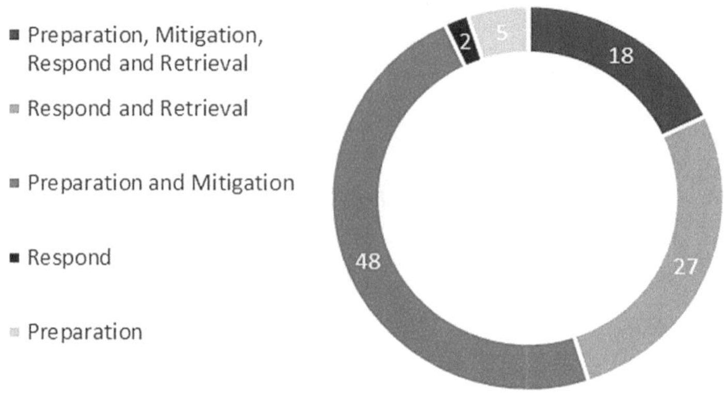

Fig. 1.13 Percentages of each of the impact phases identified in the different articles, some appear individual while they are usually shared

On the other hand, 45% of the articles analyzed are based on case studies that have to do with experiences obtained by hydrometeorological disasters such as storms, severe rains, floods, landslides and hurricanes, earthquakes and their variations are also contemplated within this percentage. However, they do not correspond precisely to this type of condition. The remaining 55% propose solutions to specific conditions such as inventory tracking systems or transport networks to optimize the flow of supplies to the points of the disaster.

In general, the case studies were developed in cities such as; Brazil, Japan, Pakistan, and the Indian Ocean, Malaysia, Uruguay, Colombia, Peru, Mexico, USA, Tehran, China, India, Amsterdam.

As regards the aspects related to in Figs. 1.12 and 1.13, they represent the other characteristics that describe humanitarian logistics. These analyzed elements seek to contextualize the components that define the actions of humanitarian aid and where the different research focuses mainly evaluated.

The impact phases were validated based on the orientation of the analyzed article, which is why a large proportion of them was linked to the preparation-mitigation followed by the response-recovery, few of them were related to a single-phase such as 2% was related only to the response and 5% to the preparation.

On the other hand, the type of approach of the article was classified in quantitative qualitative; the vast majority of them are oriented in quantitative aspects, which is observed in Table 1.8.

It is worth mentioning that the impact phases in Table 1.8 have been ordered as follows; (1) Preparation, (2) Mitigation, (3) Response and (4) Recovery, followed by the logistics decision areas listed as; (A) Transportation, (B) Inventories and (C) Location to finally add the type of approach; (+) Qualitative and (*) Quantitative.

Table 1.8 The concentration of the main characteristics and data of the 39 articles analyzed

Paper title	Impact phase				Logistics decision area			Paper approach		Authors
	1	2	3	4	A	B	C	+	*	
Improving civil-military coordination in humanitarian logistics: the challenge			x	x	x	x	x	x		Heaslip and Barber (2017)
Humanitarian logistics: How to help even more?	x	x			x	x		x		Agostinho (2013)
Supply Chains in Humanitarian Operations: Cases and Analysis	x	x			x	x		x		Costa et al. (2012)
Humanitarian logistics: Empirical evidence from a natural disaster	x	x	x	x	x		x	x		Scarpin and De Oliveira Silva (2014)
Analyzing transportation and distribution in emergency humanitarian logistics			x		x		x	x		Safeer et al. (2014)
The contribution of scientific productions at the beginning of the third millennium (2001–2014) for humanitarian logistics: A bibliometric analysis	x							x		Zary et al. (2014)
An analysis of the literature on humanitarian logistics and supply chain management: paving the way for future studies	x							x		Chiappetta Jabbour et al. (2017)

(continued)

Table 1.8 (continued)

Paper title	Impact phase				Logistics decision area			Paper approach		Authors
	1	2	3	4	A	B	C	+	*	
On the unique features of post-disaster humanitarian logistics	x							x		Holguín-Veras et al. (2012)
On the appropriate objective function for post-disaster humanitarian logistics models	x	x			x	x	x	x		Holguín-Veras et al. (2013)
Humanitarian logistics: A clustering methodology for assisting Humanitarian operations	x	x			x	x	x		x	Santos Lima et al. (2014)
The relative effects of logistics, coordination and human resource on humanitarian aid and disaster relief mission performance	x	x			x	x	x		x	Idris and Che Soh (2014)
Research in Humanitarian Logistics	x	x			x		x	x		Overstreet et al. (2013)
Humanitarian logistics and its application in Uruguay	x	x			x	x	x		x	Viera et al. (2012)
Identification and Analysis of a Humanitarian Aid Network: A Case Study			x	x	x		x		x	Gómez et al. (2017)

(continued)

Table 1.8 (continued)

Paper title	Impact phase				Logistics decision area			Paper approach		Authors
	1	2	3	4	A	B	C	+	*	
A bi-criterion model for the location of shelters, as part of a flood evacuation plan			x	x	x		x		x	Iniestra et al. (2012)
Think about collective action in the context of humanitarian logistics: the lessons of the Pisco earthquake.	x	x			x	x	x	x		Chandes and Paché (2009)
Challenges for the development of disaster support strategies in Mexico	x	x			x	x	x	x		Pilar Arroyo Lópeza et al. (2015)
Potential use of logistics focused on disaster management logistics systems. A conceptual analysis	x	x	x	x	x	x	x	x		Rodríguez et al. (2012)
Considerations for the management of post-disaster humanitarian logistics			x	x	x		x	x		Anglica and Hernández (2015)
A humanitarian logistics model for disaster relief operation considering network failure and standard relief time: A case study on San Francisco district	x	x			x		x		x	Ahmadi et al. (2015)
A bi-objective model for the location of relief centers and distribution of commodities in disaster response operations	x	x						x		Moreno et al. (2017)

(continued)

Table 1.8 (continued)

Paper title	Impact phase				Logistics decision area			Paper approach		Authors
	1	2	3	4	A	B	C	+	*	
Humanitarian Logistics: Models for Attention to Populations Affected by Natural Disasters	x	x			x	x			x	Berger Vidal et al. (2019)
A dynamic multi-objective location–routing model for relief logistic planning under uncertainty on demand, travel time, and cost parameters	x	x	x	x	x	x	x		x	Bozorgi-Amiri and Khorsi (2016)
Multiple-objective analysis of integrated relief supply and network restoration in humanitarian logistics operations			x	x	x	x			x	Ransikarbum and Mason (2016)
Optimal Decisions for Prepositioning Emergency Supplies Problem with Type-2 Fuzzy Variables	x	x			x	x			x	Bai (2016)
A Qualitative Analysis of Inventory Management Strategies in Humanitarian Logistics Operations			x	x		x		x		Mora-Ochomogo et al. (2016)
Humanitarian logistics planning for natural disaster response with Bayesian information updates			x	x	x	x	x		x	Ye and Liu (2013)

(continued)

Table 1.8 (continued)

Paper title	Impact phase				Logistics decision area			Paper approach		Authors
	1	2	3	4	A	B	C	+	*	
An agent-based model for material convergence in humanitarian logistics	x	x				x				Suárez-Moreno et al. (2017)
Inventory-Allocation Distribution Models for Post-disaster Humanitarian Logistics with Explicit Consideration of Deprivation Costs Inventory-Allocation			x	x		x			x	Pérez-Rodríguez and Holguín-Veras (2015)
No Vehicle Means No Aid—A Paradigm Change for the Humanitarian Logistics Business Model	x	x	x	x	x			x		Hartmann and The (2015)
Models, solutions and enabling technologies in humanitarian logistics	x	x	x	x	x	x	x	x		Özdamar and Ertem (2015)
Humanitarian logistics network design under mixed uncertainty	x	x			x	x			x	Tofighi et al. (2016)
Supply Chain Management in Humanitarian Relief Logistics	x	x			x	x			x	Awan and Shafiq (2015)
Demand forecasting and order planning for humanitarian logistics: An empirical assessment	x	x				x			x	van der Laan et al. (2016)

(continued)

Table 1.8 (continued)

Paper title	Impact phase				Logistics decision area			Paper approach		Authors
	1	2	3	4	A	B	C	+	*	
A path to a public-private partnership: Commercial logistics concepts applied to disaster response			x	x	x	x	x	x	x	David Swanson and Smith (2013)
A Peek into the Future of Humanitarian Logistics: Forewarned Is Forearmed	x	x	x	x	x	x	x	x		Majewski et al. (2018)
Models and metrics to assess humanitarian response capacity			x	x		x			x	Acimovic and Goentzel (2016)
Humanitarian Logistics: An Approach to Supply from Agro-Food Chains	x	x			x	x			x	Fernando et al. (2014)
Facility location optimization model for emergency humanitarian logistics	x	x	x	x			x	x		Boonmee et al. (2017)

Finally, it should be stressed that a considerable percentage of the articles evaluated focus on logistics decisions related to transport followed by inventories. In general, these approaches can relate or converge on a proposal or be an individual proposal. It depends on a considerable measure of the conditions proposed by the article and the characteristics that contextualize the disaster.

1.4 Conclusions

Once the information regarding humanitarian logistics and natural disasters in Mexico is analyzed, it is possible to summarize a set of conclusions developed during the development of this research.

First of all, the indisputable impeccability that natural disasters continuously provide to the national territory and especially those of the hydrometeorological type and specifically those derived from extreme rains, as has been pointed out previously and repeatedly, must be considered. On the other hand, in the data provided by EM-DAT, CENPRED, and CONAGUA, significant evidence is detected about a new trend in weather conditions, such as drought and extreme temperatures. However, it is clear that climate change. It is playing an essential role in generating new conditions for natural disasters. It can be seen in greater detail in the comparisons made between the states of the Republic with the highest amount of extreme rainfall and its derivatives versus those most affected by drought and high temperatures. At first, it is thought the southern states are affected by the rains and their derivatives and those of the north by drought and high temperatures. However, once the records are reviewed in detail, it can be determined that some states from the south of the republic already begin to suffer from droughts and extreme temperatures, such as in the case of Veracruz, Campeche, and the center of the country Puebla and Hidalgo.

On the other hand, the factors used in the characterization of humanitarian logistics say a lot about what is currently being developed through research focused on this line of research. For example, the phases of the disaster where the highest amount of research is concentrated on humanitarian logistics; are those related to the preparation and mitigation that is equivalent to the first and second time according to the management of risks in the national civil protection system. Some of the phases forgotten within the review or much less frequently. They are the answer and recovery, which is why post-disaster strategies do not stand out as a repetitive component within the context of humanitarian logistics research.

Another considerable aspect of the search carried out in the CENAPRED records has to do with the description of the disaster. In the case of droughts, in general, they focus on the effects on the field, animals, and heatstroke that imply the population without naming even the affectations that result in the migration of the affected populations and other problems, such as, the food shortage that the drought causes jointly with the erosion of lands and the decrease of the water deposits in the sinister regions.

In the end, hydrometeorological disasters dominate the effects in Mexico and within this context rains, and their derivations are the leading causes of problems to the population. However, it cannot be ignored that droughts are changing the face of some sectors of the national territory where this condition seemed difficult or almost impossible to happen. It must be remembered that the drought phenomenon is a passive and slow-moving agent whose impact is much more substantial once it has been consummated. Humanitarian logistics is currently highly concentrated in conditions of immediate as earthquakes, tsunamis, and floods, which have kept it distant from conditions such as droughts and its affectations that in recent years have caused great migrations and global conflicts. The idea is to focus the capabilities of humanitarian logistics on phenomena of this type. It be able to constitute a slope that reaches to analyze and confront the problems Ethics of the drought in Mexico as well as their next repercussions.

References

Abeldaño Zúñiga, R. A., & González Villoria, A. M. (2018). Desastres en México de 1900 a 2016: Patrones de ocurrencia, población afectada y daños económicos. *Revista Panamericana de Salud Pública, 42*, 1–8. https://doi.org/10.26633/rpsp.2018.55. [In Spanish].

Acimovic, J., & Goentzel, J. (2016). Models and metrics to assess humanitarian response capacity. *Journal of Operations Management, 45,* 11–29. https://doi.org/10.1016/j.jom.2016.05.003.

Agostinho, C. F. (2013). Humanitarian logistics: How to help even more? In *IFAC Proceedings Volumes* (IFAC-Papers Online) (Vol. 6). IFAC. https://doi.org/10.3182/20130911-3-BR-3021.00075.

Ahmadi, M., Seifi, A., & Tootooni, B. (2015). A humanitarian logistics model for disaster relief operation considering network failure and standard relief time: A case study on San Francisco district. *Transportation Research Part E: Logistics and Transportation Review, 75,* 145–163. https://doi.org/10.1016/j.tre.2015.01.008.

Anglica, L., & Hernández, G. (2015). Consideraciones para la gestión de la logística humanitaria postdesastre. Instituto Mexicano Del Transporte, (433), 46. Retrieved from https://www.imt.mx/archivos/Publicaciones/PublicacionTecnica/pt433.pdf. [In Spanish].

Awan, A. G., & Shafiq, M. (2015). Supply chain management in humanitarian relief logistics. *Industrial Engineering Letters, 5*(10), 12–22. http://citeseerx.ist.psu.edu/viewdoc/download?doi=10.1.1.857.6281&rep=rep1&type=pdf.

Bai, X. (2016). Optimal decisions for prepositioning emergency supplies problem with type-2 fuzzy variables. *Discrete Dynamics in Nature and Society, 2016,* 1–17. https://doi.org/10.1155/2016/9275192.

Beamon, B. M. & Kotleba, S. A. (2006). Inventory management support systems for emergency humanitarian relief operations in South Sudan. *The InternationalJournal of Logistics Management, 17*(2), 187–212.

Berger Vidal, E., Velásquez Pino, C., Huaroto Sumari, C., Zacarías Díaz, M., Núñez Ramírez, L., & Arriola Sánchez, J. (2019). Logística humanitaria: Modelos para la atención de poblaciones afectadas por desastres naturales. *Pesquimat, 21*(2), 17. https://doi.org/10.15381/pes.v21i2.1571255. [In Spanish].

Boonmee, C., Arimura, M., & Asada, T. (2017). Facility location optimization model for emergency humanitarian logistics. *International Journal of Disaster Risk Reduction, 24*(June 2016), 485–498. https://doi.org/10.1016/j.ijdrr.2017.01.017.

Bozorgi-Amiri, A., & Khorsi, M. (2016). A dynamic multi-objective location–routing model for relief logistic planning under uncertainty on demand, travel time, and cost parameters. *International Journal of Advanced Manufacturing Technology, 85*(5–8), 1633–1648. https://doi.org/10.1007/s00170-015-7923-3.

Capacci, A., & Mangano, S. (2015). Las catástrofes naturales. *Cuadernos de Geografía: Revista Colombiana de Geografía, 24*(2), 35–51. https://doi.org/10.15446/rcdg.v24n2.50206. [In Spanish].

Caunhye, A. M., Nie, X., & Pokharel, S. (2012). Optimization models in emergency logistics: A literature review. *Socio-Economic Planning Sciences, 46*(1), 4–13. https://doi.org/10.1016/j.seps.2011.04.004.

Caunhye, A. M., Zhang, Y., Li, M., & Nie, X. (2016). A location-routing model for prepositioning and distributing emergency supplies. *Transportation Research Part E: Logistics and Transportation Review, 90,* 161–176. https://doi.org/10.1016/j.tre.2015.10.011.

CENAPRED. (2019). Gestión integral del riesgo de desastres|México X. Retrieved August 28, 2019, from http://www.mexicox.gob.mx/courses/course-v1:CENAPRED+GIDR19042X+2019_04/about. [In Spanish].

Chakravarty, A. K. (2014). Humanitarian relief chain: Rapid response under uncertainty. *International Journal of Production Economics, 151,* 146–157. https://doi.org/10.1016/j.ijpe.2013.10.007.

Chandes, J., & Paché, G. (2009). Pensar la acción colectiva en el contexto de la logística humanitaria: Las lecciones del sismo de Pisco. The Bi-Annual Academic Publication of Universidad ESAN, *14* (27), 47–62. http://www.redalyc.org/articulo.oa?id=360733607003DBRedalyc. [In Spanish].

Chiappetta Jabbour, C. J., Sobreiro, V. A., de Sousa, Lopes, Jabbour, A. B., de Souza Campos, L. M., Mariano, E. B., et al. (2017). An analysis of the literature on humanitarian logistics and supply chain management: Paving the way for future studies. *Annals of Operations Research.* https://doi.org/10.1007/s10479-017-2536-x.

CONAGUA. (2019). Monitor de Sequía de México Fuente: Comisión Nacional del Agua-Servicio Meteorológico Nacional. Retrieved September 7, 2019, from https://smn.conagua.gob.mx/tools/DATA/Climatología/Sequía/MonitordesequíaenMéxico/SeguimientodeSequía/MSM20190815.pdf. [In Spanish].

Costa, S. R. A. da, Campos, V. B. G., & Bandeira, R. A. de M. (2012). Supply chains in humanitarian operations: Cases and analysis. *Procedia-Social and Behavioral Sciences, 54,* 598–607. https://doi.org/10.1016/j.sbspro.2012.09.777.

Cruz Benítez, D. P., Garzón Garnica, E. A., Chayban Abdul-Massih, J. A., & Martínez Flores, J. L. (2015). Reduction of disaster effects through localization of first aid centers in Chiapas, Mexico. In *International Congress on Logistics & Supply Chain.* https://www.researchgate.net/publication/283460700_Reduction_of_disaster_effects_through_localization_of_first_aid_centers_in_Chiapas_Mexico.

David Swanson, R., & Smith, R. J. (2013). A path to a public-private partnership: Commercial logistics concepts applied to disaster response. *Journal of Business Logistics, 34*(4), 335–346. https://doi.org/10.1111/jbl.12031.

Diéguez, E. T., Mancera, G. M., Falcón, A. C., Garibay, A. N., Valdez Cepeda, R. D., García Hernández, J. L., & Amador, B. M. (2014). Análisis de la sequía y desertificación mediante índices de aridez y estimación de la brecha hídrica en Baja California Sur, noroeste de México. *Investigaciones Geograficas, 85*(85), 66–81. https://doi.org/10.14350/rig.32404. [In Spanish].

Dufour, É., Laporte, G., Paquette, J., & Rancourt, M. È. (2018). Logistics service network design for humanitarian response in East Africa. *Omega (United Kingdom), 74,* 1–14. https://doi.org/10.1016/j.omega.2017.01.002.

Esparza, M. (2014). Drought and water shortages in Mexico: Current status and future prospects. *Secuencia*, (89), 193–219. http://ref.scielo.org/b95sv5.

Fernando, S., Judith, C., & Gabriel, V. (2014). Logística humanitaria: Un enfoque del suministro desde las cadenas agroalimentarias. *Información Tecnológica, 25*(4), 43–50. https://doi.org/10.4067/s0718-07642014000400007. [In Spanish].

Gómez, D. M., Sarache, W., & Trujillo, M. (2017). Identificación y Análisis de una Red de Ayuda Humanitaria: Un Caso de Estudio. *Información Tecnológica*, *28*(2), 115–124. https://doi.org/10.4067/s0718-07642017000200013. [In Spanish].

Hartmann, M. H. R. M. T. S. E., & The. (2015). No vehicle means no aid—a paradigm change for the humanitarian logistics business model. *Perspective from Practice*, *49*(5), 630–631. https://doi.org/10.1002/tie.

Heaslip, G. E., & Barber, E. (2017). Improving civil-military coordination in humanitarian logistics: The challenge. *The Irish Journal of Management, 35*(2), 143–158. https://doi.org/10.1515/ijm-2016-0011.

Holguín-Veras, J., Jaller, M., Van Wassenhove, L. N., Pérez, N., & Wachtendorf, T. (2012). On the unique features of post-disaster humanitarian logistics. *Journal of Operations Management, 30*(7–8), 494–506. https://doi.org/10.1016/j.jom.2012.08.003.

Holguín-Veras, J., Pérez, N., Jaller, M., Van Wassenhove, L. N., & Aros-Vera, F. (2013). On the appropriate objective function for post-disaster humanitarian logistics models. *Journal of Operations Management, 31*(5), 262–280. https://doi.org/10.1016/j.jom.2013.06.002.

Huang, K., Jiang, Y., Yuan, Y., & Zhao, L. (2015). Modeling multiple humanitarian objectives in emergency response to large-scale disasters. *Transportation Research Part E: Logistics and Transportation Review, 75,* 1–17. https://doi.org/10.1016/j.tre.2014.11.007.

Idris, A., & Che Soh, S. N. (2014). The relative effects of logistics, coordination, and human resource on humanitarian aid and disaster relief mission performance. *The South East Asian Journal of Management, 8*(2), 87–103. https://doi.org/10.21002/seam.v8i2.3928.

Iniestra, J. G., Arroyo López, P. E., & Enríquez Colón, R. (2012). Un Modelo Bi-Criterio Para La Ubicación De Albergues, Como Parte de un Plan De Evacuación a Bi-Criteria Model to Locate Shelters as Part of an Evacuation Plan in The Case of loods. *Revista Ingeniería Industrial, 11* (2), 35–56. https://dialnet.unirioja.es/servlet/articulo?codigo=4398257.

Jais, N., Lebus, F., Pfitzer, M., & Rodriques, A. (2017). Insuring shared value. Shared value initiative, 30. Retrieved from https://www.sharedvalue.org/sites/default/files/resource-files/Insuring%20Shared%20Value_7-5-17.pdf.

Magaña-Rueda, V. O., Méndez, B., Neri, C., & Vázquez Cruz, G. (2018). El riesgo ante la sequía meteorológica en México. Realidad, Datos y Espacio. *Revista Internacional de Estadística y Geografía*, *9*(1), 35–48. Retrieved July 24, 2018, from https://www.inegi.org.mx/rde/2018/04/01/riesgo-ante-la-sequia-meteorologica-en-mexico/. [In Spanish].

Majewski, B., & Heigh, I. (2010). A peek into the future of humanitarian logistics: Forewarned is forearmed. *Supply Chain Forum: An International Journal, 11*(November), 4–20. https://doi.org/10.1080/16258312.2010.11517236.

Majewski, B., Navangul, K. A., & Heigh, I. (2018). A peek into the future of humanitarian logistics: Forewarned is forearmed. *Supply Chain Forum: An International Journal, 11*(3), 4–19. https://doi.org/10.1080/16258312.2010.11517236.

Mora-Ochomogo, E. I., Mora-Vargas, J., & Serrato, M. (2016). A qualitative analysis of inventory management strategies in humanitarian logistics operations. *International Journal of Combinatorial Optimization Problems & Informatics*, *7*(1), 40–53. https://ijcopi.org/index.php/ojs/article/view/45/43.

Moreno, A., Ferreira, D., & Alem, D. (2017). A bi-objective model for the location of relief centers and distribution of commodities in disaster response operations. *DYNA (Colombia)*, *84*(200). https://doi.org/10.15446/dyna.v84n200.54810.

Ochoa Lupián, L., & Ayvar Campos, F. (2015). Migración y cambio climático en México. *Cimexus, 10*(1), 35–51. [In Spanish].

Ortega-Gaucin, D. (2013). Sequía: Causas y efectos de un fenómeno global. *Ciencia UANL, 16*(61), 8–15. Retrieved July 22, 2019, from https://www.researchgate.net/publication/260163188. [In Spanish].

Overstreet, R. E., Hall, D. J., Hanna, J. B., Overstreet, R. E., Hall, D., Hanna, J. B., & Rainer, R. K. (2013, September). Research in Humanitarian Logistics. https://doi.org/10.1108/20426741111158421.

Özdamar, L., & Ertem, M. A. (2015). Models, solutions and enabling technologies in humanitarian logistics. *European Journal of Operational Research, 244*(1), 55–65. https://doi.org/10.1016/j.ejor.2014.11.030.
Pérez-Rodríguez, N., & Holguín-Veras, J. (2015). Inventory-allocation distribution models for postdisaster humanitarian logistics with explicit consideration of deprivation costs. *Transportation Science, 50*(4), 1261–1285. https://doi.org/10.1287/trsc.2014.0565.
Pilar Arroyo Lópeza, Gaytán Iniestrab, J., & Mejía-Argueta, C. (2015). *Retos para el Desarrollo de Estrategias de Apoyo Ante Desastres en México* (Primera ed.). Cortazar, Gto: Universidad Politécnica de Guanajuato. [In Spanish].
Rancourt, M. È., Cordeau, J. F., Laporte, G., & Watkins, B. (2015). Tactical network planning for food aid distribution in Kenya. *Computers & Operations Research, 56,* 68–83. https://doi.org/10.1016/j.cor.2014.10.018.
Ransikarbum, K., & Mason, S. J. (2016). Multiple-objective analysis of integrated relief supply and network restoration in humanitarian logistics operations. *International Journal of Production Research, 54*(1), 49–68. https://doi.org/10.1080/00207543.2014.977458.
Rodríguez, L. J. G., Kalenatic, D., Velasco, F. J. R., & Bello, C. A. L. (2012). Potencial uso de la logística focalizada en sistemas logísticos de atención de desastres. Un análisis conceptual. *Revista Facultad de Ingenieria, 62,* 44–54. [In Spanish].
Rua, T. A. (2013). Refugiados Ambientales. Cambio Climático y migración forzada. (P. Fondo Editorial de la Pontificia Universidad Católica del Perú, 2014 Av. Universitaria 1801, Lima 32, Ed.) (Primera Ed.). PERÚ: Enero 2014. Retrieved July 18, 2019, from www.pucp.edu.pe/publicaciones. [In Spanish].
Rueda, V. O. M., & Neri, C. (2012). Cambio climático y sequías en México. *Ciencia,* 26–35. Retrieved September 6, 2019, from www.amc.edu.mx/revistaciencia/images/revista/63_4/PDF/sequiasMexico.pdf. [In Spanish].
Safeer, M., Anbuudayasankar, S. P., Balkumar, K., & Ganesh, K. (2014). Analyzing transportation and distribution in emergency humanitarian logistics. *Procedia Engineering, 97,* 2248–2258. https://doi.org/10.1016/j.proeng.2014.12.469.
Santos Lima, F., de Oliveira, D., Buss Gonçalves, M., & Marcondes Altimari Samed, M. (2014). Humanitarian logistics: A clustering methodology for assisting humanitarian operations. *Journal of Technology Management & Innovation, 9*(2), 86–97. https://doi.org/10.4067/S0718-27242014000200007.
Scarpin, M. R. S., & De Oliveira Silva, R. (2014). Humanitarian logistics: Empirical evidences from a natural disaster. *Procedia Engineering, 78,* 102–111. https://doi.org/10.1016/j.proeng.2014.07.045.
Serrano Vincenti, S., Zuleta, D., Moscoso, V., Jácome, P., Palacios, E., & Villacís, M. (2017). Análisis estadístico de datos meteorológicos mensuales y diarios para la determinación de variabilidad climática y cambio climático en el Distrito Metropolitano de Quito. *La Granja, 16* (2), 23. https://doi.org/10.17163/lgr.n16.2012.03. [In Spanish].
Suárez-Moreno, J. D., Osorio-Ramírez, C., & Adarme-Jaimes, W. (2017). Agent-based model for material convergence in humanitarian logistics. *Revista Facultad de Ingeniería Universidad de Antioquia, 81,* 24–34. https://doi.org/10.17533/udea.redin.n81a03.
Tofighi, S., Torabi, S. A., & Mansouri, S. A. (2016). Humanitarian logistics network design under mixed uncertainty. *European Journal of Operational Research, 250*(1), 239–250. https://doi.org/10.1016/j.ejor.2015.08.059.
Unidas, N. (2017). Junta Ejecutiva del Programa de las Naciones Unidas para el Desarrollo y del Fondo de Población de las Naciones Unidas. Policy, 1–28. [In Spanish].
van der Laan, E., van Dalen, J., Rohrmoser, M., & Simpson, R. (2016). Demand forecasting and order planning for humanitarian logistics: An empirical assessment. *Journal of Operations Management, 45,* 114–122. https://doi.org/10.1016/j.jom.2016.05.004.
Viera, O., Moscatelli, S., & Tansini, L. (2012). Logística humanitaria y su aplicación en Uruguay. *Gerencia Tecnológica Informática, 11,* 47–56. [In Spanish].

Ye, Y., & Liu, N. (2013). Humanitarian logistics planning for natural disaster response with Bayesian information updates. *Journal of Industrial and Management Optimization, 10*(3), 665–689. https://doi.org/10.3934/jimo.2014.10.665.

Zary, B., Bandeira, R., & Campos, V. (2014). The contribution of scientific productions at the beginning of the third millennium (2001–2014) for humanitarian logistics: A bibliometric analysis. *Transportation Research Procedia, 3*(July), 537–546. https://doi.org/10.1016/j.trpro.2014.10.002.

Chapter 2
An Analysis of Inventory Levels on Humanitarian Relief for Vulnerable Municipalities of Puebla, Mexico

Paola Tapia Muñoz, Diana Sánchez-Partida, Santiago-Omar Caballero-Morales, and Patricia Cano-Olivos

Abstract Recently, the concept of humanitarian relief was introduced, and there remains work to do to help people in any situation of disaster. The present research is focused on determining an inventory level for relief kits that can benefit 24 municipalities that belong to the State of Puebla, Mexico. Historically, these municipalities have been vulnerable to hydrometeorological phenomena. From 2001 to 2017, Puebla has had 1,632 emergency declarations, of which 59.7% were classified as a hydrometeorological issue. According to this historical, the disasters increased during August, September, and October, so it is proposed with this research to have an adequate inventory level of kits before the disasters happen in accordance to the months above. In the official records, there are not registered the number of affected people of these municipalities, so to determine demand, the frequency was found using the historical data regarding affected people by hydrometeorological phenomena at a national level. The lot size was calculated using the Newsboy Inventory Model, and the demand was separated in different age range and gender to make kits following the necessities.

Keywords Humanitarian relief · Newsboy inventory model · Disaster

2.1 Literature Review

The emergencies caused by natural disasters are unpredictable, an adequate and timely delivery help in goods or services, and a correct administration in the relief operations are required to minimize the human suffering and inclusive the loss of

Electronic supplementary material The online version of this chapter (https://doi.org/10.1007/978-3-030-67295-9_2) contains supplementary material, which is available to authorized users.

P. T. Muñoz · D. Sánchez-Partida (✉) · S.-O. Caballero-Morales · P. Cano-Olivos
Department of Logistics and Supply Chain Management, Universidad Popular Autónoma Del Estado de Puebla, (UPAEP University), 17 Sur 901, Barrio de Santiago, 72410 Puebla, Mexico
e-mail: diana.sanchez@upaep.mx

D. Sánchez-Partida et al. (eds.), *Disaster Risk Reduction in Mexico*,
https://doi.org/10.1007/978-3-030-67295-9_2

lives. For any disaster type, some logistics decisions have to be made and structured according to each community's needs (Ilhan, 2011).

During an emergency, the humanitarian relief consists of donor a large number of medicines, clothes, foods, and non-necessary articles or not priority articles. The donations received after a disaster, but if the received items are not necessary, can generate serious logistics problems (Pan American Health Organization, 1999). So why not foresee what is necessary before the disaster? The number of victims and damages caused primarily by natural phenomena as torrential rains, floodwater, high winds, and droughts has increased because these kinds of disasters are more frequent and appear with high intensity. According to the United Nations and Humanitarian Relief (UNHR), 70% of the natural disasters are related to climate change; and this is double than 20 years ago (ONU, 2013).

The UNISDR (United Nations Office for Disaster Risk Reduction) informs that the available information about disasters sometimes is limited because it is not taking into account the little disasters, although they caused losses and occurred very frequently (UNISDR, 2013). This omission will cause that people cannot imagine the real dimension of the problem, and sometimes the problem is passed over, and it could cause the most significant problem (UNISDR, 2013).

The experiences regarding disasters for other nations and the help of international organizations represent a valuable source of knowledge (Moore et al., 2009). This knowledge contributes to improving humanitarian relief. Useful and appropriate coordination is an essential element for successful disaster relief (Moore et al., 2009).

According to UNICEF, children are very vulnerable to crises, violence, and instability situations. From birth and until the age of six (especially the first two years) are fundamental for the intelligence, personality, and social behavior development. The arm conflicts and natural disasters have serious consequences, and the highest risk that children have in these cases is that they have to grow up in an environment that limits their creativity and their physical and mental development (UNICEF, 2009).

It is demonstrated that natural disasters and climate change are forcing children to escape from home; this puts them at considerable risk of disease, violence, and child exploitation (UNICEF, 2017). In any emergency, children required protection that could guarantee their wellness and security. The hydrometeorological phenomena like hurricanes, typhoons, and tropical storms could cause damage in the developing nations, but in places with poverty, the damage for children could be more significant (UNICEF, 2017).

Women are vulnerable too to natural disasters, climate change, and industrial hazards, because of their position in society and their susceptible moments of the life cycle. For example, women are discriminated against and excluded by culture and social norms (PNUD, 2014).

According to Pan American Health Organization, the most vulnerable groups are formed by children, pregnant women, breastfeeding mothers, and the elderly; the survival for that group is threatened by a slow recovery of services (Pan American Health Organization, 1999).

Nowadays, there is a need to deal with all the disaster′s phases: prevention, response, and recovery (Canyon, 2017). In this research, we deal with a prevention proposal because prevention is focused on identifying disaster risks and developing strategies to improve resilience. This term is the ability to adapt and react to changes in a system. When the cities are vulnerable and exposed to disturbing events caused by climate change, it becomes imperative (Da Silva Stefano et al., 2017).

According to the "Manual to Develop More Resilient Cities," published by UNISDR, a resilient city can minimize the disaster because the people and local authorities know and share information about disasters, threats, and risks. Therefore, people can anticipate disasters and mitigate their impact (UNISDR, 2012).

One of the disciplines that intervene in the case of attention to disaster is Humanitarian Logistics (HL). Its discipline is very similar to corporate logistics. The HL refers to "process and involved systems to mobilize people, resources, abilities, and knowledge to help vulnerable people that were affected during an impact of a natural disaster and complex emergencies" (Thomas, 2005). Nowadays, there is intense pressure to develop better coordination to respond to disasters; there is still some work to do, for example, standardize logistics practices regarding humanitarian relief (Canyon, 2017).

Therefore, the present research has the goal of helping 24 municipalities in the State of Puebla that is vulnerable to hydrometeorological phenomena determining an optimal inventory level for relief kits early instead to wait for goods donations because this activity can cause severe distribution problems.

2.2 Problem Description

The State of Puebla is located in the center of the Mexican Republic. Puebla has 217 municipalities, and according to the National Institute of Statistics and Geography (INEGI), it represents 1.7% of the total Mexico territory (INEGI, 2016). The National Council of Evaluation of Social Development Policy (CONEVAL) situates the State with a high percentage of the population living in poverty, between 50 and 65% (CONEVAL, 2017).

According to information provided by the Official Journal of the Federation (DOF), the State of Puebla has had 1,632 emergency declarations from 2001 to 2017. As can be seen in Table 2.1, these 1,632 declarations, 567 were declarations of contingency, 527 declarations of disaster, and 538 were declaratory of emergency. A contingency declaration is when could be some latent risk; a disaster declaration is referred to the damages caused to houses and public infrastructure, and an emergency declaration is addressed to the attention of the life and health of the population. The information divided by phenomena type indicates that the majority of declarations are due to hydrometeorological events such as tropical cyclones, floods, rains, snowfall, frost, hailstorm. Puebla is affected by 974 events of this kind. In Table 2.1 are shown the event and the emergency type during the period 2001–2017.

Table 2.1 Declarations divided by emergency and phenomenon type from 2001 to 2017

Event/emergency type	Low temperatures	Tropical cyclone	Floods	Rains	Snowfall, frost or hailstorm	Drought	Earthquake	Total
Contingency				145	123	299		567
Disaster		192	1	135	17	70	112	527
Emergency	65	189	1	116	55		112	538
Total	65	381	2	396	195	369	224	1,632

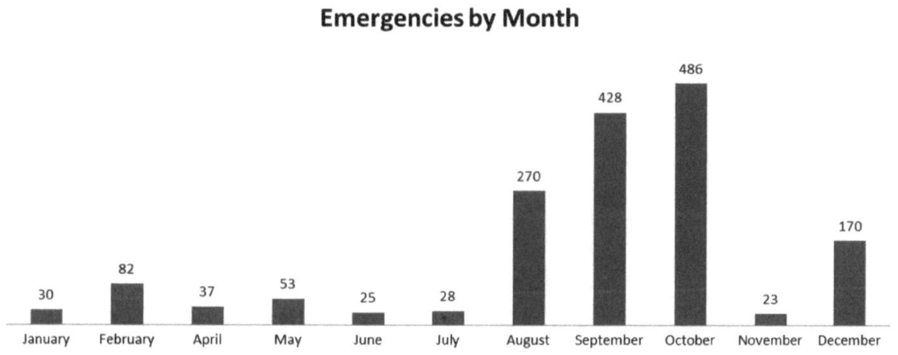

Fig. 2.1 Emergency declarations by month

Once the hydrometeorological event was identified as the most impacting event for Puebla, the emergency declarations were analyzed by month. The data during the period 2001–2017 divided by month showed a significant increase during August (24.13%), September (12.42%), and October (40.76%). This data is shown in Fig. 2.1.

The information of 2013–2017 offered by the National Water Commission (CONAGUA) shows that the months where the atmospheric precipitation of water increases in the State of Puebla are June, July, August, and September. This information agrees with the information obtained from the Emergency declarations because the precipitations start in June, and the disasters begin to reflect in August until October Fig. 2.2 refers to the data regarding rainfall for the State of Puebla for the period (CONAGUA, 2018).

Once corroborated that the hydrometeorological phenomena affect the State of Puebla during summer and part of autumn, and taking into account the declarations during those months, 24 municipalities were determined as the most affected with the most significant number of declaratory. This information is shown in Table 2.2.

Figure 2.3 refers to the location of the 24 municipalities. The most affected municipalities are located in the north of the state of Puebla, according to

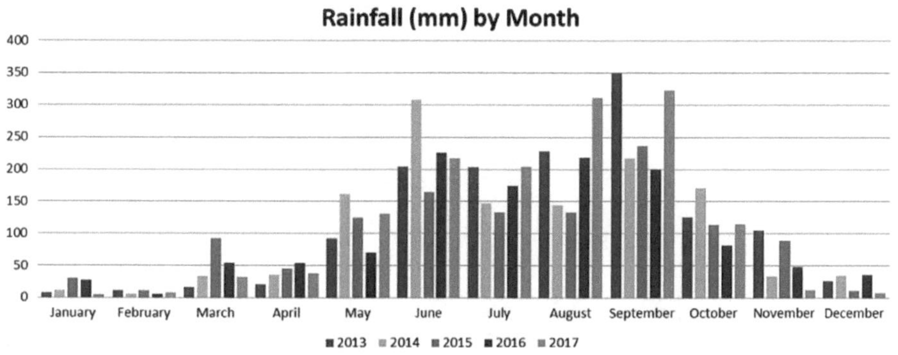

Fig. 2.2 Atmospheric precipitation of water (CONAGUA)

Table 2.2 Emergency declaration by municipality from 2001 to 2017

Municipality	Tropical cyclone	Rains	Snowfall, frost or hailstorm	Total
Tlatlauquitepec	7	6	3	16
Zacapoaxtla	7	6	1	14
Huauchinango	5	7	2	14
Xicotepec	5	7	1	13
Atempan	6	6	1	13
Zautla	3	6	3	12
Chignautla	4	7	1	12
Tetela de Ocampo	5	6	1	12
Tlaola	3	8	1	12
Tlapacoya	4	7	1	12
Chiconcuautla	4	7	1	12
Zaragoza	3	6	2	11
Chalchicomula de Sesma	3	3	5	11
Cuetzalan del Progreso	5	4	2	11
Hueyapan	4	6	1	11
Jopala	4	6	1	11
Tenampulco	6	4		10
Ayotoxco de Guerrero	7	3		10
Esperanza	2	3	5	10
Hueytamalco	7	3		10
Acateno	5	3	1	9
Huehuetla	3	5	1	9
Guadalupe Victoria	1	4	3	8
Tecamachalco	2	1	2	5

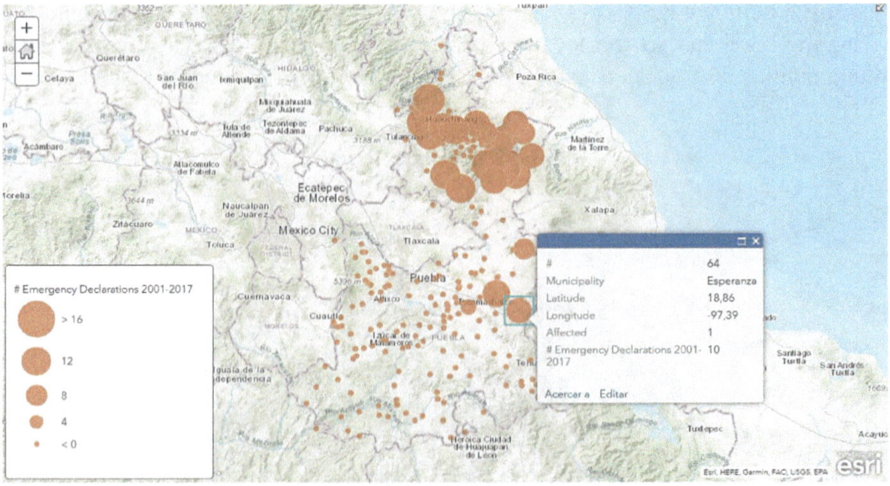

Fig. 2.3 Location of the 24 municipalities

CONEVAL (2015). Some of the municipalities belong to the 80% of the population that are in extreme poverty.

The National Risk Atlas, managed by Mexico´s National Center for Prevention of Disasters (CENAPRED), measures resilience with a 5 points scale, where 1 represents the lowest grade, and 5 represents the higher grade of resilience. A municipality is resilient when they are coordinated to understand and reduce the risk for a disaster, the population is continuously trained about risk reduction, and they have early warning systems. The opposite meaning of this measure represents the social vulnerability. The vulnerability is defined as "Lack of access to prevention and resilience." The 24 affected municipalities have a resilience average of 3, they are in the middle of the scale, so they still have some work to do to anticipate the disasters in a better way, and 37% of the municipalities have a high grade of social vulnerability, they do not have enough prevention culture (CENAPRED, 2015). In Fig. 2.4, the resilience average is shown, and social vulnerability is shown in Fig. 2.5.

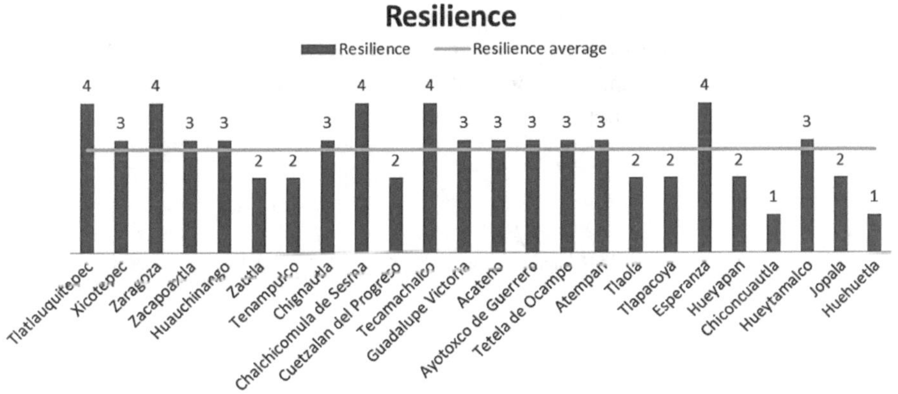

Fig. 2.4 Resilience grade for 24 municipalities of State of Puebla

Fig. 2.5 Social Vulnerability for 24 municipalities of the State of Puebla

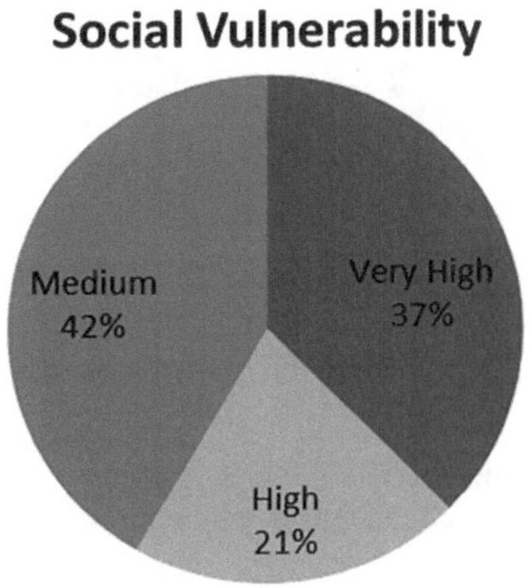

Hence, after analyzing the information obtained, the principal municipalities to help are 24 with high social vulnerability and low resilience grade.

2.3 Methods and Procedures

Even though the emergency declarations derivate by hydrometeorological phenomena, presents seasonality in the municipalities studied, the number of the affected population is uncertain. Thus the emergency magnitude cannot be predicted.

The necessary data to determine the inventory of the kits that need to be ready to help the affected population is the demand estimation. The demand is entirely uncertain, and the purchase of the kits will be only once per year. Because of these characteristics, the lot size will be calculated using a stochastic model called the Newsboy Model. This model can be used to calculate a single purchase or purchase with seasonality (Hillier & Lieberman, 2001).

It is called the Newsboy Model; also unique inventory model considers the risk that a person is willing to take to run out of inventory (Chase et al., 2009). The model has two assumptions: first, immediately, the demand occurs when the order is received, and second, there is not a preparation cost (Taha, 2012).

Equation (2.1) for the Newsboy Model, according to (Axsäter, 2015) is as follow:

$$Q = D_{\min} + (D_{\max} - D_{\min})\left(\frac{p - c}{p}\right) \tag{2.1}$$

where:

Q Unit quantity to buy
D_{min} Minimum registered demand.
D_{max} Maximum registered demand.
p Sale value for inventory unit
c Purchase cost for inventory unit

Equation (2.1) calculates the lot size to buy when the minimum and maximum demand is known.

2.4 Experimental Setting

2.4.1 Estimation of the Demand

There are no records of affected people in the State of Puebla, so it is necessary to estimate the maximum and minimum demand; thus, the data regarding the affected population that were hit by hydrometeorological phenomena in all the Mexican Republic was analyzed. This information is reported by Mexico´s Fund for Natural Disasters (FONDEN, 2018). The data regarding the population census reported in 2010 by the National of Statistics and Geography Institute (INEGI) was analyzed as well. The purpose of analyzing the complete Mexican Republic data is to approximate the minimum and the maximum demand for each of the 24 municipalities of the State of Puebla. In Appendix 2A, it is possible to see the estimated demand by age range.

First, the data for the emergency declaratory in all the Mexican Republic was obtained from 2014 to 2016, with a total of 783 declaratory. Then, the proportional relationship was calculated, taking into account the total of the affected population against 100% of the population. For each of the 514 municipalities involved made this calculation, and we obtained the percentage of affectation.

Once obtained the proportional relationship, it was divided into ranges. The frequency was calculated to verify that the limits were the significant numbers of the observations. The range where the probability to help the 84% of the population was between 0.31 and 0.4; thus, 30 and 40% were used to approximate the minimum and maximum demand of the 24 municipalities of Puebla. The Frequency and accumulated frequency is shown in Fig. 2.6

The probability to assist the 84% of the population is between 30% and 40%. Thus, the minimum percentage will be 30% to estimate the D_{min} and 40% to estimate the $D_{ma2.}$

According to the last population census executed in 2010, the involved 24 municipalities have a population of 720,764 people (INEGI, 2010). Figure 2.7 shows the population divided by age according to INEGI data.

With this probability of cover the 84% of demand, the calculations were made to obtain the optimum lot size of relief kits separating by gender and age.

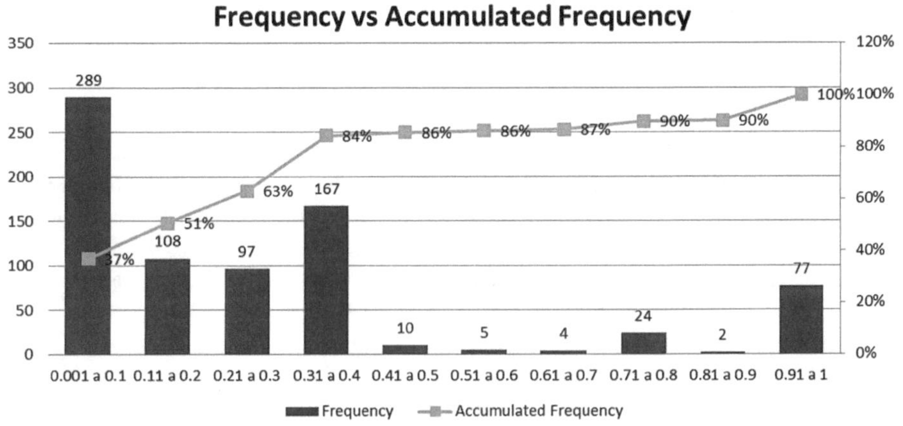

Fig. 2.6 Frequency vs. Accumulated Frequency

Fig. 2.7 Distribution of population by age (INEGI, 2010)

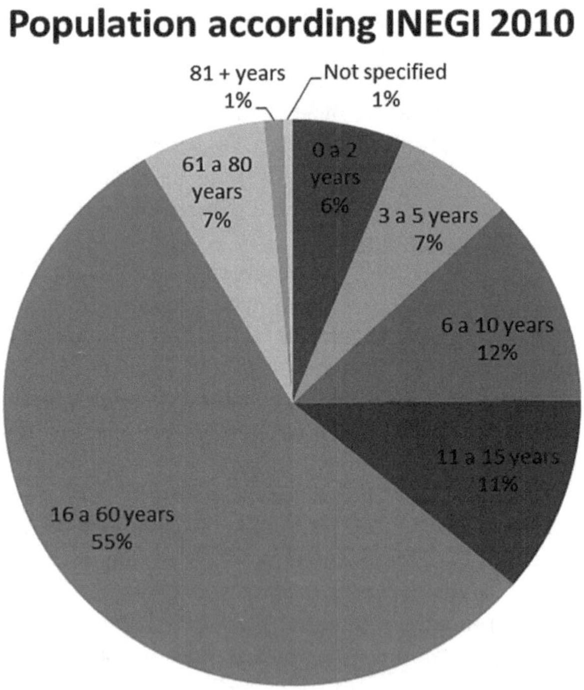

In this research, the kits were classified by age range to define an adequate relief kit for gender and age:

0–2 years
3–5 years
6–10 years
11–15 years Men
11–15 years Women
16–60 years Men
16–60 years Women
61–80 years
and more than 81 years

Table 2.3 contains the proposed articles to be included in each different relief kit, taking into account the different necessities for children, women, and older adults. The choice of articles was based on recommendations from UNICEF and ONU, where they emphasize that it is essential to make the people feel comfortable to help them to forget the tragic situation (UNICEF, 2017)

Then the cost was calculated to obtain the variable "*c*" for the Newsboy Model. The costs are as follows in Table 2.4.

How it is difficult to obtain a sale value in this research was considered the economic losses caused by a negative impact of hydrometeorological phenomena

Table 2.3 Kits descriptions

Kit's name	Kit A	Kit B	Kit C	Kit D	Kit E	Kit F	Kit G	Kit H	Kit I
Items/age range	0–2 years	3–5 years	6–10 years	11–15 years—Men	11–15 years—Women	16–60 years—Men	16–60 years—Women	61–80 years	81 and more
Shampoo	X	X	X	X	X	X	X	X	X
Baby diapers	X								
Baby wet towel	X								
Antibacterial gel	X	X	X	X	X	X	X	X	X
Toilet soap		X	X	X	X	X	X	X	X
Toothpaste		X	X	X	X	X	X	X	X
Infantile toothbrush		X							
Toothbrush			X	X	X	X		X	X
Nail clipper				X	X	X	X	X	X
Hair comb				X	X	X	X	X	X
Sanitary towels					X		X		
Deodorant				X	X	X	X	X	X
Toilet paper		X	X	X	X	X	X	X	X
Coloring book		X							
Activity book			X						
Crayons		X	X						
Adult diapers									X
	1	4		3	8				
	2	1		2	5				

Table 2.4 Kit costs separated by age range and gender

Kit´s name	Age range	Cost
Kit A	0–2 years	$ 231
Kit B	3–5 years	$ 176
Kit C	6–10 years	$ 176
Kit D	11–15 years—Men	$ 158
Kit E	11–15 years—Women	$ 178
Kit F	16–60 years—Men	$ 158
Kit G	16–60 years—Women	$ 178
Kit H	61–80 years	$ 158
Kit I	81 and more	$ 228

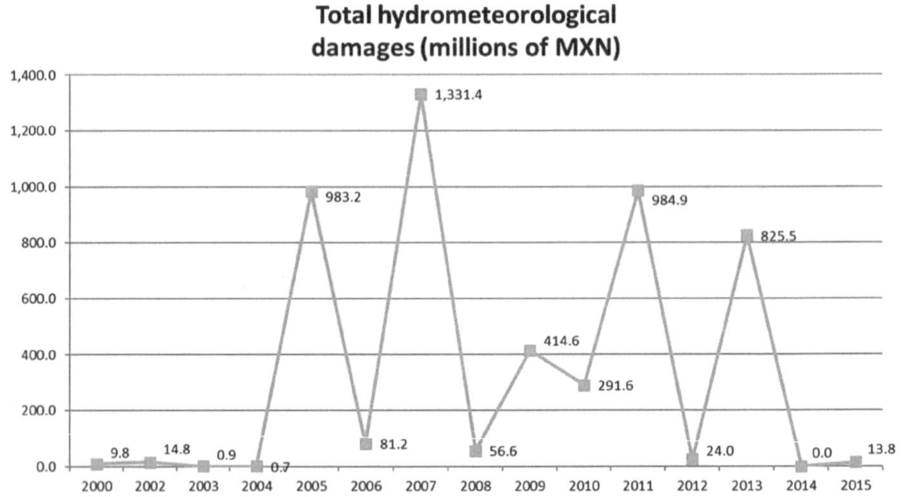

Fig. 2.8 Total damages caused by hydrometeorological emergencies

occurred. Therefore, the total amount of damage reported by the CENAPRED affected the State of Puebla during 15 years from (2000 to 2015) was calculated to obtain the variable *p*. The total damage reported sums 5,033 million MXN, the annual average was 335.5 million MXN, and the total affected population was 254,959. Calculating the relationship between the average losses and the total affected people, we obtained 1,316 MXN per person in Fig. 2.8 is shown the total damages caused by hydrometeorological emergencies.

Table 2.5 Total inventory kits (Q*) by age range and by the municipality

	0–2 years	3–5 years	6–10 years	11–15 years—Men	11–15 years—Women	16–60 years—Men	16–60 years—Women	61–80 years	81 and more	Total
Total	17,420	18,637	33,026	15,882	15,553	72,746	81,879	19,395	3,064	277,602
Tlatlauquitepec	1,211	1,304	2,223	1,112	1,058	5,258	5,960	1,530	246	19,902
Huauchinango	2,315	2,462	4,280	1,978	1,953	9,900	11,594	2,354	339	37,175
Xicotepec	1,855	1,922	3,331	1,692	1,626	7,697	8,752	2,036	307	29,218
Guadalupe Victoria	407	450	824	394	389	1,604	1,774	462	98	6,402
Zautla	498	537	906	442	411	1,875	2,129	574	109	7,481
Chalchicomula de Sesma	950	1,037	1,838	891	906	4,603	5,139	1,332	243	16,939
Tlaola	513	548	1,063	520	496	1,911	2,082	468	66	7,667
Zaragoza	348	353	630	314	317	1,584	1,946	405	74	5,971
Cuetzalan del Progreso	1,113	1,240	2,229	1,070	1,051	4,772	5,214	1,476	181	18,346
Acateno	176	203	369	189	168	945	972	373	58	3,453
Tetela de Ocampo	598	599	1,064	550	567	2,633	2,917	875	175	9,978
Esperanza	321	344	613	279	306	1,416	1,574	392	89	5,334
Chiconcuautla	482	499	975	433	397	1,361	1,520	369	50	6,086
Zacapoaxtla	1,261	1,395	2,466	1,182	1,173	5,328	6,108	1,426	222	20,561
Tenampulco	117	134	253	137	120	705	772	332	49	2,619
Chignautla	825	931	1,545	688	675	3,045	3,333	556	90	11,688
Tecamachalco	1,738	1,852	3,334	1,499	1,510	7,378	8,361	1,386	257	27,315

(continued)

Table 2.5 (continued)

	0–2 years	3–5 years	6–10 years	11–15 years—Men	11–15 years—Women	16–60 years—Men	16–60 years—Women	61–80 years	81 and more	Total
Ayotoxco de Guerrero	181	196	334	195	181	867	934	230	37	3,155
Tlapacoya	156	152	310	176	182	629	679	176	22	2,482
Jopala	307	326	607	338	326	1,248	1,398	432	48	5,030
Huehuetla	372	406	739	354	337	1,500	1,678	617	64	6,067
Hueyapan	334	346	562	279	267	1,203	1,290	276	36	4,593
Hueytamalco	623	635	1,178	580	549	2,793	3,019	824	122	10,323
Atempan	719	766	1,353	590	588	2,491	2,734	494	82	9,817

Table 2.6 Total inventory cost in MXN by age range and by the municipality

	0–2 years	3–5 years	6–10 years	11–15 years —Men	11–15 years —Women	16–60 years —Men	16–60 years —Women	61–80 years	81 and more	Total
Total	4,024,020	3,280,112	5,812,576	2,509,356	2,768,434	11,493,868	14,574,462	3,064,410	698,592	48,225,830
Tlatlauquitepec	279,741	229,504	391,248	175,696	188,324	830,764	1,060,880	241,740	56,088	3,453,985
Huauchinango	534,765	433,312	753,280	312,524	347,634	1,564,200	2,063,732	371,932	77,292	6,458,671
Xicotepec	428,505	338,272	586,256	267,336	289,428	1,216,126	1,557,856	321,688	69,996	5,075,463
Guadalupe Victoria	94,017	79,200	145,024	62,252	69,242	253,432	315,772	72,996	22,344	1,114,279
Zautla	115,038	94,512	159,456	69,836	73,158	296,250	378,962	90,692	24,852	1,302,756
Chalchicomula de Sesma	219,450	182,512	323,488	140,778	161,268	727,274	914,742	210,456	55,404	2,935,372
Tlaola	118,503	96,448	187,088	82,160	88,288	301,938	370,596	73,944	15,048	1,334,013
Zaragoza	80,388	62,128	110,880	49,612	56,426	250,272	346,388	63,990	16,872	1,036,956
Cuetzalan del Progreso	257,103	218,240	392,304	169,060	187,078	753,976	928,092	233,208	41,268	3,180,329
Acateno	40,656	35,728	64,944	29,862	29,904	149,310	173,016	58,934	13,224	595,578
Tetela de Ocampo	138,138	105,424	187,264	86,900	100,926	416,014	519,226	138,250	39,900	1,732,042
Esperanza	74,151	60,544	107,888	44,082	54,468	223,728	280,172	61,936	20,292	927,261
Chiconcuautla	111,342	87,824	171,600	68,414	70,666	215,038	270,560	58,302	11,400	1,065,146
Zacapoaxtla	291,291	245,520	434,016	186,756	208,794	841,824	1,087,224	225,308	50,616	3,571,349
Tenampulco	27,027	23,584	44,528	21,646	21,360	111,390	137,416	52,456	11,172	450,579
Chignautla	190,575	163,856	271,920	108,704	120,150	481,110	593,274	87,848	20,520	2,037,957
Tecamachalco	401,478	325,952	586,784	236,842	268,780	1,165,724	1,488,258	218,988	58,596	4,751,402
Ayotoxco de Guerrero	41,811	34,496	58,784	30,810	32,218	136,986	166,252	36,340	8,436	546,133

(continued)

Table 2.6 (continued)

	0–2 years	3–5 years	6–10 years	11–15 years—Men	11–15 years—Women	16–60 years—Men	16–60 years—Women	61–80 years	81 and more	Total
Tlapacoya	36,036	26,752	54,560	27,808	32,396	99,382	120,862	27,808	5,016	430,620
Jopala	70,917	57,376	106,832	53,404	58,028	197,184	248,844	68,256	10,944	871,785
Huehuetla	85,932	71,456	130,064	55,932	59,986	237,000	298,684	97,486	14,592	1,051,132
Hueyapan	77,154	60,896	98,912	44,082	47,526	190,074	229,620	43,608	8,208	800,080
Hueytamalco	143,913	111,760	207,328	91,640	97,722	441,294	537,382	130,192	27,816	1,789,047
Atempan	166,089	134,816	238,128	93,220	104,664	393,578	486,652	78,052	18,696	1,713,895

2.5 Results and Discussion

The model was solved once the necessary data was obtained. The results were a total of relief kits of 277,602 pieces that represent 48,225,830.00 MXN. It is suggested that this amount of relief kits can be held in August, September, and October, and being prepared before the disasters happen. The order quantity calculated and the inventory cost for each kit is shown in Tables 2.5 and 2.6, also this information can be seen in Appendix 2B.

As an example, Tlatlauquitepec, the first municipality listed below, needs a total of 19,902 kits in order to attend all range of ages, which will have a cost of 3,453,985 MXN.

2.6 Conclusions and Future Research

The paper is based on having enough inventory levels of relief kits to help the most significant possible population in the rainy season. The statistical data shows that the municipalities are affected during the summer.

It is important to emphasize that the Northern zone of the State of Puebla is in poverty and the social vulnerability is high, this is why the study considered the fact that people receive help from the government of Mexico. Remember that children, women, and the elderly are the most vulnerable in cases of disasters, so the study suggests focussing on a special relief kit for them. For example, some of the items include in the children′s kit are colors and a book; these items will help them to forget the critical situation.

As future research, using location-allocation models can be identified as the best place to suggest a distribution center with sufficient capacity to hold the inventory determined in this study and thus bring the aid closer to the endpoint. Also, this distribution can safeguard the relief kits and afterward to distribute them to the affected people. Additionally, an optimal route can be analyzed to take away this aid to the municipalities above. This route should minimize costs and maximize the delivery of relief kits through an appropriate routing model.

References

Axsäter, S. (2015). *Inventory control* (3rd ed., Vol. 225). Springer. https://doi.org/10.1007/978-3-319-15729-0.

Canyon, D. V. (2017). Military provision of humanitarian assistance and disaster relief in non-confllict crises. *Journal of Homeland Security and Emergency Management*, 1–5. https://doi.org/10.1515/jhsem-2017-0045.

CENAPRED. (2015). Mexico′s National Center for prevention of disasters. Retrieved 2018, May 31st from https://www.gob.mx/cenapred [In Spanish].

Chase, R. B., Jacobs, F. R., & Aquilano, N. J. (2009). *Administración de Operaciones, Producción y Cadena de Suministro* (Twelfth ed.) McGRAW-HILL/INTERAMERICANA EDITORES, S. A. DE C.V. [In Spanish].
CONAGUA. (2018). CONAGUA. Retrieved 2018, March 1st from http://smn.cna.gob.mx/es/climatologia/temperaturas-y-lluvias/resumenes-mensuales-de-temperaturas-y-lluvias [In Spanish].
CONEVAL. (2015). Puebla. Retrieved March 1st, 2018 from https://www.coneval.org.mx/coordinacion/entidades/Puebla/PublishingImages/municipal_2015/Pue15_ind_1.jpg.
CONEVAL. (2017). Contenido. Retrieved March 1st from https://www.coneval.org.mx/Medicion/MP/Paginas/Pobreza_2016.aspx [In Spanish].
Da Silva Stefano, G., Pacheco Lacerda, D., Veit, D. R., & Pantaleao, L. H. (2017). Identifying constraints to increase the resilience of cities : A case study of the city of Porto Alegre. *Journal of Homeland Security and Emergency Management*, 1–24. https://doi.org/10.1515/jhsem-2016-0057.
Hillier, F. U., & Lieberman, G. J. (Late of S. U. (2001)). Introduction to operations research (7th ed., Vol. 20). McGraw-Hill. https://doi.org/10.1016/0305-0548(93)90042-H.
Ilhan, A. (2011). The humanitarian relief chain. *South East European Journal of Economics and Business, 6*(2), 45–54. https://doi.org/10.2478/v10033-011-0015-x.
INEGI. (2010). Census of Population and Housing Units. Retrieved May 20th from https://en.www.inegi.org.mx/programas/ccpv/2010/ [In Spanish].
INEGI. (2016). Anuario estadístico y geográfico de Puebla 2017. Retrieved May 20th from http://internet.contenidos.inegi.org.mx/contenidos/Productos/prod_serv/contenidos/espanol/bvinegi/productos/nueva_estruc/anuarios_2015/702825077174.pdf [In Spanish].
Moore, M., Trujillo, H. R., Stearns, B. K., Basurto-davila, R., & Evans, D. K. (2009). Learning from exemplary practices in international disaster management : A fresh avenue to inform U. S. policy ? Learning from exemplary practices in international disaster management : A fresh avenue to inform U. S. Policy ? *Journal of Homeland Security and Emergency Management, 6*(1), 40. https://doi.org/10.2202/1547-7355.1515.
Natural Disasters Fund. (FONDEN). Retrieved 2018, June 31st, from https://www.gob.mx/segob/documentos/fideicomiso-fondo-de-desastres-naturales-fonden [In Spanish].
ONU. (2013). United Nations and humanitarian relief. Retrieved May 25th from http://www.un.org/es/humanitarian/overview/disaster.shtml.
Pan American Health Organization. (1999). Asistencia humanitaria en caso de desastres: guía para proveer ayuda eficaz. Retrieved from http://books.google.com/books?hl=en&lr=&id=j6GY6cVoyjcC&oi=fnd&pg=PP3&dq=Asistencia+humanitaria+en+caso+de+desastres+Guía+para+proveer+ayuda+eficaz&ots=tRlBjSK8y4&sig=rrObmNbHPelZrqIkz1cHthob_ao [In Spanish].
PNUD, P. de N. U. para el D. (2014). *Informe sobre Desarrollo Humano 2014 Equipo del Informe sobre Desarrollo Director y autor principal*, 19 [In Spanish].
Taha, H. A. (2012). Investigación de Operaciones (Ninth Edit). PEARSON EDUCACIÓN.
Thomas, A. (2005). *Humanitarian logistics : Enabling disaster response* (pp. 1–17). Fritz Institute. Retrieved from http://www.fritzinstitute.org/pdfs/whitepaper/enablingdisasterresponse.pdf.
UNICEF. (2009). UNICEF Primera Infancia. Retrieved May 20th from https://www.unicef.org/spanish/earlychildhood/index_40745.html [In Spanish].
UNICEF. (2017). UNICEF México. Retrieved May 20th from https://www.unicef.org/mexico/spanish/noticias_36267.html%0A [In Spanish].
UNISDR. (2012). *Cómo desarrollar ciudades más resilientes Un Manual para líderes de los gobiernos locales*, 103 [In Spanish].
UNISDR. (2013). *Impacto de los Desastres en América Latina y el Caribe, 1990–2013* [In Spanish].

Chapter 3
Location of a Regional Humanitarian Response Depot (RHRD) in Puebla, Mexico Using an Analytical Hierarchical Process

Diana Sánchez-Partida, Brenda López-Durán, José-Luis Martínez-Flores, and Santiago-Omar Caballero-Morales

Abstract In recent years, the number of natural disasters has been increasing due to climate change. It is why the study of these events has become necessary, and finding practical solutions and responding to them has also been crucial since human lives are at risk. Humanitarian support has been of great importance for their care and rescue, which is why a quick response to this task is paramount. The present case study focuses on Puebla, Mexico. Events of hydrometeorological and seismological types have hit this entity. Two types of studies have been made in this case, one qualitatively and the other using an Analytical Hierarchical Process (AHP) in order to obtain the best municipalities to install a Regional Humanitarian Response Depot (RHRD) that protect items of first necessity, which will be delivered to the victims in case of emergency. The state is formed by 217 municipalities in which 203 have been impacted by some natural disaster, according to the data between the years 2006 and 2017. In the first scenario, the information above was considered and the results showed that the safest sites are found in the central and southeastern areas of the state, with a total of 14 municipalities. In the second scenario, an AHP was used but considering other aspects like infrastructure, roads, and so forth, the safety sites were found in the northwest, center-west, center-east, southwest, southeastern, and central areas of the state of Puebla.

Keywords Natural disaster · Mexico · Regional Humanitarian Response Depot (RHRD) · Location problem · Analytical Hierarchical Process (AHP)

Electronic supplementary material The online version of this chapter (https://doi.org/10.1007/978-3-030-67295-9_3) contains supplementary material, which is available to authorized users.

D. Sánchez-Partida (✉) · B. López-Durán · J.-L. Martínez-Flores · S.-O. Caballero-Morales
Department of Logistics and Supply Chain Management, Universidad Popular Autónoma Del Estado de Puebla (UPAEP University), 17 Sur 901, Barrio de Santiago, CP 72410 Puebla, Mexico
e-mail: diana.sanchez@upaep.mx

D. Sánchez-Partida et al. (eds.), *Disaster Risk Reduction in Mexico*,
https://doi.org/10.1007/978-3-030-67295-9_3

3.1 Introduction

Global natural disasters have caused economic and infrastructure damage of millions of dollars, unexpected disruption of socio-economic activities, and the tragic loss of human lives each year. Zhou et al. (2009) mentioned that it is of paramount importance to collect, maintain, and manage detailed and accurate records of disasters for adequate risk assessment and mitigation operations. Recent devastating events, such as the Indian Ocean tsunami in 2004, the US hurricanes in 2005, the earthquake in Haiti in 2010, the earthquake and tsunami in Japan in 2011, they have made us acutely aware of the significant impacts that natural disasters have caused to humanity. A disaster is any occurrence that causes damage, destruction, ecological disruption, loss of human life, human suffering, or the deterioration of health and health services on a scale sufficient to warrant an extraordinary response from outside of the affected community or area. Van Wassenhove (2006) classified the disasters in natural and human-made, and in sudden-onset and slow-onset (see Table 3.1).

That is why, in recent years, one of the topics that have become relevant, in the face of such disasters, has been one aspect of logistics: humanitarian logistics. Logistics is the discipline that deals with the study of the problems related to supply, in the time, place, form, and quality required of material goods and services; it is a crucial aspect in emergency humanitarian action. Humanitarian logistics deals with the study of the problems related to the supply of articles and materials in humanitarian emergencies, either by natural catastrophes or by human-made disasters. Cozzolino (2012) identifies four primary stages in the cycle of emergency humanitarian aid management: mitigation, preparedness, response, and reconstruction. It is essential to know what are the phases of disaster management in order to identify in which stage is found and act before, during, and after the phenomenon (see Fig. 3.1).

These phases may vary; for example, in Heaslip and Barber (2016), only three phases are mentioned in humanitarian aid (preparation, immediate response, and reconstruction phases). Humanitarian actions in crises require a high level of response, reliability, and adaptation to the needs of the victims. Then, the performance of humanitarian logistics depends also on the nature of the site of the disaster, the amount and concentration of affected people, and the quality of the infrastructure, among others. Chandes and Paché (2009) recognized the great

Table 3.1 Types of disasters found in the world by Van Wassenhove (2006)

	Natural	Human-made
Sudden-onset	Earthquake	Terrorist attack
	Hurricane	Putsch
	Tornadoes	Chemical leak
Slow-onset	Famine	Political crisis
	Drought	Refugee crisis
	Poverty	

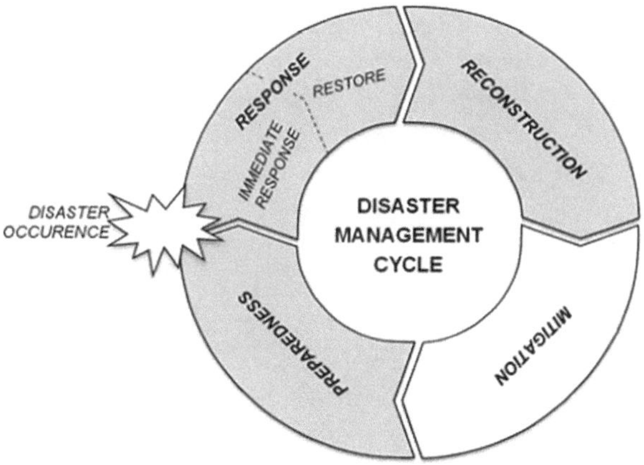

Fig. 3.1 Disaster management cycle

importance of improving global management of the humanitarian supply chain using more suitable locations for logistical resources in the event of a disaster. That is, instead of waiting passively for a crisis to occur to launch humanitarian operations, it is better to be prepared and mobilize material or non-material resources in advance. For this reason, the location of the resource storage sites should be taken into account, since a complex location would increase the risk of lengthening the reaction times and would harm the execution of humanitarian aid (Boonmee et al. 2017).

The strategy of having a prepositioned-inventory with the necessary products for the first critical hours after a disaster occurs has been studied. Because the inner necessity of people that are affected when a disaster of any nature occurs is survival, these people will need appropriate products such as water, medicine, food, hygiene articles, clothes, technological, and electronic devices, and so forth. Pre-positioned inventory has advantages such as an increase in the resilience and facilitation of high-speed and low-cost delivery (Barojas-Payán et al. 2019). In humanitarian operations, it takes relevance to the economic resources that governmental and non-governmental agencies, corporations, and the general public have. Some non-governmental agencies use the strategy of relief humanitarian depots to improve their abilities to help the affected people and reduce the cost and lead-time in their humanitarian operations (İlhan, 2011).

For example, the United Nations Humanitarian Response Depot (UNHRD) developed a network of depots around the world to store, manage, and distribute emergency supplies. These pre-positioned inventories help to respond faster to the needs of people affected in cases of disaster. The UNHRD has six Humanitarian Response Depots (HRD) in Europe, Africa, Middle East, Southeast Asia, and Latin America (see Fig. 3.2).

The HRD in Panama supports relief operations for agencies in Latin America and the Caribbean, the HRD in Spain (Las Palmas) is focused on the crisis in Africa and can provide support in less than 48 h. In Italy (Brindisi) it is the main departure point for humanitarian flights and convoys, the HRD in United Arab Emirates

Fig. 3.2 The six United Nations Humanitarian Response Depots (UNHRD)

(Dubai) can support a crisis anywhere in the world with a lead time of 24 h. Moreover, the Ghana depot (Accra) was built with the intention to support a significant part of Africa, and the HRD in Malaysia (Kuala Lumpur) is prepared to support emergencies in Asia (UNHRD, consulted April 2018).

Due to the geographical location of the state of Puebla and the natural phenomena to which it is exposed, the state has been impacted by some natural disasters. According to the government data between the years 2006 and 2017, these have meant economic losses and human lives. Thus, the present work has the objective of finding the most appropriate municipalities in the state of Puebla, in which a Regional Humanitarian Response Depot (RHRD) could be established and support crisis operations.

3.2 Literature Review

The essence of facility location problems is to determine the position of a set of facilities in a given location space to provide services to a set of entities. This implies that the availability of geographically referenced information represents the fundamental prerequisite to model and solve such problems. There has been significant research in which the problem of localization and the required solving methods have been addressed such as mathematical programming, Analytical Hierarchical Process (AHP), Geographic Information Systems (GIS), and others (Bruno and Giannikos, 2015).

Prabowo et al. (2017) developed location-allocation models for the distribution and evacuation of refugees. They mentioned that logistics distribution in the event of natural disasters, such as volcanic eruptions, earthquakes, and floods, is an important matter that needs the attention of both government and society. The logistic distribution in many disaster-prone countries is poorly planned and managed. Their research was focused on determining the characteristics of the disaster management system of a volcanic eruption in Mount Merapi, Indonesia. The mathematical model was built, and a set of solution procedures was proposed for the problem. The objective of the model was to minimize the total costs of the system and determine the optimal solutions at the three alert-levels (alert phase, danger phase, and post-disaster phase). The results were discussed, and models were analyzed to provide several managerial insights regarding the model's optimal solution. Based on their results, the proposed models were proven to give more effective solutions at lower costs, allowing a minimum waste of capacity as well as costs related to evacuation and distribution.

On the other hand, Geographic Information Systems (GIS) offer enormous possibilities for integrating, storing, editing, analyzing, sharing, and displaying spatial as well as non-spatial information. GIS can play a crucial role in supporting decision making in the field of location science. Since 1960 location science has evolved into a genuinely multidisciplinary area since it considers elements from mathematics, engineering, geography, and economics, among other disciplines (Bruno and Giannikos, 2015).

Bruno and Giannikos (2015) provided valuable information about the topic of GIS and its vital use in location problems. They discussed the relationships between location science and GIS, which are present in suitability analysis and data generation, visualization of results, formulation of new models, uncertainty and error propagation, and problem solution. Other researchers such as Rodríguez-Espíndola et al. (2016) developed a methodology for flood preparedness, including a GIS analysis to discard floodable facilities and identify road failures in the states of Veracruz and Tabasco in Mexico.

Also, the AHP technique has been used in a broad set of topics, ranging from industry to the environment. This technique is a fundamental approach to decision making. According to Saaty and Vargas (2012), AHP was designed to cope with both the rational and the intuitive aspects to select the best from some alternatives based on several criteria. In this process, the decision-maker carries out pure pairwise comparison judgments, which are then used to develop overall priorities for ranking the alternatives. These researchers mentioned that the purest form used to structure a decision problem is a hierarchy consisting of three levels: the goal of the decision at the top level, followed by a second level consisting of the criteria by which the alternatives, located in the third level, will be evaluated.

Aydi et al. (2016) presented a multi-criteria site selection tool based on the geographic information system of an oil mill wastewater (OMW) disposal site in the Sidi Bouzid region, Tunisia. The multi-criteria decision framework integrated ten constraints and six factors that were related to environmental and economic concerns and built a hierarchy model for the suitability of the OMW elimination site. The methodology was used for the preliminary evaluation of the most appropriate OMW elimination sites, combining the theory of fuzzy sets and the AHP. The fuzzy set theory was used to standardize the factors using different diffuse membership functions, while the AHP was used to establish the relative importance of the criteria. The AHP made comparisons by pairs of relative importance between hierarchies' elements grouped by environmental criteria.

Other researchers such as Daim et al. (2012) developed a decision model to select an external logistics provider (3PL) since, in international business, it can be a complex task. They defined a problem of multiple criteria, with alternative ways of weighing the importance of the chosen criteria. The document used the approaches of the hierarchical decision model. Among the results that were obtained given the criteria, weights, and attribute scores, it was found that two 3PLs were located very close in the upper position. It was also found that a simple classification of the criteria can produce very similar aggregate weights, provided that the number of experts is sufficiently high. The document contained an extensive review of the literature on the use of AHP and related methods in a logistics context.

González-Prida et al. (2014) worked with a dynamic view of the AHP, considering the criteria and alternatives as temporary variables. They obtained the right choice for a specific moment and a complete picture of the alternatives that were more important for business, according to the strategy and the passage of time. The value of their work was the description of AHP as a tool that can facilitate decision-making related to some of the critical aspects in the maintenance or after-sales area, allowing them to align the actions with the company's objectives.

Kavurmaci and Üstun (2016) investigated the spatial distribution of groundwater quality in the Sereflikochisar basin, in the central Anatolian region of Turkey, using different hydrochemical, statistical, and geostatistical methods. A total of 51 groundwater samples were collected from the observation wells in the study area to evaluate the characteristics of the groundwater quality. They made use of the AHP as a relatively simple and practical method to obtain a Groundwater Quality Index (GWQI).

Trivedi and Singh (2017) mentioned that disaster resilience is the intrinsic capacity of a hazard susceptible society to adapt and survive by changing, adapting, and rebuilding itself. This conceptualization has implications for disaster risk reduction and resilient community development efforts. They proposed a hybrid approach to preparing hazard-prone societies to become more resilient against disasters. They integrated fuzzy AHP and TOPSIS to address a case study for prioritizing emergency shelter areas. Six qualitative factors were identified and prioritized using judgments of experts and the AHP approach.

Further, a real case study of the 2015 Nepal earthquake was undertaken to evaluate a set of candidate shelter sites in Gorkha district of Nepal by adopting TOPSIS technique. Finally, the results were validated through assessment reports by the International Organization of Migration. This research offered many contributions. First, it contributed by exploring and establishing key factors to select a location for an emergency. It also successfully highlighted and applied theories of group decision support in the same context. Further, the use of hybrid multi-criteria decision approach for effective shelter planning was demonstrated through the case study.

At this point, many ways to solve the selection site or allocation problem have been reviewed, so is the versatility of the AHP technique, which is adequate in the selection of an alternative considering several objectives.

3.3 Case Study in Puebla, Mexico

Due to the geographical location of the Mexican Republic, each year, the risk of receiving disturbing phenomena such as storms, hurricanes, intense rains that cause floods, as well as earthquakes due to being in the so-called "fire belt" is latent. Where 80% of the world's seismic activity occurs so that on average, the country experiences more than 90 earthquakes a year with a magnitude of 4.0 or more on the Richter scale. Also, additionally to the risk of human losses in a disaster, there are other effects such as the change of the natural environment in different regions of the country and economic losses (González, 2015).

Taking into account the background of Mexico as a nation vulnerable to disasters, it is evident the need to improve logistics planning before an emergency response to reduce the adverse impact it causes on the population. The state of Puebla is part of the 32 states that are part of the country, it is located in the eastern region of Mexico, bounded on the north by Tlaxcala and Hidalgo, northeast by Veracruz, south by Oaxaca, southwest by Guerrero and west by Morelos and the state of Mexico (see Fig. 3.3).

In this research, two solution scenarios were developed: in the first scenario the solution was obtained qualitatively, considering the criteria of the alternatives separately; then, in the second scenario a solution was obtained using the AHP technique, analyzing all the criteria and all the alternatives at the same time.

Fig. 3.3 Location of Puebla in the country. *Source* National Institute of Statistic and Geography

3.3.1 Scenario 1. Qualitative Solution

The state of Puebla has been hit by natural disasters of the hydro-meteorological and geological (seismological) type, according to the information obtained from the emergency declarations of the National Center for Disaster Prevention (CENAPRED) and the Natural Disasters Fund (FONDEN). The primary responsibility of CENAPRED is to support the National System of Civil Protection (SINAPROC) in technical requirements. It carries out research, training, instrumentation, and dissemination activities about natural and anthropogenic phenomena that may lead to disaster situations, as well as actions to reduce and mitigate the adverse effects of such phenomena to contribute to better preparation of the population to confront them. The FONDEN is a financial instrument within the National System of Civil Protection, that through their Operating Rules and the procedures derived from them, integrates a process respectful of the competencies, responsibilities, and needs of the different levels of government, which has as its purpose, principles of co-responsibility, complementarity, opportunity, and transparency, to support the states of the Mexican Republic, as well as the dependencies and entities of the Federal Public Administration, in the attention and recovery of the effects that produce a natural phenomenon, in accordance with the parameters and conditions provided in its Operating Rules (FONDEN, consulted March 2017).

Table 3.2 contains information about the type of natural phenomena that have caused disasters in this entity. The data covers from 2006 to 2017.

On one side, Puebla has been affected by hydrometeorological phenomena many times, and on the other side, it has witnessed many seismological events. According to the National Seismological Service (2019), Puebla has been affected by five different earthquakes with magnitudes between 6.5 and up to 7.1 degrees: (1) year 1928, magnitude 6.5 degrees; (2) year 1945, magnitude 6.5 degrees; (3) year 1980, magnitude 7.1 degrees; (4) year 1999, magnitude 7.0; and (5) September 19th, 2017, magnitude 7.1. In many of these situations, the population was severely affected to the extent that the municipalities had to receive support from different institutions. Therefore, it is essential to locate a safe area where a Regional

Table 3.2 Different types of natural disasters in Puebla from (2006–2017)

Types of phenomena	Classification	Year of occurrence
Low temperature	Hydrometeorological	2006
Tropical cyclone	Hydrometeorological	2007, 2017
Floods	Hydrometeorological	2012
Frost or hailstorm	Hydrometeorological	2006, 2007, 2010 2011, 2013, 2015 2016
Drought	Hydrometeorological	2006, 2007, 2008 2009, 2011
Earthquake	Seismological	2017

Source Data obtained from the National Center for Disaster Prevention (CENAPRED, 2017) and Natural Disasters Fund (FONDEN, 2018). Own elaboration

Humanitarian Response Depot (RHRD) can be established to safeguard useful supplies and provide them with time to safeguard human lives.

To identify the safe municipalities where the warehouse can be established, it was necessary to know what kind of natural disaster has affected the region. For this reason, Appendix 3.A was elaborated. It contains the 217 municipalities that comprise the state with their respective coordinates (latitude and longitude), and these were obtained from the Google Maps® tool. The type of disaster was classified as numbers, with a range from 0 to 5. The classification is no incidents = 0, earthquake = 1, river flood = 2, severe frost = 3, severe rain = 4, undefined phenomenon = 5.

Once the information was classified, it was introduced into the GIS ESRI® tool to visualize it and have better management of it to make a better decision about the establishment of the RHRD. The results were 112 municipalities affected by the earthquake, 64 affected by the undefined phenomenon, 25 affected by severe frost, 14 without affectation, and two affected by severe rain, as can be seen in Fig. 3.4.

The 64 municipalities without an identified phenomenon, it was concluded that landslides are the most likely because there are mountains, and they are near the state of Veracruz, which frequently receives emergency declarations of hydrometeorological events. Just like on the National Risk Atlas, the system created by the Mexican government, has many records that the mountains of Puebla present many situations of softening of slopes.

As a result, 14 safe municipalities were identified for the establishment of the RHRD. In Table 3.3, the names of these safe sites are presented. These are located in the central and southeastern regions of the state. Moreover, in Fig. 3.5 it is possible to see the geographical representation. Although they are listed, they all have the same weight when they were selected.

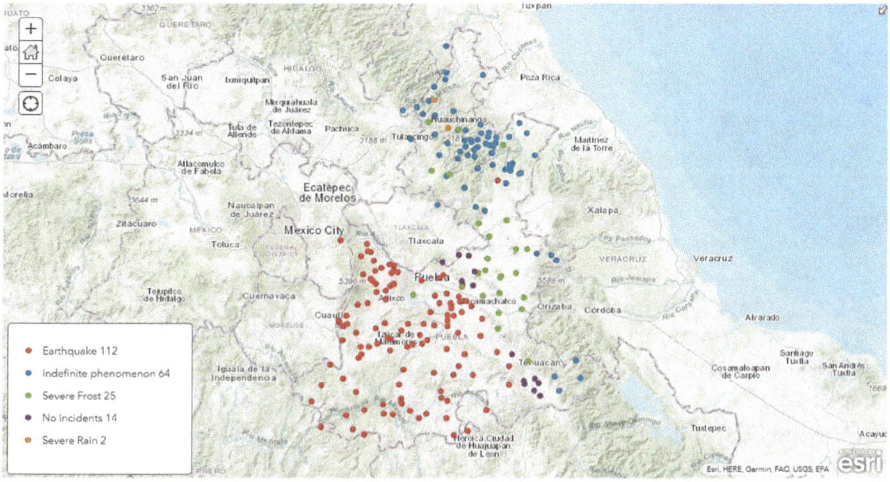

Fig. 3.4 Location of the type of natural disasters occurred in Puebla. *Source* Compiled by the author with information from CENAPRED and FONDEN

Table 3.3 Municipalities free of natural disasters in Puebla

No.	Municipality	Area
1	Acatzingo	Central
2	Ajalpan	Southeastern
3	Altepexi	Southeastern
4	Coxcatlán	Southeastern
5	General Felipe Ángeles	Central
6	Nopalucan	Central
7	Rafael Lara Grajales	Central
8	San Gabriel Chilac	Southeastern
9	San José Chiapa	Central
10	San José Miahuatlán	Southeastern
11	Santiago Miahuatlán	Southeastern
12	Soltepec	Central
13	Vicente Guerrero	Southeastern
14	Zinacatepec	Southeastern

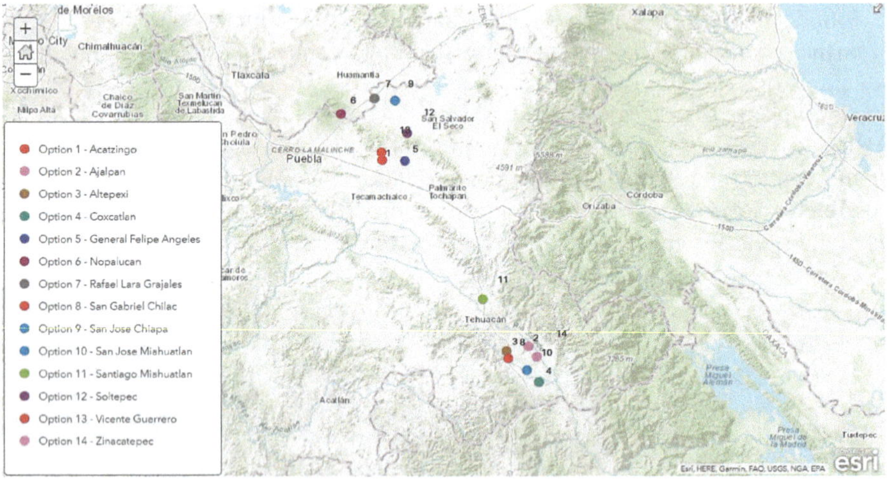

Fig. 3.5 Location of the fourteen municipalities free of natural disasters in Puebla

To ensure that the 14 municipalities in Puebla are safe, of the municipalities of the contiguous municipalities that belong to the states of Tlaxcala and Oaxaca, just the closest to the safe areas were considered. In Tlaxcala, the municipalities of Cuapiaxtla, El Carmen Tequexquitla, Huamantla, and Zitlaltepec de Trinidad Sánchez Santos; and, in Oaxaca, the municipalities of Santa Ana Ateixtlahuaca, San Francisco Huehuetlán, San Lorenzo Cuaunecuiltitla, Santa María Chilchotla, Santa María Teopoxco, and Santiago Texcalcingo, no natural disasters have been reported (see Table 3.4).

Table 3.4 Municipalities of Tlaxcala and Oaxaca near to Puebla

State	Municipality	Affected by natural disasters
Oaxaca	Santa Ana Ateixtlahuaca	No
	San Francisco Huehuetlán	
	San Lorenzo Cuaunecuiltitla	
	Santa María Chilchotla	
	Santa María Teopoxco	
	Santiago Texcalcingo	
Tlaxcala	Cuapiaxtla	No
	El Carmen Tequexquitla	
	Huamantla	
	Zitlaltepec de Trinidad Sánchez Santos	
	Cuapiaxtla	

3.3.2 Scenario 2. AHP Solution

The occurrence of natural phenomena that affect Mexico makes necessary the permanent monitoring of earthquakes, tropical cyclones, and the volcanoes of Colima and Popocatepetl. Based on information published by the National Center for Disaster Prevention (CENAPRED), the National Seismological Service, the Earth Observation Laboratory and the National Oceanic and Atmospheric Administration, a system is integrated into the National Risk Atlas which allows its dynamic visualization on a single board (CENAPRED, consulted March 2017).

For that reason, the Atlas is a significant source of information, from which the data for the preparation the AHP discussed in this chapter were obtained. From this source, data related to the state of Puebla, such as non-threatened municipalities by natural phenomena, was obtained. In this case, not only was the frequency used, but also were considered the Municipal indicators of danger, exposure, and vulnerability contained in the Atlas. Other considered indicators were the number of floods, droughts, thunderstorms, hail, warm waves, tropical cyclones, low temperatures, snowfalls, earthquakes, the susceptibility of hillsides, dangers by flammable substances, dangers due to toxic substances, number of authorized shelters, number of hospitals, declarations of contingency, declarations of disaster, and declarations of emergency. All these data were obtained from each of the 217 municipalities that make up the state. We also used the data of the current refuges in each municipality of Puebla; this information was obtained from the SINAPROC. As a result, the information considered in this case consists of 17 criteria and 217 alternatives (SINAPROC, consulted February 2018).

Ghiani et al. (2013) mentioned that the Analytic Hierarchy Process (AHP) is a structured technique for organizing and analyzing complex decisions based on mathematics and psychology. It was developed by Thomas L. Saaty in the 1970s, and it has been extensively studied and refined since then. Users of the AHP first decompose their decision problem into a hierarchy of more easily understood sub-problems, each of which can be analyzed independently. The elements of the hierarchy can relate to any aspect of the decision problem—tangible or intangible, carefully measured, or roughly estimated (Saaty, 1980).

3.3.2.1 Overview of the Principal Eigenvalue Method (EM)

Bana-e-Costa and Vansnick (2008) mentioned that it is necessary to review the principal eigenvalue method used in AHP to derive priorities vectors. Let $X = \{x_1, x_2, \ldots, xn\}$ be a set of elements and φ "a property or criterion that they have in common" (Saaty, 1996). How can we help a person J quantify the relative priority (or importance) that the elements of *X* have for her, in terms of φ*?* The EM used in AHP to derive priorities for the elements of X requires that a number—denoted w_{ij}—be assigned to each pair of elements (x_i, x_j) representing, in the opinion of *J*, the ratio of the priority of the dominant element (x_i) relative to the priority of the dominated element (x_j) (Saaty, 1997). *J* is invited to compare the elements pairwise and can express her judgments in two different ways:

- either numerically, by giving a real number between 1 (inclusive) and 10 (exclusive) (Saaty, 1989);
- or verbally, by choosing one of the following expressions: equal importance, moderate dominance, strong dominance, very strong dominance, extreme dominance, or an intermediate judgment between two consecutive expressions; each verbal pairwise comparison elicited is then automatically converted into a number w_{ij} as exhibited in Table 3.5 (Saaty, 1994).

During the elicitation process, a positive reciprocal matrix is made, in which each element $x_1, x_2, \ldots, x_n$ of *X* is assigned one row, and one column can be filled by placing the corresponding number at the intersection of the row of x_i with the column of x_i

Table 3.5 Saaty scale (1994)

Verbal scale	Numerical scale
Both criteria or elements are equally important	1
Weak or moderate importance of one over another	3
Essential or strong importance of one criterion over the other	5
Demonstrated importance of one criterion over another	7
Absolute importance of one criterion over another	9
Intermediate values between two adjacent judgments	2, 4, 6, 8
Between equally and moderately preferable	2
Between moderately and strongly preferable	4
Between strongly and extremely preferable	6
Between very strongly and extremely preferable	8

$$\begin{cases} w_{ij} & \text{if } x_i \text{ dominates } x_j, \\ 1/w_{ij} & \text{if } x_j \text{ dominates } x_i, \\ 1 & \text{if } x_i \text{ does not dominate } x_j \\ & \text{and } x_j \text{ does not dominate } x_i. \end{cases}$$

For example, assuming that for all $i, j \in \{1, 2, \ldots, n\}$ x_i dominates x_j if and only if $i < j$, the format of the positive reciprocal matrix is

$$W = \begin{pmatrix} 1 & w_{12} & \ldots & w_{1n} \\ 1/w_{12} & 1 & \ldots & w_{2n} \\ \ldots & \ldots & \ldots & \ldots \\ \ldots & \ldots & \ldots & \ldots \\ 1/w_{1n} & 1/w_{2n} & \ldots & 1 \end{pmatrix}$$

In order to assign a "priority" (or a "weight") to each element x_i—a numerical value that we will denote as $w(x_i)$—the principal eigenvalue λ_{max} of matrix **W** and its normalized eigenvector is calculated: the components of this vector are $w(x_i)$. This procedure has a very interesting property: if the judgements of J are such that $w_{ij} \cdot w_{jk} = w_{ik}$ for all $i < j < k$ (cardinal consistency condition), the derived $w(x_i)$ are such that $w_{ij} = w(x_i)/w(x_j)$ for all $i < j$.

However, cardinal consistency is seldom observed in practice. Therefore, AHP makes use of a "consistency test" that prevents priorities from being accepted if the inconsistency level is high. In order to measure the deviation of matrix **W** from "consistency", a consistency index CI is defined as λ_{max}—$n/(n-1)$ and a random index RI (of order n) is calculated as the average of the CI of many thousands reciprocal matrices (of order n) randomly generated from the scale 1 to 9, with reciprocals forced. The ratio of CI to RI for the same order matrix is called the consistency ratio CR. According to Bana-e-Costa and Vansnick (2008), "a consistency ratio of 0.10 or less is considered acceptable. That is, an inconsistency is stated to be a matter of concern if CR exceeds 0.1, in which case the pairwise comparisons should be re-examined".

Bana-e-Costa and Vansnick (2008) mentioned "if the elements are to be compared according to several φ, the AHP proposes that a hierarchy is built with the general goal on top, the elements at the bottom and the φ at intermediate levels. The procedure described above is then repeatedly applied bottom-up: to calculate a vector of priorities for the elements concerning each φ situated at the bottom intermediate level; to calculate a vector of weights for each cluster of φ at the different levels. All this judgmental information is then synthesized from bottom to top by successive additive aggregations, in order to derive a vector of overall priorities for the elements".

3.3.2.2 Power Method

Palacios (2007) provides a more detailed procedure to obtain a vector solution. It is assumed that the matrix A has a base of eigenvectors $\{\upsilon_1, \ldots, \upsilon_n\}$ associated to the eigenvalues $\lambda_1, \ldots \lambda_n$, respectively, and, furthermore, that λ_1 is the dominant eigenvalue, that is,

$$|\lambda_1| > |\lambda_2| \geq \ldots \geq |\lambda_n|$$

For any vector $v \in R^n$, since $\{v_1, \ldots, v_n\}$ is a base, scalars $\alpha_1, \ldots, \alpha_n$ must exist such that

$$x = \alpha_1 v_1 + \ldots + \alpha_n v_n$$

Multiplying repeatedly by matrix A we have:

$$\begin{aligned} Ax &= \sum_{j=1}^{n} \alpha_j A v_j = \sum_{j=1}^{n} \alpha_j \lambda_j v_j \\ A^2 x &= \sum_{j=1}^{n} \alpha_j \lambda_j A v_j = \sum_{j=1}^{n} \alpha_j \lambda_j^2 v_j \\ &\ldots \\ A^k x &= \sum_{j=1}^{n} \alpha_j \lambda_j^{k-1} A v_j = \sum_{j=1}^{n} \alpha_j \lambda_j^k v_j \end{aligned}$$

Taking common factor λ_1^k in each term of the second member, in the last expression, we obtain:

$$\mathrm{Ak\ x} = \lambda_1^k \sum_{j=1}^{n} \mathrm{aj} \left(\frac{\lambda_j}{\lambda_1}\right) \mathrm{k}\ vj$$

Taking into account that $|\lambda 1| > |\lambda j|$, j = 2, 3, .., n, you have $\lim_k \left(\frac{\lambda_j}{\lambda_1}\right)^k = 0$ and, therefore,

$$\lim_k A^k x = \lim_k \lambda_1^k \alpha v$$

Consequently, this sequence converges to 0, if $|\lambda_1| < 1$, and diverges, in another case. It can take advantage of this behavior by building the following algorithm of the method. The algorithm for the power method is:

The initial stage, k = 0

Given the arbitrary vector $x = (x^1, \ldots, x^n) = \alpha_1 v_1 + \ldots + \alpha_n v_n$, we will call

$$
\begin{aligned}
\mathrm{p}_0 &= \min\{j|\ ||x||\infty = |x^j|\} \\
x_0 &= \frac{x}{x^{P0}} \Rightarrow x_0^{P0} = 1 = ||x||\infty \\
\mu_0 &= x_0^{P0} = 1
\end{aligned}
$$

Stage k = 1

$$
\begin{aligned}
y_1 &= AX_0 \\
\mu_1 &= y_I^{po} = \frac{y_1^{p0}}{x_0^{p0}} \\
\mathrm{p}_1 &= \min\{j|\ ||y||\infty = |y_1^j|\} \\
X_1 &= \frac{y1}{y_1^{p1}} = \frac{Ax_0}{y_1^{p_1}} \Longrightarrow \quad x_f^{p1} = 1 = ||y_1||\infty
\end{aligned}
$$

Stage k = 2

$$
\begin{aligned}
y_2 &= AX_1 = \frac{1}{y_1^{p1}} A^2 X_0 \\
\mu_2 &= y_2^{p1} = \frac{y_2^{p1}}{x_1^{p1}} = \frac{\sum_{j=1}^n \alpha_j \lambda_j^2 v_j^{p1}}{\sum_{j=1}^n \alpha_j \lambda_j v_j^{p1}} = \lambda_1 \frac{\sum_{j=1}^n \alpha_j (\lambda_j/\lambda_1)^2 v_j^{p1}}{\sum_{j=1}^n \alpha_j (\lambda_j/\lambda_1) v_j^{p1}} \\
p_2 &= \min\{j|\,||y||\infty = |y_2^j|\} \\
X_2 &= \frac{y^2}{y^{p2}} = \frac{Ax_1}{y^{p2}} = \frac{A^2 x_0}{y_2^{p2} y_1^{p1}} \Longrightarrow x_2^{p2} = 1 = ||y_2||\infty
\end{aligned}
$$

Stage k + 1

$$
\begin{aligned}
Y_{k+1} &= AX_k = \frac{1}{y_k^{pk} \cdots y_2^{p2} y_1^{p1} y_1^{p1}} A^{k+1} X_0 \\
\mu_{k+1} &= y_{k+1}^{pk} = \frac{y_{k+1}^{pk}}{x_k^{pk}} = \frac{\sum\limits_{j=1}^{n} \alpha_j \lambda_j^{k+1} v_j^{pk}}{\sum\limits_{j=1}^{n} \alpha_j \lambda_j^k v_j^{pk}} = \lambda_1 \frac{\sum\limits_{j=1}^{n} \alpha_j (\lambda_j/\lambda_1)^{k+1} v_j^{pk}}{\sum\limits_{j=1}^{n} \alpha_j (\lambda_j/\lambda_1)^k v_j^{pk}} \\
\mathrm{p}_{k+1} &= \min\{j|\,||y||\infty = |y_2^j|\} \\
X_{k+1} &= \frac{y^{k+1}}{y^{pk+1}} = \frac{Ax_k}{y^{pk+1}} = \frac{A^{k+1} x_0}{y_{k+1}^{pk+1} \cdots y_2^{p^2} y_2^{p_1}} \Longrightarrow x_{k+1}^{Pk+1} = 1 = ||y_{k+1}||\infty
\end{aligned}
$$

Final stage

$$\lim \mu_k = \lambda_1, \quad \lim X_k = v_1, \quad \|v_1\|\infty = 1$$
$$|\mu_k - \lambda_1| \approx C\left|\frac{\lambda_2}{\lambda_1}\right|^k \Longrightarrow \lim\left|\frac{\mu_{k+1} - \lambda_1}{\mu_k - \lambda_1}\right| \approx \left|\frac{\lambda_2}{\lambda_1}\right| < 1$$

3.3.2.3 Solving Scenario 2

Seventeen criteria were considered for this project, which can be seen in Table 3.6; these criteria conform to matrix **W**.

This matrix was necessary because it was essential to know the weight for each one of these criteria using the Saaty scale mentioned before:

$$\begin{array}{c} \\ \textit{Criterion } 1 \\ \textit{Criterion} 2 \\ \cdots \\ \textit{Criterion } \mathrm{m} \end{array} \begin{array}{c} \begin{array}{cccc} \textit{Criterion } 1 & \textit{Criterion} 2 & \ldots & \textit{Criterion } \mathrm{m} \end{array} \\ \left(\begin{array}{cccc} P'_1 & & & \\ P'_2 & & & \\ \cdots & & & \\ P'_m & & & \end{array} \right) \end{array}$$

The weights or relationship of judgment were established based on the frequency, danger or benefit of each of the criteria. In the case of Puebla, the most frequent events which have caused the most damage in the municipalities are floods and

Table 3.6 The seventeen criteria and their identifier

Criteria identifier	Concept
C1	Floods
C2	Droughts
C3	Thunderstorms
C4	Hail
C5	Warm waves
C6	Tropical cyclones
C7	Low temperatures
C8	Snowfalls
C9	Earthquakes
C10	Susceptibility of hillsides
C11	Dangers by flammable substances
C12	Dangers due to toxic substances
C13	Number of authorized shelters
C14	Number of hospitals
C15	Declarations of contingency
C16	Declarations of disaster
C17	Declarations of emergency

earthquakes (which are hydrometeorological and seismological events). This is the reason to rated them high. Also, the criteria related to infrastructure, which are the authorized shelters and hospitals, were also rated high since they are relevant if such events occur. As an example of the resulting matrix, see the representation of the matrix of 17 criteria by 17 criteria (consult the complete matrix in Appendix 3.B).

$$\begin{array}{c} \\ Floods \\ Droughts \\ \cdots \\ Declarations\ of\ emergency \end{array} \begin{array}{c} \begin{array}{cccc} Floods & Droughts & \cdots & Declarations\ of\ emergency \end{array} \\ \begin{pmatrix} 1 & 3 & \cdots & 5 \\ 1/3 & 1 & \cdots & 1/2 \\ \cdots & \cdots & \cdots & \cdots \\ 1/5 & 2 & \cdots & 1 \end{pmatrix} \end{array}$$

The second matrix considered 217 alternatives (which are the 217 municipalities) by the normalized information of the 17 previous criteria. For m criterion and n alternatives we have:

$$\begin{array}{c} \\ Alternative\ 1 \\ Alternative\ 2 \\ \cdots \\ Alternative\ n \end{array} \begin{pmatrix} Criterion\ 1 & Criterion\ 2 & \cdots & Criterion\ m \\ P_{11} & P_{12} & \cdots & P_{1m} \\ P_{12} & P_{22} & \cdots & P_{2m} \\ \cdots & \cdots & \cdots & \cdots \\ P_{n1} & P_{n2} & \cdots & P_{nm} \end{pmatrix}$$

Where P_{ij} is the priority of alternative i with respect to criterion j, for $i = 1, 2, \ldots, n$; and $j = 1, 2, \ldots, m$. As an example of the resulting matrix, see the following representation of the matrix of 217 alternatives by 17 previous criteria (consult the complete matrix in Appendix 3.C).

$$\begin{array}{c} \\ \text{Acajete} \\ \text{Acateno} \\ \cdots \\ \text{Zoquitlan} \end{array} \begin{pmatrix} \text{Floods} & \text{Droughts} & \cdots & \text{Declarations of emergency} \\ 0.00550 & 0.00461 & \cdots & 0.00462 \\ 0.00275 & 0.00615 & \cdots & 0.00458 \\ \cdots & \cdots & \cdots & \cdots \\ 0.00413 & 0.00461 & \cdots & 0.00460 \end{pmatrix}$$

The global priority for each decision alternative is summarized in the column vector that results from the product of the priority matrix with the priority vector of the criteria. Finally, a matrix of 217 by one or solution vector was obtained, in which the final weights of each one of the municipalities are reflected, in order to select the ideal municipality or in this case less vulnerable.

$$\begin{pmatrix} P_{11} & P_{12} & \cdots & P_{1m} \\ P_{21} & P_{22} & \cdots & P_{2m} \\ \cdots & \cdots & \cdots & \cdots \\ P_{n1} & P_{n2} & \cdots & P_{nm} \end{pmatrix} \begin{pmatrix} P'_1 \\ P'_2 \\ \cdots \\ P'_m \end{pmatrix} = \begin{pmatrix} Pg_1 \\ Pg_2 \\ \cdots \\ Pg_n \end{pmatrix}$$

where Pg_i is the global priority (concerning the global goal) of the alternative i ($i = 1, 2, \ldots, n$).

In order to compare with the qualitative solution of the first scenario, the first best 14 results (the highest) of the AHP were selected, and they were represented geographically. In Fig. 3.6 these options are presented in order of importance, while in Table 3.7, appear the corresponding weights. The municipality that obtains the highest rating is Teziutlán, which is located in the northwest region of the state. In the case of Teziutlán, the greatest danger to which it is exposed, according to the National Risk Atlas (consulted August 2018), is susceptibility to hillsides and thunderstorms. It has eight hospitals, and within the services it offers are 29 banks, 12 gas stations, 23 hotels, 40 supermarkets. This place is followed by San Miguel Xoxtla, which is next to Puebla (capital of the state), which is the third option in the rank. According to the same source, San Miguel Xoxtla has one hospital, one bank, zero gas stations, one hotel, and zero supermarkets. The greatest danger to which it is exposed is flooding. Moreover, Puebla (capital of the state), has an essential infrastructure since it has a more significant infrastructure and services, such as 61 hospitals, 537 banks, 158 gas stations, 649 supermarkets. The hazards exposed to Puebla are susceptibility to hillsides and danger from flammable substances. However, the municipality is tacked in second place with events such as floods, electrical storms, hail, and earthquakes. For this reason, its rating decreases, and it is placed in third place.

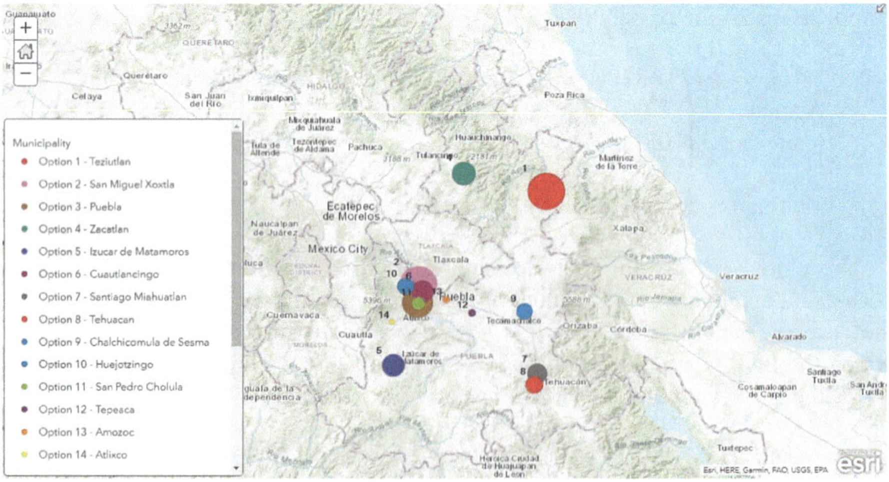

Fig. 3.6 Location of the 14 municipalities less vulnerable to disasters in Puebla according to AHP

Table 3.7 Municipalities free of natural disasters in Puebla according to AHP

No.	Municipality	Weight	Area
1	Teziutlán	0.013670939	Northwest
2	San Miguel Xoxtla	0.013616067	Center-west
3	Puebla	0.012368291	Center-west
4	Zacatlán	0.011011129	Northwest
5	Izúcar de Matamoros	0.010843039	Southwest
6	Cuautlancingo	0.010834206	Center-west
7	Santiago Miahuatlán	0.010373660	Southeastern
8	Tehuacán	0.009931276	Southeast
9	Chalchicomula de Sesma	0.009658993	Center-east
10	Huejotzingo	0.009651304	Center-west
11	San Pedro Cholula	0.008880879	Center-east
12	Tepeaca	0.007914700	Central
13	Amozoc	0.007856477	Central
14	Atlixco	0.007641258	Center-west

3.4 Discussion

3.4.1 Discussion for the First Scenario

As we can see in the results of the first scenario, two zones were identified to allocate the RHRD. One of the elements that are crucial in an emergency event is time; it is a vital factor to attend the zone, which is in danger. Thus, a distance of the RHRD to the possible affected zones is important. How far is the affected area from any of the zones (central and southeastern) that have been considered safe?

To evaluate that, two zones that frequently have been affected by natural disasters were selected (orange circles enclosed in black squares - Tlahuapan located in the west-center area of Puebla and Tlaxco located in the northwest area of Puebla), see Fig. 3.7, in order to obtain the time and distance when being treated by any of the proposed areas as safe. These selected municipalities were considered based on the characteristics that the government contemplates convenient such as infrastructure, services, among others to be viable and be considered as safe municipalities. In the case of the central zone, the municipality of Rafael Lara Grajales (green triangle) was chosen, and the municipality of Ajalpan (blue triangle) was chosen from the southeast area.

The measures are in kilometers and were obtained from the Google Maps® tool for the results see Table 3.8.

The same analysis was made but with the results of the second scenario (AHP). In this case, the municipality of Teziutlan (located in the northwest area of Puebla), which was the best of the 14 municipalities, and Puebla (located in the center-west

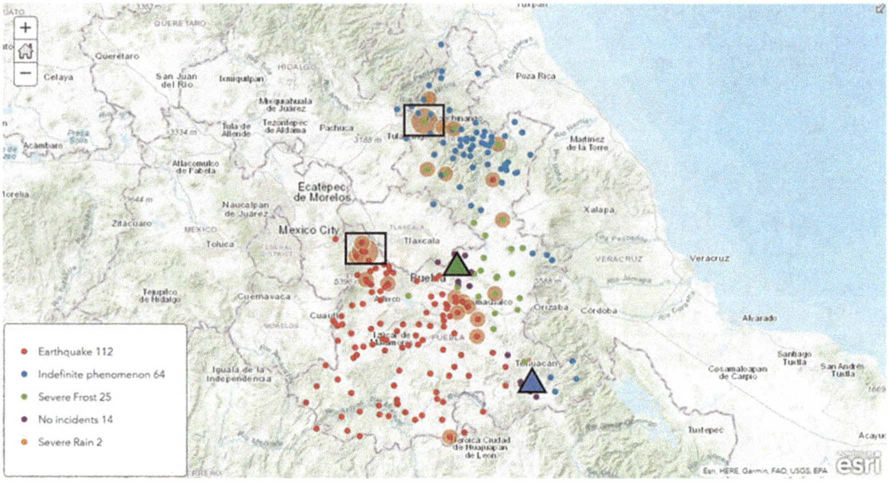

Fig. 3.7 Location of the municipalities where more disasters have occurred (frequency) and location selected in the first scenario. *Source* Compiled by the author with information from CENAPRED and FONDEN

Table 3.8 Time and distance between safe municipalities of the first scenario and vulnerable municipalities

From Rafael Lara Grajales to Tlahuapan	
Time	**Distance**
1 h 28 min	114 km
From Rafael Lara Grajales to Tlaxco	
Time	**Distance**
3 h 12 min	209 km
From Ajalpan to Tlahuapan	
Time	**Distance**
2 h 53 min	202 km
From Ajalpan to Tlaxco	
Time	**Distance**
5 h 26 min	375 km

zone) which is the capital of the state, were selected. The same two zones that frequently have been affected by natural disasters were selected (orange circles enclosed in black squares - Tlahuapan and Tlaxco), see Fig. 3.8, in order to obtain the time and distance when being considered by any of the proposed areas as safe. The municipality of Teziutlan is marked with a yellow triangle and the municipality of Puebla with a pink triangle.

The measures are in kilometers and were obtained from the Google Maps® tool for the results see Table 3.9.

The numbers obtained in this analysis show that times and distances are short, and in an emergency caused by a natural disaster, these will not be considered an

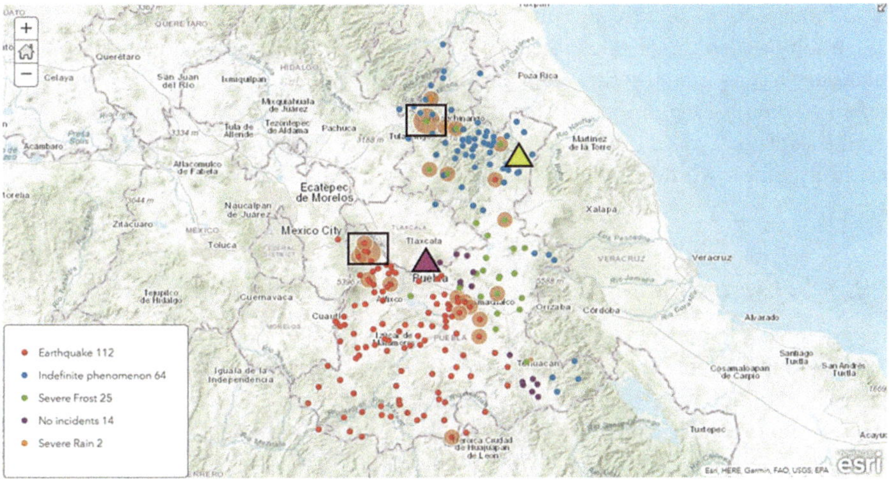

Fig. 3.8 Location of the municipalities where more disasters have occurred (frequency) and location selected in the second scenario—AHP

Table 3.9 Time and distance between safe municipalities of the second scenario (AHP) to vulnerable municipalities

From Teziutlan to Tlahuapan	
Time	**Distance**
2 h 30 min	194 km
From Teziutlan to Tlaxco	
Time	**Distance**
4 h 13 min	232 km
From Puebla to Tlahuapan	
Time	**Distance**
57 min	56.7 km
From Puebla to Tlaxco	
Time	**Distance**
3 h 46 min	267 km

obstacle to acting quickly and effectively. Do not forget that the primary goal is to safeguard human integrity. Only in the cases of Ajalpan-Tlaxco (first scenario) and Teziutlan-Tlaxco (second scenario), the help will take longer to arrive.

As we have already seen, the state of Puebla is prone to be affected by natural disasters. It should be remembered that on September 19th, 2017, it was severely affected by an earthquake of magnitude 7.1 on the Richter scale, and it caused considerable damage in the state. In this situation, the establishment of an RHRD is necessary for the entity. This establishment must be sponsored by the three levels of government that exist in the country, who must be aware of the vital aspects of its construction.

Of course, civil society and other organizations (NGO's) would participate in the construction of the RHRD, either by supporting with financial donations or with materials for its construction. Not only the construction of this RHRD will bring benefits when a natural disaster occurs, and the affected people are attended, but it would also be a source of employment for all those people in charge of administering it, which would bring a positive development in the region.

3.4.2 Discussion for the Second Scenario

As previously mentioned, for this work it was necessary to elaborate the second scenario that let to consider various criteria, with the purpose of knowing the suitable municipality so that the depot can be built. With the criteria and alternatives mentioned above, the results were the municipality of Teziutlán, obtaining a weight of 0.013670939, this being the largest of the 217 municipalities.

Both scenarios, solved by qualitatively and AHP, have one municipality in common, that is Santiago Miahuatlán, shown in Fig. 3.9, which is in the southeastern of the state of Puebla.

This municipality limits the north with the municipalities of Tepanco de López, Chapulco and Nicolás Bravo; to the east with the municipalities of Nicolás Bravo and Tehuacán; to the south with the municipality of Tehuacán; to the west with the municipalities of Tehuacán and Tepanco de López (INEGI, 2000) (consulted January 2019). According to information obtained from the National Institute of Statistics and Geography (INEGI) of the year 2000, 36.9% of the population is engaged in primary activities (including agriculture, livestock, hunting, and fishing)

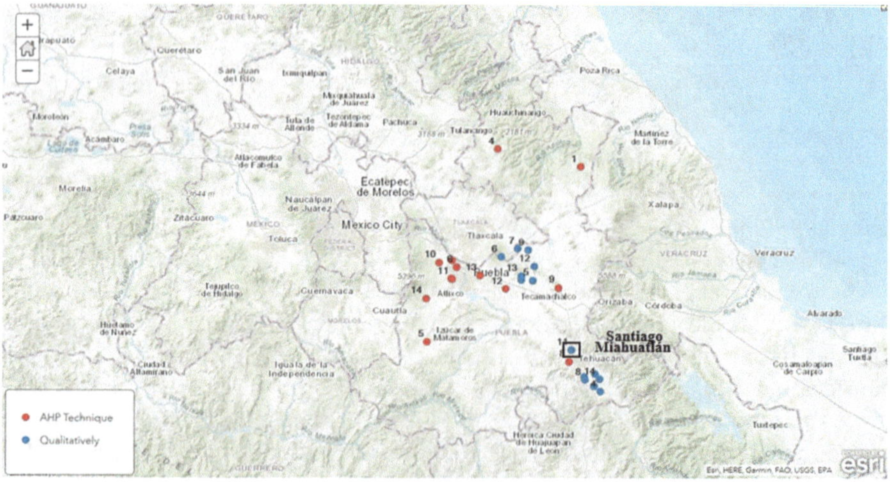

Fig. 3.9 Santiago Miahuatlán location (blue point enclosed in black square)

Table 3.10 General Information Santiago Miahuatlán in accordance with INAFED (2010)

Number of towns of the municipality	14
Area of the municipality in km^2	94
% of the surface that represents concerning the state	0.27
Municipal head	Santiago Miahuatlán
The population of the municipal head	13,909
Men	6649
Women	7260
Geographical coordinates of the municipal seat	
Length	97°26′31″ W
Latitude	18°33′16″ N
Altitude	1793 meters
Classification of the municipality according to the size of localities	Semi-urban

and 35.2% is dedicated to the tertiary sector (trade and services). It has only had one declaration of emergency and one declaration of disaster, both in the year 2005 and mainly was susceptibility to hillsides. Relevant information of this municipality is shown in Table 3.10. According to the National Institute for Federalism and Municipal Development (INAFED, consulted January 2019) and National Atlas of Risks (consulted January 2019).

3.5 Concluding Remarks

Disaster is a situation or an event that overwhelms the capacity of the community, needing the help of national or international levels of external assistance. It is an unforeseen and often sudden event that causes considerable damage, destruction, and human suffering. For that reason, humanity needs to be prepared to face those situations.

Undoubtedly natural disasters have increased in the state of Puebla; for this reason, we must be prepared to act effectively in such situations. The results obtained in this analysis are essential, because known which municipalities are safe in the entity is relevant for the establishment of a secure RHRD and that it works optimally. Additional aspects such as the security or the active volcano Popocatepetl must be considered before, during, and after the construction of the RHRD.

Puebla is a state that has been affected by many natural disasters; for that reason, we have to try being resilient. In this case, all the applied knowledge plays a fundamental role because we know what places in the state are good candidates to establish a depot, which is aimed at helping people in disaster situations.

This study can serve as an example for other states in the country or other places in the world. Also, we consider that this work can be applied to the whole country in order to supply not only the state of Puebla but also any other state of the country in a disaster situation.

The analysis developed in this study can be further extended by including other criteria such as the Popocatepetl volcano and the security in the municipalities. Maybe other criteria or factors concerning the humanitarian supply chain can be considered in the context of disaster management as unpredictable demand, timing, short lead time, lack of primary resources, and so forth.

References

Aydi, A., Abichou, T., Hamdi-Nars, I., Louati, M., & Zairi, M. (2016). Assessment of land suitability for olive mill wastewater disposal site selection by integrating fuzzy logic, AHP, and WLC in a GIS. *Environmental Monitoring and Assessment, 188,* 59. https://doi.org/10.1007/s10661-015-5076-3.

Barojas-Payán, E., Sánchez-Partida, D., Martínez-Flores, J. L., & Gibaja-Romero, D. E. (2019). Mathematical Model for Locating a Pre-Positioned Warehouseand for Calculating Inventory Levels. *Journal of Disaster Research, 14*(4), 649–666. https://doi.org/10.20965/jdr.2019.p0649.

Bana-e-Costa, C. A., & Vansnick, J. C. (2008). A critical analysis of the eigenvalue method used to derive priorities in AHP. *European Journal of Operational Research, 187*(3), 1422–1428. https://doi.org/10.1016/j.ejor.2006.09.022.

Boonmee, Ch., Arimura, M., & Asada, T. (2017). Facility location optimization model for emergency humanitarian logistics. *International Journal of Disaster Risk Reduction, 24,* 485–498. https://doi.org/10.1016/j.ijdrr.2017.01.017.

Bruno, G., & Giannikos, I. (2015). Location and GIS. In *Location science* (1st ed.). Cham: Springer.

CENAPRED. (2017). *Declaratorias sobre emergencia, desastre y contingencia climatológica a nivel municipal entre 2000 y 2017.* National Center for Disaster Prevention. Retrieved March 10, 2017 from https://datos.gob.mx/busca/dataset/declaratorias-sobre-emergencia-desastre-y-contingencia-climatologica/resource/41444ebe-6a35-4631-8f91-9237d5114488 [in Spanish].

Chandes, J., & Paché, G. (2009). Pensar la acción colectiva en el contexto de la logística humanitaria: las lecciones del sismo de Pisco. *Journal of Economics, Finance & Administrative Science, 14*(27), 47–61. [in Spanish].

Cozzolino, A. (2012, July). Humanitarian logistics and supply chain management. In *Humanitarian logistics* (pp. 5–16). Berlin, Heidelberg: Springer.

Daim, T., Adbye, A., & Balasubramanian, A. (2012). Use of analytic hierarchy process (AHP) for selection of 3PL Providers. *Journal of Manufacturing Technological Management, 24*(1), 28–51. https://doi.org/10.1108/17410381311287472.

FONDEN. (2017). *Recursos autorizados por declaratoria de desastre.* National System of Civil Protection. Retrieved March 15, 2018, from http://www.proteccioncivil.gob.mx/es/ProteccionCivil/Recursos_Autorizados_por_Declaratoria_de_Desastre [in Spanish].

Ghiani, G., Laporte, G., & Mussmano, R. (2013). *Introduction to logistics systems management* (2nd ed.). Hoboken: Wiley.

González, L. E. (2015). *Localización de un Almacén de Abastecimiento para la Atención de Desastres en el Estado de Guerrero.* Master Tesis. Universidad Nacional Autónoma de, México [in Spanish].

González-Prida, V., Viveros, P., Barberá, L., & Crespo-Márquez, A. (2014). Dynamic analytic hierarchy process: AHP method adapted to a changing environment. *Journal of Manufacturing Technology Management, 25*(4), 457–475. https://doi.org/10.1108/JMTM-03-2013-0030.

Heaslip, G. E., & Barber, E. (2016). Improving civil-military coordination in humanitarian logistics: The challenge. *The Irish Journal of Management, 35*(2), 143–158. https://doi.org/10.1515/ijm-2016-0011.

İlhan, A. (2011). The humanitarian relief chain. *South East European Journal of Economics and Business, 6*(2), 45–54. https://doi.org/10.2478/v10033-011-0015-x.

Kavurmaci, M., & Üstun, A. K. (2016). Assessment of groundwater quality using DEA and AHP: a case study in the Sereflikochisar region in Turkey. *Environmental Monitoring and Assessment, 180*(4), 258. https://doi.org/10.1007/s10661-016-5259-6.

National Risk Atlas. Retrieved August 31st, 2018, from http://www.atlasnacionalderiesgos.gob.mx.

National Institute for Federalism and Municipal Development (INAFED). Retrieved January 29, 2019, from https://www.gob.mx/inafed.

National Institute of Statistics and Geography (INEGI). Retrieved January 29, 2019, from https://www.inegi.org.mx.

Natural Disasters Fund (FONDEN). Retrieved August 31, 2018, from https://www.gob.mx/segob/documentos/fideicomiso-fondo-de-desastres-naturales-fonden.

National Seismological Service. Retrieved February 10, 2019, from http://www.ssn.unam.mx/.

Palacios, M. (2007). Aproximación numérica de valores y vectores propios. Centro Politécnico Superior. Universidad de Zaragoza [in Spanish].

Prabowo, A. R., Dwicahyani, A. R., Jauhari, W. A., Aisyati, A., & Laksono, P. W. (2017). Development and application of humanistic logistics models for optimizing location-allocation problem solutions to volcanic eruption disaster (Case study: Volcanic eruption of Mount Merapi, Indonesia). *Cogent Engineering, 4*(1). https://doi.org/10.1080/23311916.2017.1360541.

Rodríguez-Espíndola, O., Albores, P., & Brewster, C. (2016). GIS and optimisation: Potential benefits for emergency facility location in humanitarian logistics. *Geosciences, 6*(18), 1–34. https://doi.org/10.3390/geosciences6020018.

Saaty, T. L. (1980). *The analytic hierarchy process*. New York: McGraw-Hill.

Saaty, T. L. (1989). Decision making, scaling, and number crunching. *Decision Sciences, 20*(2), 404–409. https://doi.org/10.1111/j.1540-5915.1989.tb01887.x.

Saaty, T. L. (1996). Decision making with dependence and feedback: The analytic network process. Pittsburgh, PA: RWS Publications.

Saaty, T. L. (1994). *How to make a decision: the analytic hierarchy process*. Pittsburgh: University of Pittsburgh.

Saaty, T. L. (1997). That is not the analytic hierarchy process: what the AHP is and what it is not. *Journal of Multi-Criteria Decision Analysis, 6*(6), 324–335. https://doi.org/10.1002/(SICI)1099-1360(199711)6:6%3c324::AID-MCDA167%3e3.0.CO;2-Q.

Saaty, T. L., & Vargas, L. G. (2012). *Models, methods, concepts & applications of the analytic hierarchy process* (2nd ed.). New York: Springer, The United Nations Humanitarian Response Depot. Retrieved April 14, 2018, from https://unhrd.org/.

The United Nations Humanitarian Response Depot. Retrieved April 14, 2018, from https://unhrd.org/.

Trivedi, A., & Singh, A. (2017). Prioritizing emergency shelter areas using hybrid multi-criteria decision approach: A case study. *Journal of Multi-Criteria Decision Analysis, 24*, 133–145. https://doi.org/10.1002/mcda.1611.

Van Wassenhove, L. N. (2006). Humanitarian aid logistics: Supply chain management in high gear. *The Journal of the Operational Research Society, 57*(5), 475–489.

Yx, Z., Liu Gj, F., & Ej, Z. K. (2009). An object-relational prototype of GIS-based disaster database. *Procedia Earth and Planetary Science, 1*(1), 1060–1066. https://doi.org/10.1016/j.proeps.2009.09.16.

Chapter 4
Selection of Humanitarian Response Distribution Centers (HRDC) in Puebla, Mexico

Martha Bello-Garduño, Diana Sánchez-Partida, José-Luis Martínez-Flores, and Santiago-Omar Caballero-Morales

Abstract In today's world, crises related to natural phenomena are being generated with a higher frequency and more significant impact than before. Mexico, is not an exception, speaking specifically of the state of Puebla, it has suffered substantial impacts due to natural phenomena, such as frosts, heavy rains, tropical cyclones, low temperatures, reflecting many destruction, casualties, damages, and economic losses. So the importance of improving humanitarian aid is evident in order to safeguard lives and reduce the impacts. The adequate preparation of the different elements involved in the rapid response to natural disasters is the success factor of disaster management. The purpose of this paper is to find the best location of Humanitarian Response Distribution Centers (HRDC) to provide aid kits immediately after the occurrence of the natural phenomena in order to reduce health impacts and achieve the basic needs of the affected population. For this study, first, a statistical analysis was made for identifying the most vulnerable municipalities. After, the P-median model was used for assigning the municipalities to attend. Moreover, the results obtained with the experimentation of two scenarios are presented and reflect the best locations for the selection of HRDCs.

Keywords Humanitarian Response Distribution Center (HRDC) · Humanitarian Logistics (HL) · P-median problem · Vulnerability

---

Electronic supplementary material The online version of this chapter (https://doi.org/10.1007/978-3-030-67295-9_4) contains supplementary material, which is available to authorized users.

M. Bello-Garduño (✉) · D. Sánchez-Partida · J.-L. Martínez-Flores ·
S.-O. Caballero-Morales
Department of Logistics and Supply Chain Management,
Universidad Popular Autónoma del Estado de Puebla (UPAEP University),
17 Sur 901, Barrio de Santiago, CP 72410 Puebla, Mexico
e-mail: martha.bello@upaep.edu.mx

D. Sánchez-Partida
e-mail: diana.sanchez@upaep.mx

D. Sánchez-Partida et al. (eds.), *Disaster Risk Reduction in Mexico*,
https://doi.org/10.1007/978-3-030-67295-9_4

4.1 Introduction

Climate change has become the primary author of the natural disasters that have occurred around the world. The consequences have been, in some cases, devastating. The world has experienced a notable increase in the frequency and intensity of natural disasters. Between 1980 and 2012, about 42 million were lost every year; years of life in disasters recorded in the international arena. If these are currently responsible for the loss of 100,000 annual lives, it is estimated that they will be 300,000 in the year 2050. Economic losses are reaching an average of between 250,000 million and 300,000 million US dollars a year (GAR, 2015).

The Latin American and Caribbean region not only concentrates a large number of threats of natural origin, such as earthquakes, floods, volcanic eruptions, and so forth, but they have a high vulnerability index, which far from being reduced has increased in recent years due to the combination of factors such as poverty, environmental deterioration or development policies.

The Centre for Research on the Epidemiology of Disasters (CRED) has been active for more than 40 years in the fields of international disaster and conflict health studies. CRED promotes research, training, and technical expertise on humanitarian emergencies with a particular focus on relief, rehabilitation, and development.

According to the annual statistical disaster study of the CRED (2015), in the last decade was estimated that 82 million people in Latin America and the Caribbean had been affected by disasters. It due to threats of natural origin, causing death to more than 247,000 people. The number of people at risk is growing every year, and the challenge for Public health and development is immense.

As mentioned in the previous chapter, the national territory is subject to a great variety of phenomena that can cause emergencies or disasters for being part of the so-called "Fire Belt of Pacific," so it is affected by intense seismic and volcanic activity.

Also, our country is located in an intertropical region, and this does vulnerable to hurricane attacks generated in the Pacific coastal areas of the Gulf and the Caribbean, causing damage from swells and winds. The heavy rains that these originate phenomena can cause landslides and floods. Of the 25 cyclones that on average arrive every year to the seas near the country, four or five usually penetrate the territory (CENAPRECE, 2018)

According to the Evaluation Global Disaster Risk Reduction (GAR) 2015, natural disasters that have occurred in the last five years in the country have generated economic losses of more than 2 thousand 942 million dollars, with eight million victims and around two thousand direct deaths.

An aggressive nature has been present in the last decade in Puebla, Mexico, leaving a balance of deaths and dangerous affectations. The majority is concentrated in the vulnerable municipalities of the north of the state, with declarations of emergency and disaster. During the last years, the municipalities of Puebla have suffered more forceful impacts due to natural phenomena, such as frosts, heavy rains, tropical cyclones, low temperatures, snowfalls. Reflecting numerous figures of destruction, casualties, economic losses, which have caused delays in economic, social, and cultural development; as well as they, have weakened the quality of life of the settlers.

So it has become essential to be prepared with an efficient humanitarian response, a way to shorten distances and help time is through prior planning of the best location of the humanitarian response distribution centers. Currently, in Puebla, the distribution process begins once there is an emergency declaration, without a planning definition of delivery points to the victims and, thus, the aid does not arrive in the most efficient time.

The objective of this work is to find the most efficient location of the Humanitarian response distribution centers (HRDC) to accelerate the distribution of humanitarian aid kits immediately after the occurrence of a disaster and reduce the health impacts of the affected population.

In this research, the P-median model was using, and two solution scenarios were developed; in the first scenario the solutions were obtained considering the distances, and it had experimentation assignment of 1, 2, 3 and 4 HRDC's; then, in the second scenario the solutions were obtained considering the traveling times, and it had an assignment of 1, 2, 3 and 4 HRDC's.

4.2 Description of the Problem

Puebla is one of the main cities of Mexico; its capital is the Heroic Puebla of Zaragoza. It has a total population of 6,168,883 that represents 5.2% of the total of the country. The distribution of the population is divided into 72% urban and 28% rural areas. Its participation in the Gross Domestic Product (GDP), giving the eighth-place in the nationwide (INEGI, 2016).

Puebla has 217 municipalities with a territorial extension of 33,919 square km, represents 1.75% of the national territory (see Fig. 4.1).

National Center for Disaster Prevention (CENAPRED) is a decentralized administrative and hierarchically subordinated body of the Ministry of the Interior. Its primary responsibility consists of supporting the National System of Civil Protection (SINAPROC) of the technical requirements its operational demands.

CENAPRED Carries out research, training, instrumentation, and dissemination activities about natural and anthropogenic phenomena that can lead to disaster situations, as well as actions to reduce and mitigate the adverse effects of such phenomena, in order to contribute to aid in a better-prepared population to face said situations.

According to open access information from CENAPRED, there are three kinds of declarations, which are a contingency, emergency, and disaster, for this work, emergency declarations, are considered.

On the one hand, the specific objective of the contingency declaration is to support low-income agriculture, fishing, and aquaculture producers, to reincorporate them in their activities in the shortest possible amount of time before the occurrence of atypical, relevant, non-recurrent and unpredictable weather contingency.

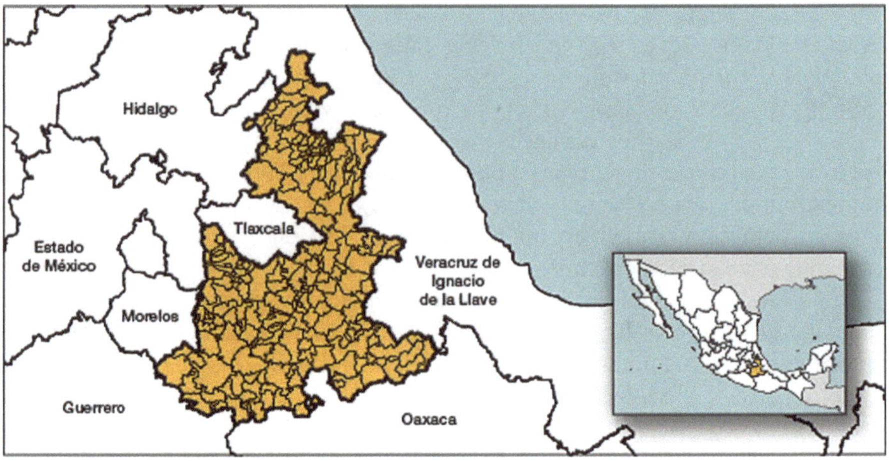

Fig. 4.1 The geographic location of Puebla, México. *Source* National Institute of Statistics and Geography

On the other hand, the emergency declaration is the recognition of the Ministry of the Interior that one or several municipalities or political delegations of a federative entity are faced with the imminence or high probability of presenting a disturbing phenomenon of natural origin, which can cause an excessive risk to the security and integrity of the population. The emergency declaration is directed to the attention of the life and health of the population. Moreover, the disaster declaration aims to provide resources for the reconstruction of the damages suffered in the homes and public infrastructure.

In summary, the total of natural phenomena presented in 18 years in Puebla is 1679, of which 38% were emergency declarations, 34% contingency declarations, and 28% disaster declarations with a whole event of 644, 561, and 474 respectively (see Fig. 4.2).

Fig. 4.2 Total natural phenomena per type of declaration in Puebla 2000–2018. Own elaboration based on open access data from CENAPRED

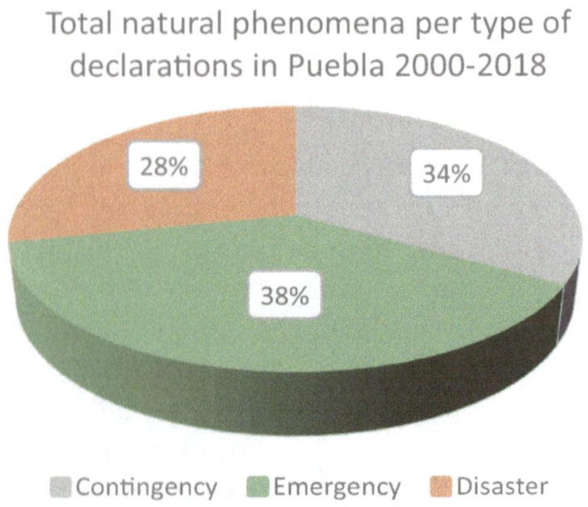

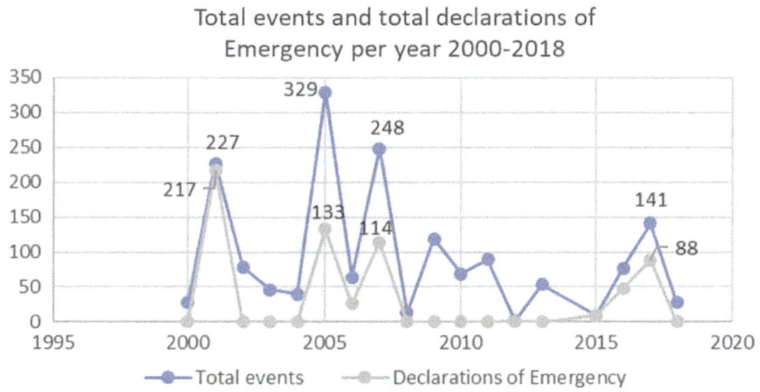

Fig. 4.3 Total events and total declarations of emergency per year from 2000 to 2018. Own elaboration based on open access data from CENAPRED

According to open access information obtained from (CENAPRED, 2006–2017), the natural phenomenon has occurred year after year. The years 2001, 2005, and 2007 are the most representative because many types of disaster were present like snowfall, tropical cyclones, heavy rain, low temperatures, and frost). Two hundred twenty-seven events were recorded at the end of 2001, of which 217 were emergency declarations that represented 95% of the total events in that year. In 2005 there were 329 events registered, which 133 were emergency declarations representing 40% of the total events in 2005, at the end of 2007 there were 248 events, of which 114 were emergency declarations that represented 45% of the total of events in 2007 (see Fig. 4.3).

Six hundred forty-four emergency declarations were recorded in the last 18 years, of which 226 were snowfalls, 192 tropical cyclones, 116 heavy rains, 64 low temperatures, and 45 touches of frost, which indicates that snowfalls occurred in greater quantity followed by tropical cyclones and heavy rains (see Fig. 4.4).

According to official information, Natural Disasters Fund (FONDEN) is a public financial instrument that aims to provide aid and assistance to the population that is facing the imminence of a disturbing natural phenomenon. This fund is activated through an Emergency Declaration or a Disaster Declaration.

According to open access information from FONDEN, in past years, at least 88 of 217 municipalities have been affected by some natural phenomenon; therefore, there have been figures of victims, according to the emergency declarations registered by FONDEN, in 2015 there were 40,416 victims and in 2016; 52,112, which indicates that there was an increase in the total number of victims (FONDEN, 2017).

Currently, in Puebla, there are no HRDC or strategic points that the distribution reaches as quickly as possible; all the municipalities affected by a natural disaster.

After analyzing the current situation of Puebla, it can see the importance of having humanitarian response centers at strategic points in order to reduce distances and times to achieve a supply of aid kits efficiently and in the shortest time possible.

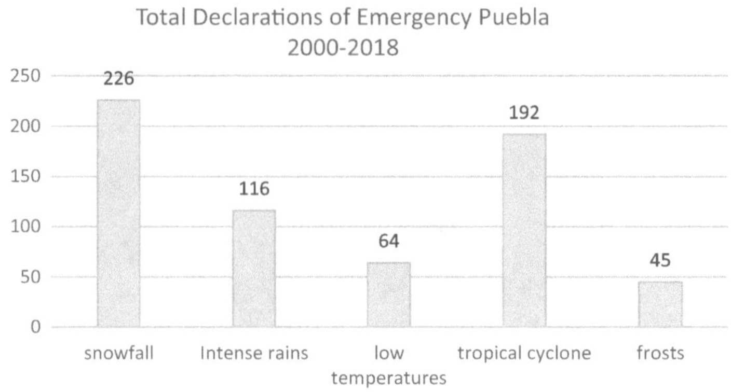

Fig. 4.4 Total declarations of emergency (2000–2018). Own elaboration based on open access data from CENAPRED

This work is the continuation of Chap. 3 of this book, which considers the results obtained from the research carried out by the authors Diana Sánchez-Partida, Brenda López-Duran, José-Luis Martínez-Flores, and Santiago-Omar Caballero-Morales, this research consisted of a case study focused on Puebla, Mexico. His work consisted of two types of studies; one qualitatively and the other using a Hierarchical Analytical Process (AHP) to obtain the best municipalities to install a Regional Humanitarian Response Deposit (RHRD) that protects the elements of first necessity, which will be delivered to the victims in case of emergency.

According to the second scenario, an AHP was used considering aspects such as infrastructure, roads, among others, safety sites were found in the northwest, center-west, center-east, southwest, southeast and center of the state of Puebla.

The 14 municipalities free of natural disasters identified in the aforementioned research are used for the application of the P-median model in the development of this work (see Table 4.1).

4.3 Literature Review

In the literature on Humanitarian Logistics (HL), several works have used models and methods of solution for the location of distribution centers in recent years.

Tavana et al. (2018) Based on a comparison between two methods, (NSGA-II) Non-dominated sorting genetic algorithm and a modified NSGA-II called reference point based non-dominated sorting genetic algorithm-II (RPBNSGA-II) they propose a network within humanitarian logistics in which central warehouses and the inventory of perishable products are located in a pre-disaster event as well as the routes of aid vehicles in a post-disaster event, finally they determine that for minor-scale problems the NSGA-II method is more useful and for large-scale problems the RPBNSGA-II method works better.

Table 4.1 Municipalities free of natural disasters in Puebla according to AHP

No..	Municipality	Weight	Area
1	Teziutlán	0.013670939	Northwest
2	San Miguel Xoxtla	0.013616067	Center-west
3	Puebla	0.012368291	Center-west
4	Zacatlán	0.011011129	Northwest
5	Izúcar de Matamoros	0.010843039	Southwest
6	Cuautlancingo	0.010834206	Center-west
7	Santiago Miahuatlán	0.010373660	Southeastern
8	Tehuacán	0.009931276	Southeast
9	Chalchicomula de Sesma	0.009658993	Center-east
10	Huejotzingo	0.009651304	Center-west
11	San Pedro Cholula	0.008880879	Center-east
12	Tepeaca	0.007914700	Central
13	Amozoc	0.007856477	Central
14	Atlixco	0.007641258	Center-west

Dufour et al. (2018) in their work present the simulation techniques, network optimization and statistical analysis to determine the total costs of installation and operation of a distribution center in order to attend the humanitarian crises in East Africa.

Albareda-Sambola et al. (2015) the objective of their work is to minimize the distance from the victims to the backup distribution centers in the assumption that the already defined distribution centers fail, the authors propose and develop different formulations executing the solution of several cases and comparing them with each other, the results show that instances with up to 50 nodes can be solved systematically.

Roh et al. (2018) present a methodology that uses fuzzy AHP (the Analytic Hierarchy Process) to identify the location points of feasible distribution centers and disseminates TOPSIS (the Technique for Order of Preference by Similarity to Ideal Solution) to evaluate the location points of distribution centers previously placed for international humanitarian aid organizations.

Maharjan and Hanaoka (2017) In their work use a simplex algorithm with branching to solve a modified version of the maximum coverage location problem where additional constraints are introduced that represent a real situation in Nepal, the purpose is to find the optimal locations of warehouses that are recommended to be installed to generate a chain of humanitarian aid ready to respond to possible disasters that occur unexpectedly.

Boonmee et al. (2017) Based on situations examined before and after a disaster in relation to the location of facilities such as warehouses, distribution centers, medical centers, debris disposal centers and shelters, the authors present the analysis of The four types of facility location problems, which are deterministic, dynamic, stochastic and fixed, also present results related to Humanitarian Logistics, finally they found that these can be solved by different techniques, such as decision theory, queuing theory, and fuzzy methods in addition to the use of advanced algorithms which are more efficient in large-scale problems.

Loree and Aros-Vera (2018) present a mixed integer non-linear programming (MINLP) mathematical model in Humanitarian Logistics to determine the location of distribution centers as well as the inventory allocation after a natural disaster with the purpose to minimize location and transportation costs that guarantee to meet the required demand by giving a profitable and equitable response.

López-Sánchez et al. (2019) present a hybrid algorithm as a solution method to the p-median model, which consists of combining two metaheuristics (GRASP) which is a greedy random adaptive search procedure, and a variable neighborhood search algorithm (VNS), finding that when solving a complete set of instances with different sizes, the hybrid GRASP_VNS algorithm obtains better solutions than the VNS algorithm or GRASP algorithm separately.

Hu and Dong (2019) present a stochastic programming model for the location of facilities with inventory and selection of suppliers considering the different discounts offered according to the order quantity and the required delivery time, in order to generate different alternatives to help humanitarian, highlighting the importance of establishing a close relationship with suppliers for the supply of first aid items.

Rezaei-Malek and Tavakkoli-Moghaddam (2014) present a bio-objective mathematical programming model in order to find a solution to determine the location of warehouses and the number of first aid products to later be distributed in logistics operations Humanitarian, the first objective is minimize the average response time and the second contemplates minimizing the costs incurred by these operations.

Xi et al. (2013) present a case study in western China, in which they use a linear programming model of the p-median considering the limitations of the rescue time with an algorithm based on the variable neighborhood search (VNS), showing a efficient performance.

Jia et al. (2007) propose a solution to determine the location of medical supply facilities, in order to minimize the loss of life and economic losses, using the p-median model.

Jia et al. (2014) present a linear programming model of p-mediana in which they use exact and approximate strategies in conjunction with the simulated annealing technique in order to find all the possible facilities to be used in a certain problem.

Zhu et al. (2013) analyze different scenarios to predict the possible communities in a region that may be at risk, using the p-median model to determine the facilities that will be used to provide care for affected communities.

Daskin and Dean (2004) examine the p-median model and the maximum coverage model to locate health care facilities, considering in the model the characteristics of accessibility, adaptability and availability to serve disaster areas or regions.

Gutierrez and Mutuc (2018) In order to optimize the delivery of supplies for medical care, they propose a mathematical model to optimally locate a temporary or fixed distribution facility or center considering a specific region.

Bean et al. (2016) They optimize the location of distribution centers for the supply of medical materials for casualties in a disaster in the Democratic Republic of the Congo with the application of a mathematical programming model.

Balcik and Beamon (2008) The objective of their work is to present a solution to the problem of the needs presented by casualties in a disaster with the implementation of a variant in the maximum coverage location model. Lin et al. (2012) propose a heuristic solution that consists of two phases to locate temporary warehouses around an area affected by natural disasters, considering the necessary resources to improve logistics planning.

Gutjahr and Dzubur (2016) in their model, they search minimize the total costs to open distribution centers and satisfy the required demand, for which they used a two-level model selecting in the first instance the distribution centers capacited and in the second instance the users choose a distribution center considering the distance that exists to arrive until there and also considering their capacity. Manopiniwes et al. (2015) propose a solution model when a case of flood or other catastrophe occurs in a specific area in Thailand, with the aim of minimizing the total costs of Humanitarian Logistics operations, locating the necessary distribution centers and inventories.

4.4 Selection of HRDC Using P-Median Model

It is essential to deliver humanitarian aid kits promptly, tents, blankets, supplies, heaters nearby to reduce losses caused by disasters. Therefore, time and distance are critical factors in emergency management. Given the importance of the selection of distribution centers. This document considers a P-median model that selects the HRDC by minimizing the total distance and the total time experimenting in each case with 1, 2, 3 and 4 HRDC.

What does the P-median model mean? Mathematically, the P-median model can be formulated in several ways.

Rosing et al. (1999) mentioned that the P-median model consists of selecting p sites of facilities in a network between the n vertices or nodes of the demand of that network. Choose those sites that reduce the weighted sum of the distances from each demand node to the facility site closest to its lowest possible value.

Erkut and Tansel (1992) mentioned that the P-median model is arguably the most popular in the literature location of the facility that treats multiple facilities. Given n demand points in some space such as the Euclidean plane or the road network, the goal of the model is to locate service facilities and assign the n demand points to the service facilities in a way that minimizes the total distance of travel for service.

Bowerman et al. (1999) mentioned that the objective of the P-median model is to choose p sites from a set of candidate sites of the facility to minimize the cost of attending to a set of existing locations of user demand.

Zhao and Batta (1999) mentioned that the P-median model is presented as follows:

P = {P_1, P_2, ..., P*n*}, a set of *n* (*n* may be larger) demand points or existing locations of ease in a skipped area A with positive weights {W_1, W_2, ..., W*n*} Assume that *p* new facilities with locations in the set X = {X_1, ..., X*n*} should be located in P. Denote as F(x) the total trip of each demand point to its nearest facility.

Hansen and Mladenovic (1997) mentioned that it is considered a set L of *m* facilities (or points of location), a set U of *n* users (clients or points of demand), and a $n \times m$ matrix D with the distances (or costs) of travel to satisfy the demand of the user located at point *i* from the localized facility to point *j*, for all $j \in$ L and $i \in$ U. The objective is to minimize the sum of the distances (or transport costs).

The primary objective of the model is to achieve the minimum time and distance in the whole system. A product of the allocation of demands to servers.

The model raises the problem of finding the location of P new centers, within *n* possible locations, concerning a series of demands to be met (*m* demand points).

The P-Median model

$$Min = \sum_{i \in N} \sum_{j \in j} C_{ij} X_{ij} \tag{4.1}$$

Constrains

$$\sum_{j \in J} X_{ij} = 1 \quad \forall i \in N \tag{4.2}$$

$$\sum_{j \in J} Y_j = p \tag{4.3}$$

$$X_{ij} \leq Y_j \quad \forall i \in N,\ j \in J \tag{4.4}$$

$$X_{ij} \in \{0, 1\},\ Y_j \in \{0, 1\} \forall i \in N, j \in J \tag{4.5}$$

i $\in N$ = index and set of demand nodes; $j \in J$ = index and set of potential facility nodes; W_i = total demand to cover the node *i*; d_{ij} = distance from node *i* to node *j; p* = number of facilities to locate; X_{ij} = take the value 1 if node i is assigned to facility j and 0 otherwise; Y_i = it takes the value 1 if a facility is opened in *j* and 0 otherwise.

The objective function (4.1) minimizes the weighted sum of the distances (or transport times) associated with the demand nodes that are assigned to the facility sites. The set of constraints (4.2) ensures that all demand nodes are assigned precisely to one facility. The constraints (4.3) sets the number of facilities to *p*. The set of constraints (4.4) prohibits any assignment to a site that does not have the ease; that is, only nodes with ease will supply the product. The set of constraints (4.5) reinforces the binary nature of the decisions of locating a faculty to a node and assigning a demand node to a facility.

With the analysis made on the historical information of the emergency declarations that each municipality had in the state of Puebla during 18 years, it was determined a municipality presents three emergency declarations on average during this period; it means that is a data to consider so that preventive measures are taken.

In this work, it is considered that municipalities that present a higher or equal number of emergency declarations than the average are the most vulnerable to a natural phenomenon. As a result, there are 100 of the 217 municipalities, which are used in the P-median model.

Also, it is considered to integrate the 14 municipalities free of natural disasters that resulted from the AHP study seen in the previous chapter and contained in Table 4.1. In Appendix 4.A, it is possible to see the names and number of the declaration of each municipality considers in this case.

The following map shows the geographical location of 100 municipalities with three or more Emergency Declarations to make more visual the geographical location of the 114 municipalities. These are represented in red color circles. Also, the map shows the geographical location of 14 municipalities free of natural disasters, which are represented in the green color circle (see Fig. 4.5).

Once the 114 municipalities are integrated, the distance matrix and a time matrix are created.

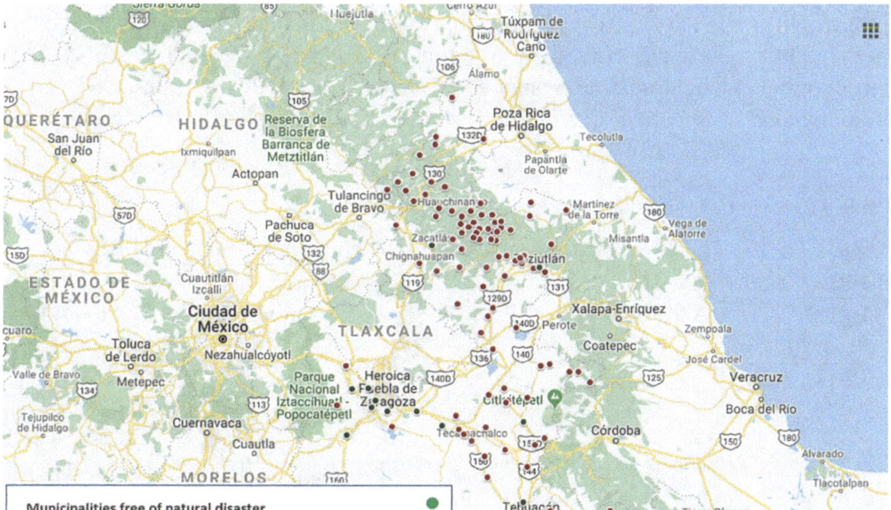

Fig. 4.5 One hundred municipalities with two or more emergency Declarations & 14 municipalities free of natural disasters. Own elaboration using the Google maps tool

4.4.1 Scenario 1, Solving Using Distances

The data matrix is generated to develop the first scenario, the P-median model is used assigning 1,2,3, and up to 4 HRDCs, the results obtained are generated and presented.

It was necessary to obtain the distances between each of the 100 municipalities to generate the data matri4. These were determined to be the most vulnerable to natural disasters, as well as the 14 municipalities free of natural disasters. For collecting these data was used the Google maps tool, in this tool, each place of origin (municipalities) and each place of destination (municipalities) was entered, several routes are shown with the corresponding distances. Those distances that reflected the smallest number were taken to realize the matri4. The result is a matrix with dimensions of 114 by 114, with a total of 12,996 data; the number of emergency declarations per municipality is considered as weight *w* to use the mathematical model, the resulting matrix is presented in Appendix 4.B.

The P-median model is solved using the software Lingo version 18. According to the results about the first scenario, we can see in the run one the municipality chose was Amozoc with a minimum distance of 16,260 km; in number 2 the municipalities have chosen are Chalchicomula de Sesma and Zacatlán with a minimum distance of 10,805 km; and in number 3 the municipalities are Teziutlán, Zacatlán, and Tepeaca with a minimum distance of 8423 km. Moreover, in number 4, the municipalities selected as HRDC are Puebla, Teziutlán, Zacatlán, and Santiago Miahuatlán with a minimum distance of 7166 km (see Table 4.2).

When these different scenarios are solved by assigning an HRDC, the chosen municipality is Amozoc, which must supply the humanitarian aid kits to the 114 municipalities. When is solved by assigning 2 HRDC's, the municipalities have chosen are Chalchicomula de Sesma and Zacatlán with a distribution of 64 and 49 municipalities, respectively. When it is resolved to assign 3 HRDC's, the selected

Table 4.2 Result of P-median model, considering distances

Runs	Distance minimum (Km)	Municipality chose as HRDC
1 HRDC	16,260	Amozoc
2 HRDC	10,805	Chalchicomula de Sesma
		Zacatlán
3 HRDC	8423	Teziutlán
		Zacatlán
		Tepeaca
4 HRDC	7166	Puebla
		Teziutlán
		Zacatlán
		Santiago Miahuatlán

Own elaboration

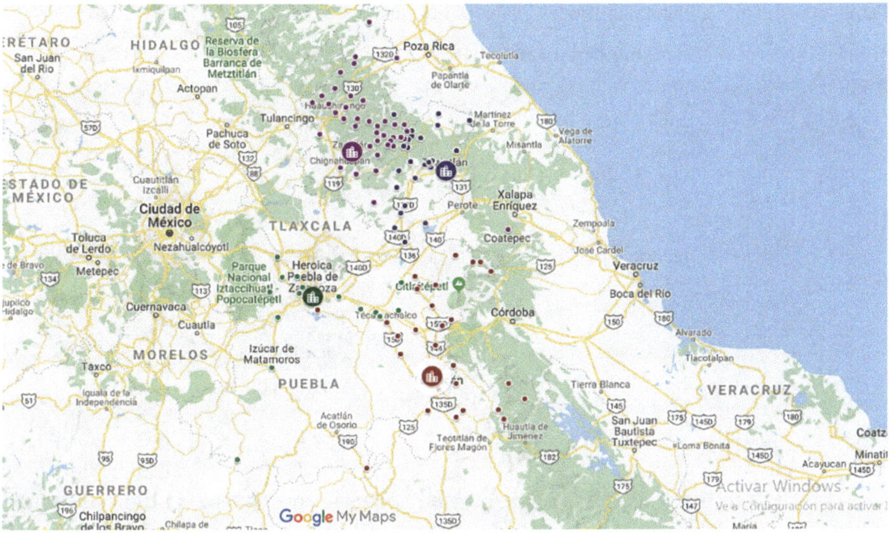

Fig. 4.6 Assignment of the 114 municipalities to 4 HRDC's (using distance). Own elaboration using the Google maps tool

municipalities are Teziutlán, Zacatlán, and Tepeaca with a distribution of 28, 40, and 46 municipalities respectively, and finally, when it is solved to assign 4 HRDC's, the elected municipalities are Puebla, Teziutlán, Zacatlán, and Santiago Miahuatlán with a distribution of 17, 28, 40, and 29 municipalities respectively all this information can be seen in Appendices 4.C, 4.D, 4.E, and 4.F.

To make the assignment more visual, we exemplified with the results when there are 4 HRDC in Fig. 4.5. The icons with the image of the building represent each RHCD, and the points represent the municipalities that are an assignment to each HRDC (See Fig. 4.6).

In the first scenario, we can see in the run number four the municipalities chosen as HRDC, so these results confirm that they are the most efficient points to generate a better humanitarian response to the population damaged by a natural disaster if the distance is considered.

4.4.2 Scenario 2, Solving Using Times

Now, the time matrix was generated to develop the second scenario, the P-median model is used assigning 1, 2, 3, and up to 4 HRDCs, the results obtained are generated and presented. The resulting matrix is presented in Appendix 4.G.

Table 4.3 Result of P-median model, considering times

Runs	Time minimum (minutes)	Municipality chose as HRDC
1 HRDC	15,743.85	Amozoc
2 HRDC	12,478.70	Amozoc
		Zacatlán
3 HRDC	10,699.72	Teziutlán
		Zacatlán
		Tepeaca
4 HRDC	9867.97	Chalchicomula de Sesma
		Teziutlán
		Zacatlán
		Cuautlancingo

s

According to the results about the second scenario, we can see in the run number one the municipality chosen, Amozoc, with a minimum time of 15,743.85 min. In run two, the municipalities chosen were Amozoc and Zacatlán with a minimum time of 12,478.7 min. In number 3, the municipalities selected were Teziutlán, Zacatlán, and Tepeaca, with a minimum time of 10,699.72 min. Moreover, the number 4 the municipalities have chosen as HRDC are Chalchicomula de Sesma, Teziutlán, Zacatlán, and Cuautlancingo with a minimum time of 9867.9 min (see Table 4.3).

From the results generated in Lingo, the assignment of the municipalities to each HRDC is obtained. That is, when the run is made with one HRDC, the municipality chosen is Amozoc, so 114 municipalities are assigned to it. When it is decided by 2 HRDC, the municipalities chosen are Amozoc and Zacatlán with an assignment of 41 y 73 municipalities, respectively. When 3 HRDCs are chosen, the municipalities chosen are Teziutlán, Zacatlán, and Tepeaca with an assignment of 35, 33 and 46 municipalities respectively, and finally when it is decided by 4 HRDC, the municipalities chosen are Chalchicomula de Sesma, Teziutlán, Zacatlán, and Cuautlancingo with an assignment of 32, 34, 33, and 15 municipalities respectively (see Appendices 4.H, 4.I, 4.J, and 4.K).

To make the assignment more visual, we exemplified with the results of run number 4 when there are 4 HRDC, and the figure shows the municipalities are the ones that humanitarian aid should give each of the HRDC. The icons with the image of the building represent each HRDC, and the points represent the municipalities that are an assignment to each HRDC (see Fig. 4.7).

In the second scenario, we can see in the run number four the municipalities chosen as HRDC, so these results confirm that they are the most efficient points to generate a better humanitarian response to the population damaged by a natural disaster if the time is considered.

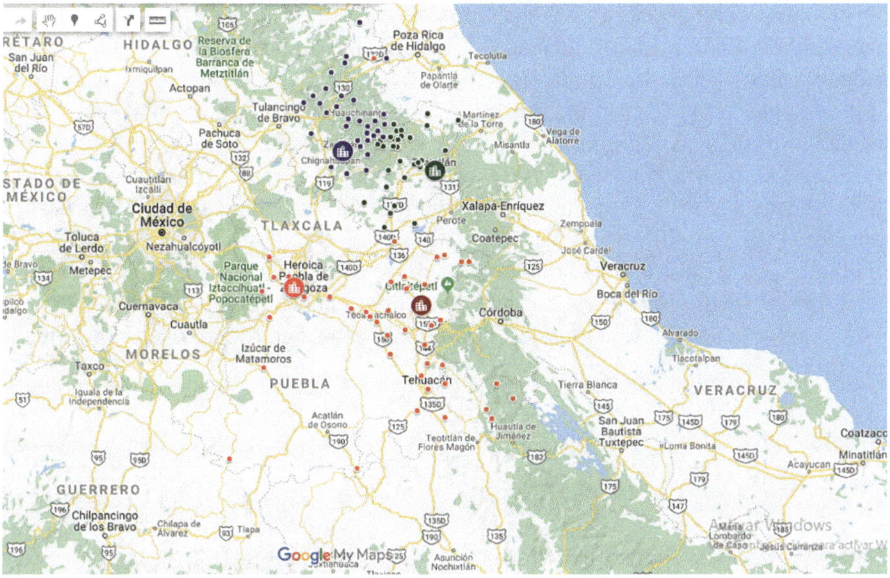

Fig. 4.7 Assignment of the 114 municipalities to 4 HRDC's (using times). Own elaboration using the Google maps tool

4.5 Concluding Remarks

After applying the P-median model in the first scenario, it was obtained the minimum distance of 7166 km; Puebla, Teziutlán, Zacatlán, and Santiago Miahuatlán were the municipalities selected as HRDC. Moreover, the minimum time was of 9867.97 min; Chalchicomula de Sesma, Teziutlán, Zacatlán, and Cuautlancingo were the municipalities selected as HRDC. The locations of those municipalities are around the Puebla city, so we can propose the installation of 4 HRDC´s because it had the minimum total distance and the minimum time.

The utmost importance to locate the HRDCs is to determine the safe place to supply aid kits to those municipalities most susceptible to natural disasters. The importance of using research techniques or methods lies in finding exact solutions to complex problems based on mathematical modeling where restrictions must be satisfied, in this case, distances and times have been considered as part of the model.

The P-median model in the application of this problem gave us satisfactory results in the sense of finding where the HRDC's should be established. It is essential to mention that some HRDCs have more municipalities assigned than others, this is due to the total sum of distances, without losing sight of the satisfaction of all the municipalities.

Having planning about the locations of the humanitarian response centers reduces the possibilities of increasing the damage to the health of the victims, safeguard lives and reduce impacts.

It is suggested to issue the results to the government as support in making decisions in order to safeguard lives from natural disasters.

Also, it must be considered the characteristics of the aid kits to hold in each HRDC because each region impacts different types of phenomenon, also must be considered the type of means of transport.

Finally, we would like to conclude by presenting some suggestions for further development of this study. This kind of research can be replicated at the national level. Another focus is to simulate the probable events that may occur and thus to support decision making.

References

Albareda-Sambola, M., Hinojosa, Y., Marín, A., & Puerto, J. (2015). When centers can fail: A close second opportunity. *Computers & Operations Research, 62,* 145–156. https://doi.org/10.1016/J.COR.2015.01.002.

Balcik, B., & Beamon, B. M. (2008). Facility location in humanitarian relief. *International Journal of Logistics, 11*(2), 101–121.

Bean, W. L., Yadavalli, V. S. S., & Mpita, S. N. (2016). Integrated facility location planning and demand assessment for humanitarian logistics: A case study in the Democratic Republic of the Congo. *Management Dynamics: Journal of the Southern African Institute for Management Scientists, 25*(1), 34–50.

Boonmee, Ch., Arimura, M., & Asada, T. (2017). Facility location optimization model for emergency humanitarian logistics. *International Journal of Disaster Risk Reduction, 24,* 485–498. https://doi.org/10.1016/j.ijdrr.2017.01.017.

Bowerman, R. L., Calamai, P. H., & Brent Hall, G. (1999). The demand partitioning method for reducing aggregation errors in p-median problems. *Computers & Operations Research, 26*(10–11), 1097–1111. https://doi.org/10.1016/S0305-0548(99)00020-9.

CENAPRECE Report. (2018). *Manual de atención a la salud ante desastres*. Retrieved May 23, 2019, from http://www.cenaprece.salud.gob.mx/programas/interior/emergencias/descargas/pdf/ManualPresentacion.pdf.

CENAPRED. (2006–2017). *Declaratorias sobre emergencia, desastre y contingencia climatológica a nivel municipal entre 2000 y 2018*. National Center for Disaster Prevention. Retrieved March 15, 2019, from https://datos.gob.mx/busca/dataset/declaratorias-sobre-emergencia-desastre-y-contingencia-climatologica/resource/41444ebe-6a35-4631-8f91-9237d5114488 [in Spanish].

CRED Report. (2015). Annual disaster statistical review 2015. Retrieved May 25, 2019, from on http://www.cred.be/sites/default/files/ADSR_2015.pdf.

Daskin, M. S, & Dean, L. K. (2004). *Location of health care facilities BT operations research and health care: A handbook of methods and applications* (M. L. Brandeau, F. Sainfort, & W. P. Pierskalla, eds.). https://doi.org/10.1007/1-4020-8066-2_3

Dufour, É., Laporte, G., Paquette, J., & Rancourt, M. (2018). Logistics service network design for humanitarian response in East Africa. *Omega, 74,* 1–14. https://doi.org/10.1016/J.OMEGA.2017.01.002.

Erkut, E., & Tansel, B. C. (1992). On parametric medians of trees. *Transportation Science, 26*(2), 149–156. https://doi.org/10.1287/trsc.26.2.149.

FONDEN. (2017). *Recursos autorizados por declaratoria de desastre. National System of Civil Protection*. Retrieved March 19, 2019, from http://www.proteccioncivil.gob.mx/es/ProteccionCivil/Recursos_Autorizados_por_Declaratoria_de_Desastre [in Spanish].

GAR Report. (2015). *Evaluación Global sobre la Reducción de Riesgo de Desastre*. Retrieved May 21, 2019, from https://www.preventionweb.net/english/hyogo/gar/2015/en/gar-pdf/GAR15_Pocket_ES.pdf [In Spanish].

Gutierrez, M. T. E., & Mutuc, J. E. S. (2018). A model for humanitarian supply chain: An operation research approach. *Procedia Engineering, 212,* 659–666. https://doi.org/10.1016/J.PROENG.2018.01.085.

Gutjahr, W. J., & Dzubur, N. (2016). Bi-objective bilevel optimization of distribution center locations considering user equilibria. *Transportation Research Part E: Logistics and Transportation Review, 85,* 1–22. https://doi.org/10.1016/J.TRE.2015.11.001.

Hansen, P., & Mladenović, N. (1997). Variable neighborhood search for the p-median. *Location Science, 5*(4), 207–226. https://doi.org/10.1016/S0966-8349(98)00030-8.

Hu, S., & Dong, Z. S. (2019). Supplier selection and pre-positioning strategy in humanitarian relief. *Omega, 83,* 287–298. https://doi.org/10.1016/J.OMEGA.2018.10.011.

INEGI. (2016). *Anuario estadístico y geográfico de Puebla 2017*. Retrieved March 15, from http://internet.contenidos.inegi.org.mx/contenidos/Productos/prod_serv/contenidos/espanol/bvinegi/productos/nueva_estruc/anuarios_2015/702825077174.pdf [In Spanish].

Jia, H., Ordóñez, F., & Dessouky, M. (2007). A modeling framework for facility location of medical services for large-scale emergencies. *IIE Transactions, 39*(1), 41–55. https://doi.org/10.1080/07408170500539113.

Jia, T., Tao, H., Qin, K., Wang, Y., Liu, C., & Gao, Q. (2014). Selecting the optimal healthcare centers with a modified P-median model: A visual analytic perspective. *International Journal of Health Geographics, 13*(1), 42. https://doi.org/10.1186/1476-072X-13-42.

Lin, Y.-H., Batta, R., Rogerson, P. A., Blatt, A., & Flanigan, M. (2012). Location of temporary depots to facilitate relief operations after an earthquake. *Socio-Economic Planning Sciences, 46* (2), 112–123. https://doi.org/10.1016/J.SEPS.2012.01.001.

López-Sánchez, A. D., Sánchez-Oro, J., & Hernández-Díaz, A. G. (2019). GRASP and VNS for solving the p next center problem. *Computers & Operations Research, 104,* 295–303. https://doi.org/10.1016/J.COR.2018.12.017.

Loree, N., & Aros-Vera, F. (2018). Points of distribution location and inventory management model for post-disaster humanitarian logistics. *Transportation Research Part E: Logistics and Transportation Review, 116,* 1–24. https://doi.org/10.1016/J.TRE.2018.05.003.

Maharjan, R., & Hanaoka, S. (2017). Warehouse location determination for humanitarian relief distribution in Nepal. *Transportation Research Procedia, 25,* 1151–1163. https://doi.org/10.1016/J.TRPRO.2017.05.128.

Manopiniwes, W., Nagasawa, K., & Irohara, T. (2015). Humanitarian relief logistics with time restriction: Thai flooding case study. *Industrial Engineering and Management Systems*. https://doi.org/10.7232/iems.2014.13.4.398.

Natural Disasters Fund. Retrieved March 20, 2018, from https://www.gob.mx/segob/documentos/fideicomiso-fondo-de-desastres-naturales-fonden.

National Institute of Statistics and Geography. Retrieved April 3, 2019, from https://www.inegi.org.mx.

Rezaei-Malek, M., & Tavakkoli-Moghaddam, R. (2014). Robust humanitarian relief logistics network planning. *Uncertain Supply Chain Management*. https://doi.org/10.5267/j.uscm.2014.1.002.

Roh, S. Y., Shin, Y. R., & Seo, Y. J. (2018). The pre-positioned warehouse location selection for international humanitarian relief logistics. *The Asian Journal of Shipping and Logistics, 34*(4), 297–307. https://doi.org/10.1016/J.AJSL.2018.12.003.

Rosing, K., ReVelle, C., & Schilling, D. (1999). A gamma heuristic for the p-median problem. *European Journal of Operational Research, 117*(3), 522–532. https://doi.org/10.1016/S0377-2217(98)00268-9.

Tavana, M., Abtahi, A.-R., Di Caprio, D., Hashemi, R., & Yousefi-Zenouz, R. (2018). An integrated location-inventory-routing humanitarian supply chain network with pre- and post-disaster management considerations. *Socio-Economic Planning Sciences, 64,* 21–37. https://doi.org/10.1016/J.SEPS.2017.12.004.

Xi, M., Ye, F., Yao, Z., & Zhao, Q. (2013). A modified p-median model for the emergency facilities location problem and its variable neighbourhood search-based algorithm. *Journal of Applied Mathematics, 2013,* 10. https://doi.org/10.1155/2013/375657.

Zhao, P., & Batta, R. (1999). Analysis of centroid aggregation for the Euclidean distance p-median problem. *European Journal of Operational Research, 113*(1), 147–168. https://doi.org/10.1016/S0377-2217(98)00010-1.

Zhu, J., Huang, J., & Liu, D.-G. (2013). Robust p-median model for facility location problem based on scenario analysis in emergency management. 11th International Symposium on Operations Research and Its Applications in Engineering, Technology and Management 2013 (ISORA 2013), pp. 1–4. https://doi.org/10.1049/cp.2013.2251.

Chapter 5
Findings in Medicine Forecast in Cases of Hydrometeorological Phenomenon in Chiapas, Mexico

Paola Jiménez-Alonso, Diana Sánchez-Partida, Patricia Cano-Olivos, and José-Luis Martínez-Flores

Abstract Incorporating forecasts for situations of natural disasters may be of great use for knowing ahead of time about the requirements of the actual demands of the victims. It makes it possible to show a more efficient capacity of response when facing the negative impacts of a natural disaster. This study case was developed in the state of Chiapas, México, because it is a location where a natural disaster has a significant adverse impact due to the existing social vulnerability. This research integrates the use of time series in order to predict the demand for medicine des-tined to treat acute respiratory infections; a disease which originates from the impact of hydrometeorological events in that state. This study was developed to estimate the demand to provide the population in advance during the following period and by that be able to avoid diseases and even the loss of human lives caused by them. In the course of the research, it was detected that the consumption of medicine for the treatment of acute respiratory diseases shows a decreasing trend, which is probably related to the fact that there are more emergency declarations caused by heatwaves.

Keywords Humanitarian logistics · Natural disasters · Medicine forecasts · Time series

---

Electronic supplementary material The online version of this chapter (https://doi.org/10.1007/978-3-030-67295-9_5) contains supplementary material, which is available to authorized users.

P. Jiménez-Alonso · D. Sánchez-Partida (✉) · P. Cano-Olivos · J.-L. Martínez-Flores
Department of Logistics and Supply Chain Management, Universidad Popular Autónoma Del Estado de Puebla (UPAEP University), 17 Sur 901, Barrio de Santiago, 72410 Puebla, Mexico
e-mail: diana.sanchez@upaep.mx

D. Sánchez-Partida et al. (eds.), *Disaster Risk Reduction in Mexico*,
https://doi.org/10.1007/978-3-030-67295-9_5

5.1 Introduction and Literary Review

It is crucial to start with a clear definition of what a natural disaster is. The International Disaster Database (2019) defines a natural disaster as a situation or event which exceeds the local capacity, requiring external support at the national or international level. It is an unforeseen and frequently sudden event that causes considerable damage, destruction, and human suffering.

Natural disasters can be classified into:

- Geophysical.
- Meteorological.
- Hydrological.
- Climatological.
- Biological.

Linnerooth-Bayer et al. (2009) mentioned that these five types of natural disasters are very different from each other, depending on their frequency, unpredictability, and devastating impact. Some may occur as a consequence of or simultaneously with others. What all-natural disasters have in common, regardless of the type, is that they are all beyond human control.

Several studies have considered the economic and social aspects, as well as the psychological effect on the affected population in the face of hurricanes, earthquakes, floods, tornados, and other natural disasters (Capacci & Mangano 2015). However, few have considered their effects on human health (Kovács and Spens, 2009).

The majority of natural disasters, as well as the demand for goods for its victims, are unpredictable. Therefore, in the event of a natural disaster, Jahre (2017) state that it is of utmost importance to count on immediate response, social coordination, as well as the organization of government authorities for the correct supply of different resources such as food, medicine and the necessary material for humanitarian aid.

5.1.1 Humanitarian Logistics

Humanitarian Logistics (HL) is a combination of different disciplines, focused on two main tasks: 1) Help the people to face disaster and 2) Improve the conditions of the people who cannot cover their basic needs. These two activities do not conflict with each other because the first tries to resolve emergencies, which occur suddenly due to natural causes, or because the human being provoked them, and the second is a constant fight for helping people to achieve higher life quality. Thomas and Kopczak (2005) define HL as "The process of planning, implementing and controlling efficiently the effective flow of costs and warehousing of goods and materials, as well as related information, from the point of origin until the point of consumption, to give relief to the suffering of vulnerable people."

HL faces particular and complicated circumstances during natural disasters. Usually, it has to be carried out in places where the infrastructure of transport and energy has been affected, which makes it even more challenging to carry out the activities. Also, Kunz and Reiner (2012) mentioned it has to be performed as fast as possible, optimizing the limited resources and making the supply chain more efficient, because countless human lives depend on it. Since most of the natural disasters are unpredictable, the demand for goods required by the victims is also unpredictable (Kovács and Spens, 2009).

Smalley (2003) considers that the main challenges of HL, and hence the problems that humanity is going to face during the next 50 years are shelter, water, food, education, access to primary health care, and safety. Unfortunately, according to the National Climatic Data Center (2018), along the same 50 years, the number of disasters per year will increase.

Resolving emergencies in disastrous situations or continuously fighting to help people reach a better quality of life, are variables, which boost each other. Unfortunately, the most vulnerable communities or those with more scarcity suffer the ravages of disasters more critically. Also, Holguín-Veras and Van Wassenhove (2014) state that the communities that are affected by disasters diminish their possibilities to cover the basic needs on a long-term basis.

5.1.2 Importance of a Proper Planning of Demand

"When you fail to prepare, you are preparing yourself for failure," Benjamin Franklin.

The administration of the supply chain is a concept, which originated and was boosted by the manufacturing industry, and has been adapted successfully in other areas. Humanitarian Logistics, for instance, has applied predictive methods for the demand for medicine in case of natural disasters.

One of the significant benefits of a reasonable administration of supply chains is "helping to resolve the disparity between demand and supply" (Waters, 2003). Creating inventory throughout the chain can control growth in demand. Nevertheless, any mistake in the prediction of the demand throughout the chain may fail the whole logistic operation. When a disaster happens, various logistic decisions have to be made. These have to be structured by the needs of the community.

When a disaster occurs, the distribution of supplies, for example, medication, to the victims, is vital. Any delay in the acquisition and distribution may complicate logistic operations and therefore accelerate the number of diseases and even, in the event they worsen, the number of deaths.

The efficient operation of the supply chain for medicine requires the optimization of financial and human resources, in order to maintain the purchasing activities before and during the disaster. Murat (2019) states that it is also essential to keep in mind that each crisis is unique and requires a response that is adequate for the demand.

During a natural disaster, mistakes in the supply chain not only pose a threat in terms of scarcity or excessive stock; mistakes are paid for with lives here. If resources are not delivered in the shortest time possible and under the necessary circumstances, entire communities may suffer different diseases, infections, and contagions. More so, if these diseases are not controlled, they may lead to epidemic outbreaks, which will result in a massive loss of human lives. That is the reason for the importance of efficient planning for the adequate supply of medication.

As Syahrir et al. (2015) mention, during a natural disaster, the demand for medical and health services increases because of that, it is imperative to optimize the services mentioned above in the supply chain of medical services. The success of a supply chain during a natural disaster is highly influenced by how the health system can work during the process.

Many types of research focus their studies on the creation of survival kits. Badillo-Valenzuela et al. (2016) determined the quantity, the type, and the turnover of the necessary essential elements which should be contained in the survival kits, as well as adequate logistics for the delivery of groceries to the settlements that suffer the effects of natural disasters. However, only a few approaches to the delivery of medicine in time and quantity. Apte et al. (2017) explore the considerations and recommendations for massive vaccine campaigns in response to natural disasters and their side effects, specifically cholera epidemics and the reserve of the vaccines necessary to treat the disease effectively. On their behalf, Shrestha et al. (2018), taking into account the importance of having access to reliable and right quality medication in low-income-countries following any disaster, based their investigation on the practices for the acquisition of medicine in pharmacies of Nepal.

5.1.3 Use of Time Series Forecasting for the Estimation of the Demand

As previously lined out, all supply process that is not planned appropriately implies related costs, and in the case of the negative impact of natural disasters, in this case, it is about human lives.

Erossa (2016) mentioned that one appropriate tool for planning is forecasting. Its purpose is to enable decision making for the future and providing a prediction of the risk involved in the decision.

One way of getting to estimations is through the method of time series forecasting. According to Chatfield (1988) forecasts with time series consist of adjusting a model of time series to historical data, and then putting it in contrast to information in the future. One of the main disadvantages of the time series methods is that the model with the best adjustment for historical data does not necessarily provide the best forecast. Masini-Vázquez (2014) state that the time series methods predict future values for the variable that is being evaluated, based exclusively on the historical pattern of that variable and supposing that this historical pattern will continue.

There is a wide variety of time series methods such as the simple single moving average, weighted moving average, exponential smoothing, exponential smoothing adjusted to trend, stationary multiplicative method, time series with stationary influences and trends, among others (Contreras-Juárez et al. 2016).

In this research, the information of 20 quarters is available. It is only a little information, which suggests the use of the simple average method. Nevertheless, the data is representing certain stationarity. Due to this fact, it is essential to assign different weights to each period, and therefore, the method of the weighted moving average will be used.

5.1.4 Used Time Series

Hanke and Wichern (2006) mentioned that the method of simple average is an adequate technique when the data that generates the series to be predicted has stabilized and when the environment of the series does not change. This method uses recent data in order to reduce the effects of random fluctuations and also to respond to changes in the process much faster. The result is the moving average given by the sum of the last N data, as shown in the Eq. (5.1).

$$M_T = \frac{1}{N}(d_{T-N+1} + d_{T-N+2} + \ldots + d_T) = \frac{1}{N} \tag{5.1}$$

where M_T is the value of the moving average; N refers to the number of periods to be considered in the moving average; d_T is the historical demand in the period T and T is the process in which the period happens (Contreras-Juárez et al., 2016).

On the other hand, Masini-Vázquez (2014) affirms that the technique of weighted moving average employs the ponderations with the subjective evaluation of the analysis concerning the importance of the most recent and oldest data when formulating a prediction. The weighted moving average of the demand is calculated by multiplying each period by a weighting factor, and dividing the result by the sum of the weighting factors; these factors may take values from 0 to 1, and the sum of them has to be equal to 1, as can be observed in Eq. (5.2)

$$X_t = \frac{\sum_{t=1}^{n} C_i * X_{t-1}}{\sum_{t=1}^{n} C_i} \tag{5.2}$$

where X_t refers to the weighted moving average of the demand in units in the period t; C_i is the weighting factor, and X_{t-1} is the real demand in units for periods before t. (Contreras-Juárez et al. 2016).

This research pursues the purpose of finding a prediction close to reality where the Mean Absolute Error (MAE) obtained from the prediction will be considered as Safety Stock (SS).

5.2 Study Case of Chiapas, Mexico

5.2.1 Introduction

Due to its geographical location, Mexico is a country, which is exposed to a high number of natural phenomena, such as earthquakes, hurricanes, floods, ice, volcanic eruption, and others. The intensity of these phenomena may provoke disasters whose effects present a high risk for human lives, turning into a test for government institutions regarding the protection against natural disasters.

Natural disasters occur in the whole world and affect an increasing number of people. According to Hugelius (2017) and Kovats et al. (2003), the effects of natural disasters in public health matters, depending on a series of factors that are present before, during, and after the event.

Unfortunately, natural disasters can neither be controlled nor predicted. However, in order to prevent them from inflicting more harm, it is vital to carry out campaigns of prevention and response in the face of the negative impact of natural disasters. It means that it is crucial to work on action plans and resilience.

Hijar et al. (2016) state that the possibilities of an effective response in the presence of sanitary risks are increased by reliable information of the needs regarding the health of the population and the needs of the rehabilitation of the Health, Water and Sanitation systems which were affected, as well as the communication of the appropriate information to the population.

In Mexico, disasters with most casualties are earthquakes and hydrometeorological disasters. According to Rodríguez (2004), the latter makes up 86% of the total number of affected people because the appearance of this phenomenon extends over a vast area and causes severe damage, mainly to housing, infrastructure, and agricultural areas.

With this context in mind, this article focuses its investigation on the planning of medication for victims in the state of Chiapas, by the declarations of emergency registered for hydrometeorological phenomena from 2013 through 2017. Its purpose is to estimate the demand in order to supply the population in advance before the next period and by this to avoid diseases and the loss of human lives, which they imply.

5.2.2 Analysis of Government Databases

The Mexican Federal Government, through its National Center for Preventive Programs and Disease Control (CENAPRECE), counts on a database with the information about declarations of emergency and disasters. By the Fund for Natural Disasters (FONDEN), a declaration of emergency is the acknowledgment that the population is encountering the high probability of the appearance of a disturbing phenomenon of natural origin, which may result in an excessive risk to safety and integrity. For its part, Civil Protection affirms that a declaration of disaster is the statement of any federal entity that a natural phenomenon has occurred and that it has caused damage to housings, as well as services and public federal, state, and/or municipal infrastructure. This research is based on reference taken from all the records of emergency declarations in the state of Chiapas during the years 2013 until 2017 retrieved from (Catalog of Health-Related Inputs for Emergency Care, 2018, and Inputs Authorized by Emergency Declarations per Year, 2018).

Through the analysis of the CENAPRECE database, the following statistics were identified (see Table 5.1).

Based on this analysis, it is concluded that hydrometeorological phenomena are those with the most significant effect in Mexico, and those can be classified as hailstorms, heavy rain, ice, hurricanes, tornados, and floods. On the other hand, it could also be determined that in the period of analysis (2013–2017), the only state that issued at least one declaration of emergency of this kind in each year was the state of Chiapas.

It is essential to point out that Chiapas is a state that has a high grade of vulnerability. According to the Council of Evaluation of Politics for Social Development (CONEVAL) in 2016, 77.1% of the population lives in conditions of poverty. 28.1% of them are considered to live in conditions of extreme poverty. This fact and the frequent impact of disturbing phenomena were taken as reference for the investigation (CONEVAL, 2018).

The study consists of planning the supply of medicine for the treatment of acute respiratory infections.

Table 5.1 Hydrometeorological events in the state of Chiapas from 2013 to 2017

Year	Number of emergency declarations	Hydrometeorological events (%)	Declarations originated in the state of Chiapas (%)
2013	77	97	4
2014	63	97	10
2015	96	92	2
2016	81	96	5
2017	63	75	5

Source National Center for Preventive Programs and Disease Control (CENAPRECE)

It is essential to mention that in case of a declaration, delivery, and supplies of medication depend on the seriousness and intensity of the phenomenon. The FONDEN establishes the following process for the request of medication

1. The municipality faces a declaration of emergency.
2. The state committee for public safety evaluates the effects caused by the phenomenon.
3. The Secretary of State sends the request for the medicine needed to control the emergency to CENAPRECE.
4. CENAPRECE controls that the request fulfills the current regulations.
5. The request for medicine complies with the norm.
6. CENAPRECE sends its ruling to the General Administration of FONDEN, establishing the necessary amount of medicine.
7. The General Administration of FONDEN authorizes the established amount of medicine and sends the request for acquisition to the General Administration of Material Resources and General Services.
8. The General Administration of Material Resources and General Services determines the supplier and carries out the acquisition and delivery of the medicine.
9. The supplier to the state's health authorities performs the delivery of the medicine.
10. The state's health authorities send CENAPRECE the report of the utilization of the medicine in a period of no longer than 60 natural days after the emergency has been declared to come to an end.

Table 5.2 Medicine for the treatment of acute respiratory diseases

Medicine	Packaging
Amoxicillin	Oral suspension
Amoxicillin	Tablets
Erythromycin	Tablets
Erythromycin	Oral suspension
Trimethoprim-Sulfamethizole	Tablets
Trimethoprim-Sulfamethizole	Oral suspension
Benzylpenicillin procaine - Benzylpenicillin crystalline	Injection
Benzylpenicillin compound benzathine: Benzathine	Injection
Chlorphenamine	Tablets
Chlorphenamine	Syrup
Fifenhidramina	Syrup
Ambroxol	Oral suspension
Salbutamol	Syrup
Salbutamol	Aerosol

Source Catalogue of health-related supplies for emergency attention

11. CENAPRECE informs the General Administration of FONDEN about the approval of the report and details the distribution in each of the municipalities that were declared to be an emergency.

Fifteen reports for utilization of medicine for acute respiratory diseases issued by the state of Chiapas in the years 2013-2017 could be obtained. The historical demand of each medication was obtained from these reports,

According to FONDEN, the following medicines were assigned for acute respiratory diseases (Table 5.2).

According to the reports, a total of 210 historical pieces of information was obtained, corresponding to these fourteen medicines. The reports indicate the amount of requested medicine as well as the number of medicines delivered. It is essential to point out that the requested amount does not always correspond to the delivered amount of medication. Of the 210 historical pieces of information, only 73.33% of the requested medication shows the same amount as the delivered medication. No justification, behavior, or pattern was found to explain the difference. For that reason and the effects of this investigation, only the historical demand for the requested medication will be considered.

5.3 Methodology, Analysis, and Results

5.3.1 Methodology and Analysis of the Demand

In this research, techniques of quantitative forecasts are applied, based on the analysis of historical data from 20 periods or quarters, with the purpose of predict-ting the use of medicine for the year 2019.

Figure 5.1 shows the demand behavior for the medicine called Ambroxol. As can be observed, the period with higher demand is the 2nd quarter (April–June 2013), while quarters 1, 4, 5, 9, 10, 12, 13, 14, 17, 19, and 20 have no records. For

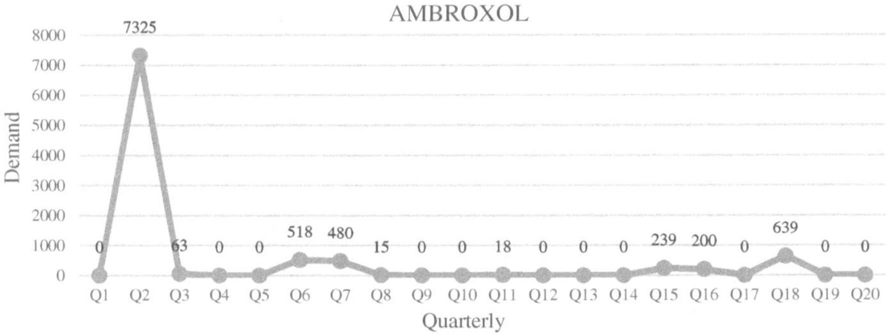

Fig. 5.1 The behavior of the demand for medicine Ambroxol

Table 5.3 Demand for the 14 medicines by period

Year	Demand	
	Quarterly 2	Quarterly 3
Ambroxol		
2013	7325	63
2014	518	480
2015	0	18
2016	0	239
2017	639	0
Amoxicillin suspension		
2013	7301	62
2014	516	450
2015	0	18
2016	0	1629
2017	2376	0
Amoxicillin Tablets		
2013	12929	110
2014	728	720
2015	0	32
2016	0	1779
2017	6082	80
Erythromycin Tablets		
2013	12929	62
2014	728	450
2015	0	18
2016	0	1629
2017	424	0
Erythromycin Suspension		
2013	7301	62
2014	516	450
2015	0	18
2016	0	1629
2017	2186	0
Trimethoprim Tablets		
2013	12929	110
2014	914	750
2015	0	32
2016	0	1779
2017	10543	800
Trimethoprim Suspension		
2013	7301	62
2014	516	450

(continued)

Table 5.3 (continued)

Year	Demand	
	Quarterly 2	Quarterly 3
2015	0	18
2016	0	1629
2017	2186	0
Bencil procaine		
2013	6739	55
2014	477	400
2015	0	17
2016	0	1629
2017	692	0
Benzathine Benzyl		
2013	12929	110
2014	914	750
2015	0	32
2016	0	1853
2017	969	80
Chlorphenamine Tablets		
2013	24000	220
2014	1830	1500
2015	0	65
2016	0	292
2017	2486	800
Chlorphenamine Syrup		
2013	7582	65
2014	530	500
2015	0	19
2016	0	239
2017	643	0
Diphenhydramine		
2013	2710	23
2014	190	180
2015	0	0
2016	0	0
2017	552	0
Salbutamol Syrup		
2013	1095	9
2014	431	100
2015	0	0
2016	0	239
2017	0	0

(continued)

Table 5.3 (continued)

Year	Demand	
	Quarterly 2	Quarterly 3
Salbutamol Suspension		
2013	2586	22
2014	183	180
2015	0	0
2016	0	239
2017	186	0

the quarters with no data about the use of medicine, the value 0 was assigned to the demand, without considering those cases contained in the report, where the requested amount was equal to zero. It is essential to mention that total units were considered for the demand (oral suspension, a box of tablets, and/or syrup), but it does not consider the dosage established for the use of these medicines.

Through the analysis of the demand, it can be determined that the quarters with a higher request for medicines are the 2nd quarter (April–June) and 3rd quarter (July–September) of each year. It indicates that there is stationarity in the "rain season." This way, the predictions for each quarter were performed independently.

Table 5.3 shows the value of demand for all the medicine for Quarter 2 (Q2) and Quarter 3 (Q3) of each year.

5.3.2 Development and Results

Once the demand was analyzed, it was determined that the time series present certain stationarity. Hence the importance of assigning a weight to the quarterly periods. Through the method of moving weighted averages, the prediction for the periods of the year 2018 was carried out.

The weighting factor C_i was obtained through the analysis of the demand. For each case, the frequency (K) presented by the demand, was determined. For the 2nd quarter, all medicines presented a frequency equal to 1. Therefore, a weighting value of 1 was assigned. For quarters 3, 13, and 14, medicines showed frequencies equal to 2, and in this case, the weighting value was obtained by calculating the difference between the maximum and minimum value of each frequency and, finally, the average.

Table 5.4 shows the frequency K, as well as the weighting factor C_i for the analysis of quarters 2 and 3 for all the medicines. In the same way, the predicted values are shown, as well as the mean absolute error (MAE), where the latter is representing the Safety Stock (SS).

Table 5.4 Results for all medicines for quarters 2 and 3

Quarterly	Frequency (K)	Weighting factors C_i	Forecast	MAE/SS
Ambroxol				
Q2	1	1, 0	0	2827
Q3	2	0.90, 0.10	215	107
Amoxicillin Suspension				
Q2	1	1, 0	0	3398
Q3	2	0.93, 0.07	1515	470
Amoxicillin Tablets				
Q2	1	1, 0	0	6580
Q3	2	0.92, 0.08	1643	444
Benzathine Benzylpenicillin				
Q2	1	1, 0	0	4937
Q3	2	0.92, 0.08	1711	462
Procaine Benzylpenicillin				
Q2	1	1, 0	0	2636
Q3	2	0.93, 0.07	1515	483
Chlorphenamine Syrup				
Q2	1	1, 0	0	2918
Q3	2	0.90, 0.10	215	114
Chlorphenamine Tablets				
Q2	1	1, 0	0	9439
Q3	2	0.82, 0.18	382	676
Diphenhydramine Syrup				
Q2	1	1, 0	0	1151
Q3	1	1, 0	0	68
Erythromycin Suspension				
Q2	1	1, 0	0	3334
Q3	2	0.93, 0.07	1515	470
Erythromycin Tablets				
Q2	1	1, 0	0	8129
Q3	2	0.92, 0.08	1643	436
Salbutamol Syrup				
Q2	1	1, 0	0	509
Q3	2	0.96, 0.04	229	55
Salbutamol Suspension				
Q2	1	1, 0	0	985
Q3	2	0.94, 0.06	225	39
Trimethoprim Suspension				
Q2	1	1, 0	0	334
Q3	2	0.93, 0.07	1515	470

(continued)

Table 5.4 (continued)

Quarterly	Frequency (K)	Weighting factors C_i	Forecast	MAE/SS
Trimethoprim Tablets				
Q2	1	1, 0	0	8129
Q3	2	0.92, 0.08	1701	615

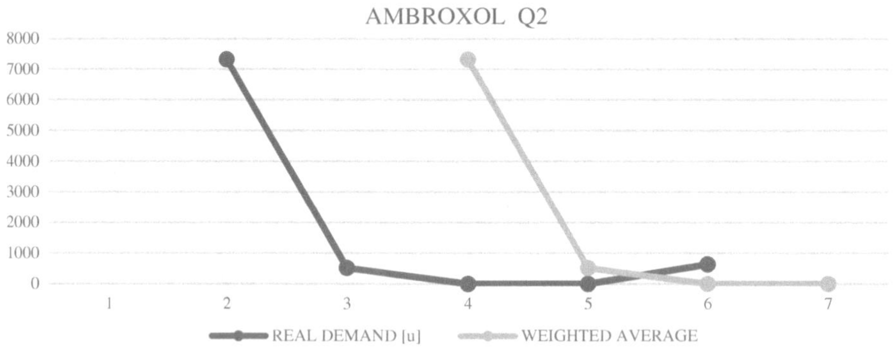

Fig. 5.2 The behavior of Ambroxol in Q2

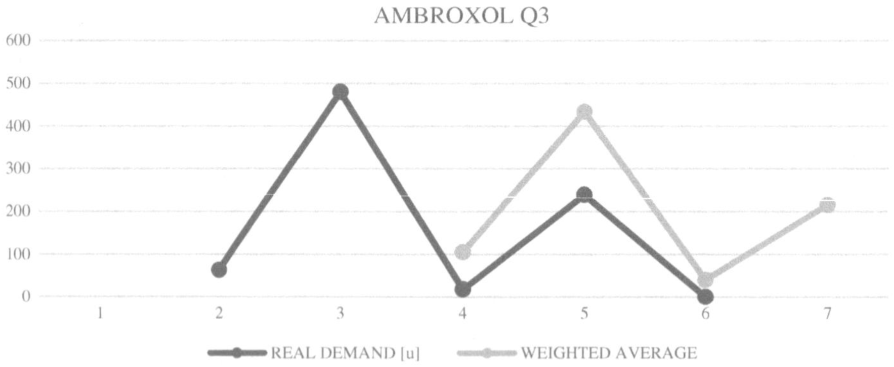

Fig. 5.3 The behavior of Ambroxol in Q3

As can be observed in Table 5.4, MAE is relatively high. It is because time series has the following disadvantage: the model that provides the best adjustment for the historical data not necessarily provides the best forecast. Although the error is high, this can be covered with an SS in order to guarantee the supply of medicine to the affected population.

In Figs. 5.2 and 5.3 the behavior of the real demand and the weighted averages for Q2 and Q3 respectively can be observed.

The weighted averages can be adjusted to the behavior of the demand, as can be observed in Figs. 5.2 and 5.3. Nevertheless, in both graphics, an imbalance can be observed. It is due to a trend in the data before carrying out the prediction. The moving averages are only based on the last prediction of the level and do not take into account a prediction of the trends.

5.4 Discussion and Conclusions

According to the results obtained, it can be concluded that the demand for medicine destined to the treatment of acute respiratory infections in Chiapas, Mexico, has a trend towards decline. It is possible because of different reasons such as:

a) The appearance and magnitude of climatic changes make it harder to predict natural phenomena. That means that it is getting more complex over time to determine if there is certain stationarity of the events. There are core data that shows that rains and hydrometeorological phenomena occur less and less. In fact, in the short term, droughts will occur, and the world will have to face food shortage. In Mexico, until July 2018, there have been a total of 37 declarations of emergency. 75.68 correspond to declarations of emergency caused by heat waves, and only 18.92% correspond to events caused by hydrometeorological phenomena. Chiapas only issued one declaration of emergency, which corresponds to a heatwave that affected a population of 124,000 people.
b) The Government has propelled different care and disease prevention programs, such as vaccine campaigns, which might be working. It is conceivable that the population is better prepared to respond to this kind of disaster.

In order to obtain predictions with a better margin of error, it is essential to count on a good number of historical data. In this research, it was impossible to obtain more information because the reports for utilization of medicine are only available for the years from 2013 to 2017.

Another critical issue is the reliability of the information. As mentioned before, the requested vs. the delivered medicine does not match entirely. There are cases where the difference exceeds 90%. Specific questions arise from this fact: What is the reason for this difference? Are the reports elaborated correctly? If the population is not receiving the medicines, it requests: are there victims who do not receive medical care and no supply of medicine? Were the needs miscalculated? Alternatively, is it a problem of the system, originated by the issuing institution, entrusted with preparing the reports?

The Mexican Government has established a percentage for the utilization of medicine based on the population, taking into account women, children, and the elderly. However, in none of the reports that were analyzed, this percent-target is fulfilled. This percentage is established in the catalog for health-related supplies for a response to emergencies of the FONDEN. There are cases where small amounts of medicine are delivered, and the report states a high number of beneficiaries. However, there are cases where the total amount of medicine was considerably high, but the number of beneficiaries was few. Because of the points stated before, it is suggested to develop a more efficient system for the request and delivery of medicine, as well as more accurate reports.

Different researches focus on the planning, establishment of warehouses, and distribution of food. However, medication is equally important because when anticipating the need, the spreading of diseases and epidemics can be avoided, which in the long run can become the cause of death in the aftermath of the disaster.

It is essential to emphasize the trend towards drought and its different effects on the population. Therefore, for future research work, it is considered crucial to perform an analysis of the medicine needed for the treatment of diseases caused by this kind of phenomenon.

References

Apte, A., Gregory, J., Hudgens, B., & Taranto, C. (2017). Inventory management of cholera vaccinations in the event of complex natural disasters. Western Decision Sciences Institute (WDSI 2017) Conference, USA.

Badillo-Valenzuela, D., Báez, O., Chancey, M., Gamica, E., Quevedo, E., Sanchez-Partida, D., & Flores, J. L. (2016). Rapid response center for disasters, inventory management in Southeastern Mexico. In *The global conference on business & finance proceedings* (Vol. 11, No. 1, p. 73). Institute for Business & Finance Research.

Capacci, A., & Mangano, S. (2015). Las catástrofes naturales. *Cuadernos de Geografía: Revista Colombiana de Geografía, 24*(2), 35–51. https://doi.org/10.15446/rcdg.v24n2.50206.

Catalog of Health Related Inputs for Emergency Care (FONDEN). Retrieved 2018, February 10th from http://cenaprece.salud.gob.mx/programas/interior/emergencias/descargas/pdf/Catalogo FONDEN2016_Version1.pdf [In Spanish].

Contreras-Juárez, A., Zuñiga, A. C., Martínez-Flores, J. L., & Sánchez-Partida, D. (2016). Análisis de series de tiempo en el pronóstico de la demanda de al macenamiento de productos perecederos. *Estudios Gerenciales, 32*(141), 387–396. https://doi.org/10.1016/j.estger.2016.11.002 [In Spanish].

Council of Evaluation of Politics for Social Development (CONEVAL). https://coneval.org.mx [In Spanish].

Chatfield, C. (1988). The future of the time-series forecasting. *International Journal of Forecas ting, 4*(3), 411–419. https://doi.org/10.1016/0169-2070(88)90108-2.

Emergency Declarations. http://www.cenaprece.salud.gob.mx/programas/interior/emergencias/declaratorias_emergencia.html [In Spanish].

Erossa. (2016). *Proyectos de Inversión en Ingeniería: su metodología* (1st ed.). Grupo Noriega Editores [In Spanish].

Hanke, J., & Wichern, D. W. (2006). *Pronósticos en los negocios* (6th ed.). México: Pearson [In Spanish].

Hijar, G., Bonilla, C., Munayco, C. V., Gutierrez, E. L., & Ramos, W. (2016). Fenómeno El Niño y desastres naturales: intervenciones en salud pública para la preparación y respuesta. *Rev Peru MedEx Salud Publica, 33*(2), 300–310. https://doi.org/10.17843/rpmesp.2016.332.2205.

Holguín-Veras, J., & Van Wassenhove, L. (2014). *Strategies to manage material convergence to disaster sites, conference on health and humanitarian logistics*, Mexico.

Hugelius, K. (2017). *Disaster response for recovery : Survivors experiences, and the use of disaster radio to promote health after natural disasters* (Ph.D. dissertation). Örebro: Örebro university. http://urn.kb.se/resolve?urn=urn:nbn:se:oru:diva-52653.

Inputs Authorized by Emergency Declarations per Year FONDEN. Retrieved 2018 February 27 from http://www.proteccioncivil.gob.mx/es/ProteccionCivil/Insumos_autorizados_por_declaratoria_de_emergencia [In Spanish].

Jahre, M. (2017). Humanitarian supply chain strategies—A review of how actors mitigate supply chain risks. *Journal of Humanitarian Logistics and Supply Chain Management, 7*(2), 82–101. https://doi.org/10.1108/jhlscm-12-2016-0043.

Kovács, G., & Spens, K. (2009). Identifying challenges in humanitarian logistics. *International Journal of Physical Distribution & Logistics Management, 39*(6), 506–528. https://doi.org/10.1108/09600030910985848.

Kovats, R. S., Bouma, M. J., Hajat, S., Worrall, E., & Haines, A. (2003). El Niño and health. *Lancet, 362*(9394), 1481–1489. https://doi.org/10.1016/s0140-6736(03)14695-8.

Kunz, N., & Reiner, G. (2012). A meta-analysis of humanitarian logistics research. *Journal of Humanitarian Logistics and Supply Chain Management, 2*(2), 116–147. https://doi.org/10.1108/20426741211260723.

Linnerooth-Bayer, J., Warner, K., Bals, C., Höppe, P., Burton, I., Loster, T., & Haas, A. (2009). Insurance, developing countries and climate change. *The Geneva Papers on Risk and Insurance-Issues and Practice, 34*(3), 381–400. https://doi.org/10.1057/gpp.2009.15.

Masini-Vázquez, A. (2014). *Compendio de Modelos Cuantitativos de Pronósticos: El primer paso en las decisiones tácticas, es predecir la demanda.* México: Advanced Value Group. [In Spanish].

Murat, A. (2019). The Humanitarian Relief Chain. *South East European Journal of Economics and Business, 6,* 45–54.

National Climatic Data Center. Retrieved 2018, August 2nd from https://www.gob.mx/inecc/acciones-y-programas/efectos-del-cambio-climatico [In Spanish].

Rodríguez, J. (2004). Los desastres de origen natural en México: el papel del FONDEN. *Estudios Sociales. Revista de alimentación contemporánea y desarrollo regional, 12*(23), 74–96 [In Spanish].

Shrestha, M., Moles, R., Ranjit, E., & Chaar, B. (2018). Medicine procurement in hospital pharmacies of Nepal: A qualitative study based on the Basel Statements. *PLoS ONE, 13*(2), e0191778. https://doi.org/10.1371/journal.pone.0191778.

Smalley, R. (2003). *Top Ten Problems of Humanity for the Next 50 Years, in Energy and Nanotechnology: Strategy for the Future Conference.* Houston, TX: Rice University.

Syahrir, I., Suparno, & Vanany, I, (2015). Healthcare and disaster supply chain: Literature review and future research. *Procedia Manufacturing, 4*(29) https://doi.org/10.1016/j.promfg.2015.11.007.

The International Disaster Database. Retrieved 2019, February 10th from http://www.emdat.be/glossary [In Spanish].

Thomas, A. S., & Kopczak, L. (2005). From Logistics to Supply Chain Management: The Path Forward in the Humanitarian Sector. *Fritz Institute, 15,* 1–15.

Waters, D. (2003). *Logistics, an introduction to supply chain management* (2nd ed.). Great Britain: Palgrave MacMillan.

Chapter 6
Strategic Location of a Logistics Cluster for Exporting Humanitarian Aid and Distributing Internally in Case of Emergency

Daniel-Alejandro Fernandez-Barajas, Diana Sánchez-Partida, Patricia Cano-Olivos, and Santiago-Omar Caballero-Morales

Abstract The purpose of this document is to better understand the opportunities involved in exports for Mexico in the field of humanitarian aid and to have a safe place to settle a logistics Cluster that can distribute at the national level. The potential of the export industry could be improved through the development of a logistics cluster with the support of first aid providers, the Mexican government, and global non-governmental entities, such as the United Nations, the Red Cross, and so forth. This research proposes a new export cluster of humanitarian aid that can support and help in an emergency caused by natural disasters, giving priority to the disasters located mainly in the country, the north and center of the American continent, and if it is necessary to attend emergencies globally. The technique of Analytic Hierarchy Process (AHP) was used to determine the strategic locations of the logistics cluster, considering five criteria such as security in the supply chain, vehicular traffic, the index of natural disasters, access routes to possible locations, and finally the importance of the airports by operations.

Keywords Exports · Logistics · Humanitarian aid · Humanitarian cluster · Natural disasters

6.1 Introduction

During the last twenty years, four billion people have been affected, which represents that natural disasters have damaged 64% of the global population. However, in the last decade, there has been the highest number of victims compared to those

---

D.-A. Fernandez-Barajas · D. Sánchez-Partida (✉) · P. Cano-Olivos · S.-O. Caballero-Morales
Department of Logistics and Supply Chain Management, Universidad Popular Autónoma del Estado de Puebla (UPAEP University), 17 Sur 901 Barrio de Santiago, CP 72410 Puebla, Puebla, Mexico
e-mail: diana.sanchez@upaep.mx

D. Sánchez-Partida et al. (eds.), *Disaster Risk Reduction in Mexico*,
https://doi.org/10.1007/978-3-030-67295-9_6

counted from the beginning of the century until 2010, according to the report of the United Nations Office for Disaster Risk Reduction (UNISDR, 2015).

According to the Global Assessment Report on Disaster Risk Reduction—GAR UNISDR (2015), in several countries, the losses from disasters have been increased and come to a significant percentage of their gross domestic product annually. Also, an even higher percentage if these losses are compared with the figures of social expenditure they could be reaching in some cases between 30 and 50% of said expenditure. This situation is even more extreme for the Small Island Developing States.

At the global level, the different governmental and non-governmental agencies are dedicated to gathering information on natural disasters and, from this hard data published to allow the countries to establish metrics towards resilience. The reality is that in Mexico, the information permitted is limited, and only some government agencies collect it even though this information has been used in the making of decisions about humanitarian logistics.

According to Thomans and Kopczak (2005), humanitarian logistics is "the process of planning, implementing, and controlling the efficiency, cost-effective flow and storage of goods and materials, as well as related information, from the point of origin to the point of consumption to alleviate the suffering of vulnerable people. The function encompasses a range of activities, including preparedness, planning, procurement, transport, warehousing tracking, and tracing, and customs clearance".

Sheffi (2013) states that the existence of logistics clusters offers advantages for the country where it is located, mainly at the transport and operational levels. In the first category, the scope, scale, density, and frequency economies are included; while, on the other hand, in the operational issue, assets and labor force are shared.

Logistics clusters are groups of humanitarian organizations and other stakeholders committed to commonly address logistics needs during humanitarian crises on a broad partnership basis (global and country-level) (Merminod et al., 2014).

According to the Inter-Agency Standing Committee (IASC) "Global cluster leads have agreed to be accountable to the Emergency Relief Coordinator for ensuring system-wide preparedness and technical capacity to respond to humanitarian emergencies, and for ensuring greater predictability and more effective interagency responses in their sector or area of responsibility" (IASC, 2013).

Clusters can be formed for different industries, for example, in Mexico, the main ones that have been established are for the automotive industry, in several states, such as Puebla, Nuevo León, Guanajuato, San Luis Potosí, and Querétaro. There are also find others of great importance like the aerospace, steel, manufacturing, and one closely related to the topic, a cluster of medical devices. Mexico has become the first Latin American exporter in restorative healing material and the fifth in medical equipment and instruments, according to (PROMEXICO, 2017). However, these exports are made only for private initiative. Generally, the private initiative obtains benefits to establish like reductions in fiscal matters or infrastructure support.

Table 6.1 Governmental restrictions imposed on relief supply chains

Type of restrictions	Examples
Import barriers	Delays at customs clearance
	The extreme complexity of clearance procedures
	Ban of import on humanitarian aid supplies
Control of activities	Extreme governmental control of NGO activities
	Extreme governmental control of people's movements
Bureaucracy	Many authorizations needed a detailed catalog for all available products
	Complex administrative procedures (car registration, labor law, and so forth)

The studies that addressed the issues of disaster has been done extensively in approximately 20 years ago. These researches focused on disaster management typically on the management of relief operations, which include the planning and distribution of relief material needs of disaster victims, evacuation, coordination, and collaboration among the parties involved in humanitarian operations. However, still few have focused on the role of the Healthcare Supply Chain (HSC) in natural disasters (Syahrir & Vanany, 2015).

The governmental restrictions imposed on relief supply chains, such as those mentioned by Kunz and Reiner (2013) in Table 6.1, play a significant role in the failure of many humanitarian interventions.

The present article intends to carry out a study through hierarchical analysis to weigh the specific criteria that will be considered with information from the government and official institutions. The reason for this is to know the optimal location of an export logistics cluster within the Mexican Republic to supply national and international demand in cases of natural disasters and reduce the adverse conditions that exist in the humanitarian supply chain.

6.2 Literature Review

Natural disasters have always existed and have been the subject of study for a long time; however, humanitarian crises experienced by societies have made research specialize in solving problems and be prevented for future eventualities, making that response times are shorter, and fatalities may be at a minimum.

There were 6,637 natural disasters between 1974 and 2003 worldwide, with more than 5.1 billion affected people, resulting in more than 182 million homeless, more than 2 million deaths, and with reported damage of $1.38 trillion USD in accordance with the Center for Research on the Epidemiology of the Disasters (CRED). In 2005 alone, over 180,000 deaths and $200 billion USD economic losses have occurred due to disasters, according to the Disaster Resource Network Humanitarian Relief Initiative (HRI). The September 11 attacks (2001), the tsunami in South Asia (2004), Hurricane Katrina (2005), earthquakes in Pakistan (2005) and Java (2006), are just some examples of the deadliest disasters witnessed by humankind in the past few years (Ergun et al., 2009).

From the methodological point of view, humanitarian logistics is considered as the set of logistic activities that prevent the suffering of vulnerable people, considering some of them as the most important planning, implementation, control, and storage of information and goods (Kovacs & Spens, 2007).

Logistics plays a crucial role in disaster response operations; it serves as a link between disaster preparedness and response, between procurement and distribution, and between headquarters and the field, and is crucial to the effectiveness and responsiveness to major humanitarian programs such as health, food, shelter, water, and sanitation (A. S. Thomas, 2003).

Various subject matter experts point out that humanitarian supply chains are unstable, rigid, and unpredictable to help the most vulnerable. If there were more institutions involved in the issue, more lives could be saved by potentially increasing efficiency and effectiveness in prevention and response during disasters.

It has been found from literature and discussion with humanitarian experts that the effectiveness of relief aid is the function of various criteria. Problems Analytic Hierarchy Process (AHP) is used as it is capable of quantifying both the experts' objective and subjective judgments to analyze such qualitative criteria. Various authors have also been used AHP successfully in areas like supplier selection, green supply chain management, reverse logistics, project risk management. The AHP starts by decomposing a complex, multi-criteria problem into a hierarchy where each level consists of a few manageable elements which are then decomposed into another set of elements. Also, the AHP is a systematic procedure for representing the elements of any problem in the form of a hierarchy. Owing to its simplicity, flexibility, and logical consistency, this research also used AHP to study the various criteria and sub-criteria of a humanitarian supply chain. In this hierarchy, a responsive humanitarian supply chain as a goal occupying the top level and various criteria and sub-criteria occupy positions at the subsequent levels (Argollo et al., 2012).

According to Sennaroglu and Varlik Celebi (2018), the extensive use of AHP can be attributed to its easy applicability to large-scale and complex decision problems involving multiple criteria and subjective evaluations, as well as its successful integration with other MCDM methods.

Most natural disasters could not be predicted when it happened, including the demand for goods to disaster victims are also unpredictable. Usually, the demand for material needs such as food, beverages, clothing, medicines, medical equipment for disaster victims will increase sharply during the disaster is still ongoing. In humanitarian operations for disaster relief, nearly 60–80% of the costs associated with supply chain activities in the operational logistics of humanitarian aid (Van Wassenhove, 2006).

A combination of supply chain responsiveness and resilience opens the door to a rapid response to environmental changes (Charles et al., 2010; Miao & Banister, 2012). In a crisis context, demand uncertainty is much higher than for final consumers in a commercial context (Liberatore et al., 2013).

Ballis (2003) uses both engineering judgment and AHP approaches for a new airport site selection based on the criteria of land use, the existence of historical or archaeological resources, wind characteristics, cost of airport construction and earthworks, and socio-economic impacts due to expropriation of land and buildings; and finds the same site as the best alternative by both approaches.

6.3 Description of the Real-World Instance

Humanitarian aid has become a trend in recent years, as natural disasters have increased in magnitude and frequency. The United Nations Organization (ONU), through its Department of Natural Hazard Risk Prevention, has created different bodies to help the affected population. It helps with rapid response inputs through five regional offices at a global level; Bangkok in Asia, Nairobi in Africa, Brussels in Europe, Cairo in the Arab States, and Panama in America and the Caribbean. In addition to maintaining its corporate office in New York, United States (United Nations Development Program [UNDP], 2014).

The clusters of humanitarian aid have been divided by commodities to solve all the problems of countries in disaster situations. All have been managed by specialized teams of logistics, coordinators, engineers, and more people to have all the supplies on time. There are different instances for each cluster, levels of hierarchy which administrates the processes in the arrival of products to the cluster until the full delivery to affected people.

The Inter-Agency Standing Committee (IASC) is the primary mechanism for inter-agency coordination of humanitarian assistance. It is a unique forum involving the key UN and non-UN humanitarian partners. The IASC was established in June 1992 in response to United Nations General Assembly Resolution 46/182 on the strengthening of humanitarian assistance. According to the IASC, at the first level exists the cluster members, formed by the companies that supply the national market and decide to consolidate themselves to grow the industry. Internally, they establish their guidelines on the operation and share supplies, transport, information, software, among other things, to make processes and their logistics chains more efficient (IASC, 2013).

Subsequently, there must be a cluster coordinator in charge of mediating and solving problems that may generate some disagreement among the members, always respecting the guidelines of the country where they are established and those of the Office of the Coordination of Humanitarian Affairs (OCHA, 2019).

There must be an agency that directs and administers the clusters dedicated to humanitarian aid in each country. It must be part of the government or with a strong relationship between governmental and non-governmental organizations. In the case of Mexico, the National Civil Protection System (SINAPROC) could be one who coordinates all the activities related to the cluster.

In the international scope of the organizations, it is necessary that all respond to the Humanitarian Coordinator (HC), who is in charge of whether or not the aid should be used in case of disaster; besides coordinating the response efforts so that the highest amount of help arrives in the shortest possible time.

At the highest level of this scale is the Office of the Coordination of Humanitarian Affairs (OCHA), which works closely with the clusters and governmental and non-governmental organizations to ensure the functioning of humanitarian aid. However, even if the efforts try to reach all places and correct all its shortcomings when occurring natural disasters (OCHA, 2019).

Within the humanitarian aid in 2017, according to the Procurement Division of the United Nations (PDUN), the following budgets were awarded for the items that have to do with humanitarian aid, see Table 6.2.

As shown in the table above, the United Nations (UN) purchasing department spends a large percentage of its resources allocated to humanitarian aid through its five hubs. The ten categories in which more inputs are spent are health, infrastructure, transportation, and food.

According to the Annual Statistics Report on the PDUN (2017), the overall procurement volume of United Nations increased for a sixth consecutive year, reaching USD 18.6 billion in 2017—an increase of 5.1% or USD 910 million compared to 2016. The increase was mainly driven by a 9.1% or USD 823 million increase in procurement of services, while procurement of goods rose by a more modest 1.0% or USD 87 million.

Table 6.2 Top ten categories of procurement by volume, UNSPSC segment level, 2017. Information retrieved from https://www.un.org/Depts/ptd/commodity_data_pdf/2017

Category (UNSPSC segment level)	Total 2017 (USD m)	Total change 2016–2017 (%)	Share of procurement from developing countries, countries with economies in transition and LDCs 2017 (%)
Pharmaceuticals, contraceptives, vaccines	$2,641.40	−7.3%	39.8%
Transportation, storage, mail services	$2,323.10	−3.5%	67.7%
Food and beverage products	$1,893.80	0.1%	79.7%
Management and administrative services	$1,809.30	8.5%	62.3%
Building and maintenance services	$1,303.00	9.9%	76.9%
Engineering and research services	$1,279.40	28.2%	45.0%
Medical equipment	$735.70	19.5%	40.8%
IT and communications equipment	$626.10	36.4%	50.2%
Fuels and lubricants	$611.70	3.7%	93.8%
Travel, food and lodging	$559.00	110.2%	64.3%
Top 10 total	$13,782.50		

Table 6.3 Top 10 Countries Supplying the UN, UNSPSC segment level, 2017. Information https://www.un.org/Depts/ptd/commodity_data_pdf/2017

Country	Number of purchase orders	Total USD (m)
United States of America	30,883	$1,741
India	18,882	$906
United Arab Emirates	4,936	$796
Belgium	3,286	$718
France	10,583	$692
United Kingdom	12,728	$555
Switzerland	19,402	$547
Netherlands	6,006	$511
Denmark	21,198	$507
Kenya	17,290	$502
United States of America	30,883	$1,741

In Table 6.3, the UN reports global numbers of purchase orders per country in the categories mentioned above. The considerable number of purchases was for the United States, followed by Denmark, Switzerland, and India, but if it is taken into account the USD, the principal vendor still being the United States, followed by India, United Arab Emirates, and Belgium.

In order to participate in the tenders of the United Nations Secretariat, companies must first register on the website of the Global Portal for United Nations Suppliers. The Secretariat maintains business relationships with suppliers around the world and has actively increased its sources of supply to developing countries and countries with economies in transition. The United States of America was the country number one supplying the United Nations in all areas like World Food Programme (WFP), United Nations International Children's Emergency Fund (UNICEF), World Health Organization (WHO), United Nations Procurement Division (UNPD) among others, being medicines, transportation, and food the most relevant concepts.

Figure 6.1 below shows how important are the countries as suppliers of the UN. The most significant marks were showing the United States, India, and the United Arab Emirates as the countries which generate more purchase orders and invoice significant quantities of dollars to the organization.

Latin America and the Caribbean had the most substantial increase in percentage terms, with procurement from suppliers in the region increasing by 39.0% to USD 1.2 billion in 2017. Primarily, to an additional USD 247 million in procurement by United Nations Office for Project Services (UNOPS) from suppliers in Argentina (up from USD 0.6 million in 2016), mainly on computer equipment and accessories and computer services. UN procurement from Argentina increased from USD 21 million in 2016 to USD 279 million in 2017. Pan American Health Organization (PAHO) also increased its procurement in the region, by USD 123 million, or 145%. The most significant decrease in the region was in Peru, where the procurement volume fell by USD140 million to $54 million following the (Annual statistical report UNPD, 2017).

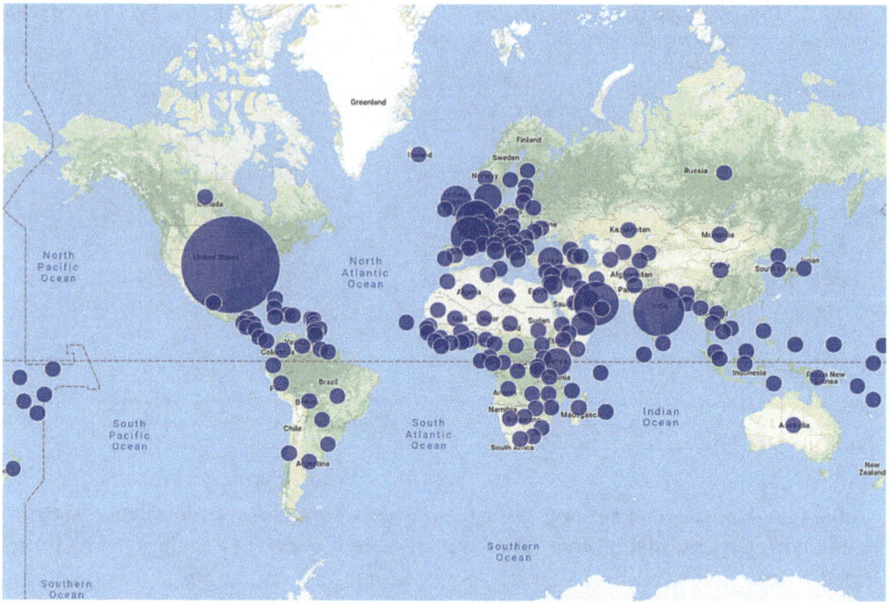

Fig. 6.1 Main suppliers of the UN at the country level

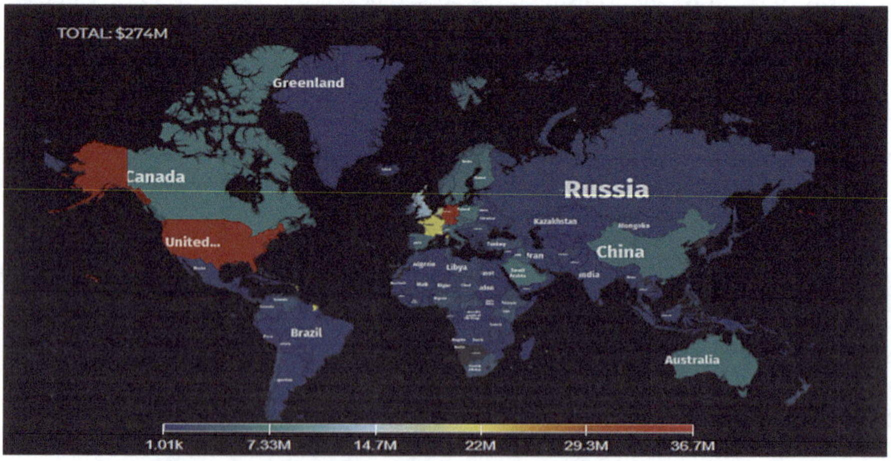

Fig. 6.2 First aid exporters (The Atlas of Economic Complexity, 2014). Available at https://atlas.media.mit.edu/en/

At a global level, 274 billion dollars are moved in first aid supplies (The Atlas of Economic Complexity, 2014), with the United States and Germany being the countries with the most considerable amount of imported goods entering their country as shown in Fig. 6.2, with approximately USD 37 billion each.

Even being the country that more considerable humanitarian aid needs due to the great disasters that year after year affects with a hydrometeorological event, devastating wildfires, and gelid waves; the United States has a broad distribution network for humanitarian aid, as well as a consolidated industry that allows it to be resilient and help developing countries through the various organisms. However, it is not able to produce all the necessary resources and needs to import a large number of goods to help its citizens.

Global exporters, led by China, followed by Germany and the United States, represents almost sixty percent of the global commerce, which was based on 274 billion dollars. As the Massachusetts Institute of Technology (MIT) shows on their Global Atlas of Goods, the Asian markets control more than half of the market in humanitarian aid supplies. However, the exchange of goods between most of these countries and the world is through trade agreements such as the BRICS (Brazil, Russia, India, China, and South Africa), the seven dragons or any other sign in the world because their protectionist policies do not allow them to be suppliers of the global NGOs. As a sample, China goes from representing 40% of the global market to only selling 1.18% of the annual budget of the United Nations, according to information from the Procurement Division. While other countries such as the United Kingdom, Denmark, Arabia, and even Kenya are essential providers of the UN. However, they are not recognized as highly exporting countries or with an industry developed to support the commercial operations of a global market.

Within the American continent, mainly in the Northern Hemisphere, the leader in exports is the United States of America, with 76% of the market, while Mexico only contributes 21% corresponding to 6.06 million dollars.

According to information PROMEXICO (2017), the supplies for health have a potential market of 26.82 MDD to the United States of America, where they do not have any tax, being the leading business partner in many aspects. Nevertheless, Mexico has a network of 12 Free Trade Agreements (FTAs) with 46 countries, 32 Agreements for the Promotion and Reciprocal Protection of Investments (APPRIs) with 33 countries, and nine agreements of limited scope Economic Complementation Agreements and Partial Scope Agreements (ECA and PSA) within the framework of the Latin American Integration Association (ALADI). They are the most important the Trans-Pacific Partnership (TPP), the United States-Mexico-Canada Agreement (USMCA), and the Mexico-EU Free Trade Agreement (Mexico-EU FTA) having the ability to link the country.

Mexico's capabilities related to the production of goods and services have been vital to detonate and expand a vast number of industries connected to crucial global chains, with a clear focus on North America. The ability to link the country has complemented these capabilities; it is a commercial opening, several international agreements, and the volume of business with the world are an enormous strength. However, it has been wasted given the lack of conditions to achieve agile production, trade, and people flows.

During the natural disasters of recent years, the most affected Mexican states have had the least development, such as Veracruz, Oaxaca, and Chiapas, according to information from the National Center for Disaster Prevention (CENAPRED,

2018). The main problem lies in the allocation of property by the State. The supplies are distributed by the Natural Disasters National Fund (FONDEN), which it is an interinstitutional instrument whose purpose is to authorize resources to mitigate the effects produced by disturbing natural phenomena in the infrastructure of federally competitive sectors, or of competition from federal entities, municipalities or political-administrative bodies of the Federal District. Also, mitigate the damage to the homes of the low-income population affected by such phenomena. The institute designates different companies for the purchase of supplies, which is why the cluster could be gathering them and having the necessary capacity in times of response and optimal inventory level to get the goods to the affected populations. Mainly, they supply such as non-perishable food pantries, covers, mattresses, zinc sheets, first aid kits, personal cleaning kits, and cleaning kits.

6.4 Methodology and Methods

In this study, the Analytic Hierarchy Process (AHP) technique was used in order to determine the criteria and alternatives to consider for the solution of the problem and be able to establish the necessary weights of each in order to standardize the units of measurement and thus make a decision without deviations or biases for the investigation.

The AHP arises at the end of the seventies from the hand of Saaty. This method is framed within multicriteria analysis methods since it is conceived for the study of discrete decisional problems. That is, the decision is made on a finite number of possible alternatives. It is, therefore, a matter of identifying the most appropriate one based on a series of criteria. The AHP proposes a basic structure of three levels. In the superior, the objectives are placed, in the next level, the criteria and the last level the alternatives. This basic structure could be complicated by incorporating new levels of sub-criteria, always linked to those of the immediately superior level, although it reflects the general idea of the operation of the method. Efforts should be directed to identify the appropriate criteria thorough knowledge of the problem, clearly define the objectives, and know the groups affected by the decision, not to reach more complex structures than necessary (Gil & Ángel, 2017).

In Fig. 6.3, it is possible to observe the aforementioned by Gil, where the primary objective of the location of the logistics cluster is determined. In the next level the criteria to be considered to reach the objective, such as security in the supply chain, vehicular traffic, the index of natural disasters, access routes to possible locations, and finally the importance of the airports by operations, all this must be evaluated for each alternative that in this case are the 32 states of the Mexican Republic.

The judgment relationship of each of the criteria used in this case was determined through the literary review of various authors regarding the influence of the topic within the supply chain. In this case, security and traffic are the most important within the classification according to the explanation below.

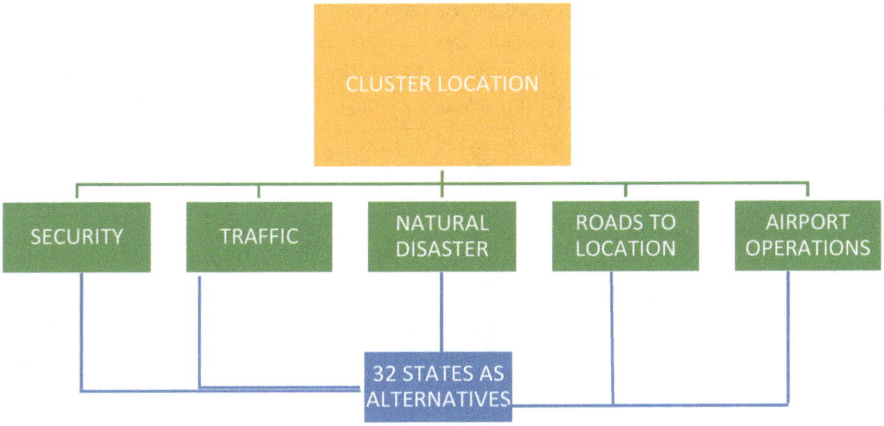

Fig. 6.3 Illustrative model to determine the optimal alternative

6.4.1 Security Criterion

Security within the supply chain is of vital importance for the merchandise in question, arrives at the indicated destination, and does not suffer any damage during its transfer from the factory or warehouse where they will be safeguarded until their commercialization.

In accordance to A. Thomas (2010), factors that affect are the timing of the shipment of certain goods from the manufacturer, their availability in the market place or transit, the level of protection given to the products when unattended, the level of market demand, and the profit associated with the sale of that commodity. These components round out the characteristics of what is called the product risk factor, while the higher the product risk factor, the greater the need to address protection of that product while in transit and at the retail level. It allows assuming that at the time of the need for the goods, the protection to them must be total in the door-to-door deliveries, thus avoiding the robbery and delinquency to the carriers, ensuring that there will be no problems in the supply chain and the full supply of the required supplies.

Rarely, the organized thieves play in the speculative theft arena, because their marketing and supply chain do not support this platform. The smaller the organization, the more random the theft and product base, but the large the organization, the more item-specific its targets are.

This paper intends to analyze the incidence of crimes on Mexican roads to cargo transport, from platforms to dry boxes and containers, as these have increased in all states according to information from the Federal Police and the Government from Mexico during 2010–2018 (Secretariado Ejecutivo del Sistema Nacional de Seguridad Publica, 2018).

For this case study, it was decided to analyze the impact of the crime on the transit of goods by land on the roads exclusively because it is the most effective means in Mexico for the internal transfer of goods. The lack of rail infrastructure means that the internal trade of the country is carried out with trucks.

6.4.2 Traffic Criterion

Traffic congestion is a condition of traffic delay because the number of vehicles trying to use a road exceeds the design capacity of the traffic network to handle it" (Weisbrod et al., 2001). Most transportation literature and transportation impact models treat congestion as a cost factor, comprised of time delay and operating expense (Cambridge Systematics, 2008; Short et al., 2010).

As traffic volumes and congestion grows on highways and urban roadways, freight and delivery service operators become increasingly challenged to maintain dependable and reliable schedules. It affects supply chains and truck-dependent businesses, both of which are of increasing importance for both public policy and private sector operators (Weisbrod & Stephen, 2011).

The most basic impact is that congestion delay and uncertainty increase requirements for, and hence, costs of product inventory (Disney et al., 1997; Mason-Jones et al., 1997).

Delivery times in logistics are of the utmost importance. Not only allow to evaluate the service level of the companies, in some cases, but they also compromise the supply chain and above all in humanitarian logistics; failures or delays cannot be allowed since in most cases it is already committed to the place that just suffered a natural disaster (Frazelle, 2018).

Currently, Mexico is a country in exponential demographic growth, to the last Intercensal Survey 2015 conducted by the INEGI, 119,53,753 population were counted, increasing 6.40% concerning 2010, it directly impacts the vehicle registration. Vehicular traffic was calculated using the Communications and Transportation Secretariat (SCT, 2017) databases on the vehicle fleet registered by the State and consider the mileage of roads within it, to determine the level of traffic that could be generated in the case that all the vehicles of the state.

6.4.3 Roads to Location Criterion

The ability of nations to achieve efficient flows of goods and people is essential for their growth and prosperity. Geographic distances have less impact on global supply chains and international trade, than those generated by inefficiencies that obstruct business connections between countries. That is why it is necessary that the transportation, infrastructure, management, storage, information exchange, and other elements that make up the logistics, operate in sync to articulate local, regional and international networks, through which the mobilization of production, trade, and people, favor the development of business, cultural exchange and the welfare of societies.

Mexico has made a significant effort for the development of infrastructure. The national territory hosts 76 airports, of which 64 are international; 117 ports and terminals enabled, 58 in the Pacific Ocean and 59 in the Gulf of Mexico and the Caribbean; 63 border crossings, 378 thousand kilometers of land routes and almost 27 thousand kilometers of railways that are complemented by 49 customs, of which 19 are on the northern border, two on the southern border, 11 are interior and 17 maritime. The national load that is mobilized through this network is concentrated in the auto transport of cargo (56.2%), followed by maritime transport (30.9%), rail transport (12.8%), and finally, air transport (0.1%). The use of different modes of transport is different for export, although the order of importance is preserved: the most important is the motor transport of cargo (63.5%), second is the transportation by sea (17.8%), followed by rail (13.8%), air (4.6%) and other means (0.3%) (SCT, 2017).

This criterion was considered the number of official roads within the Mexican Republic and monitored by the Communications and Transportation Secretariat. However, there are certainly more used roads, they are unsafe and without the corresponding infrastructure so that the cargo trucks arrive in time and form to the cluster, airport or distribution center. In a subsequent study, the exact location within the selected state can be determined, considering the fluidity of the road networks.

6.4.4 Airport Operations Criterion

The characterization of the airport supply chain is a network of multiple organizations and their relationship. It encompasses suppliers, airlines, airport operators, government agencies such as customs, security, and immigration, and other service providers. The airport is a service system which is a vital part of the air transportation of today's society. The airports in the 21st-century make an exciting long-term growth industry with excellent prospects. The industry is undergoing quick change and becoming much more challenging for airport managers (Neufville, 2012).

Supply chain management (SCM) in the airport business involves overseeing relationships among all members in the chain, controlling inventory, demand forecasting, and getting constant feedback on what is happening at every link to add value to all members down the supply chain. SCM decisions have a direct impact on the airport's operation and financial performance. In turn, this influences all members of the chain (Jain & Hailemariam, 2010).

For this study, airports and aerodromes located in the thirty-two states were considered, considering only those that can accommodate aircraft with a capacity higher than 19 m^3 or 1.5 tons. Moreover, it is considered the number of operations per state; this is the sum of the operations that all airports and aerodromes have had, which is why Mexico City and Jalisco were the most valuable in the inde6. This information was obtained from the Secretariat of Transportation and Communications (SCT, 2017).

6.4.5 Natural Disasters

There has been an increasing demand for humanitarian relief operations in line with more and more disasters happening around us. There are about 500 disasters every year, killing about 75,000 people and affecting some 200 million others (Van Wassenhove, 2006).

The various effects of disasters on the population and its surroundings generate different kinds of needs and require different approaches to meet those needs. It is therefore essential to have a general sense of what these effects are, and which systems are most commonly affected. However, experience shows that the effects in question cannot be taken as absolute, since the impact and form a disaster takes on the specifics of the affected region (WHO, 2002).

Natural disasters not only affect the territory where they originate or the place where they pass, the collateral damage usually causes affectations to the population that increases the mortality of the same. In the same way, there are health implications, communication, migration, nutrition, infrastructure, among others, which increase the risks of the population (PAHO, 2001).

For the event index per year, only those disasters that could complicate the logistics during the development of the supply chain were considered, such as earthquakes, floods, hurricanes, landslides, and subsidence during the last ten years (2008–2018). The information was obtained through the National Center for Disaster Prevention (Ortiz, 2006).

6.4.6 Instances

Three different instances were used to carry out the analysis, which was classified in scenarios for their interpretation at the time of obtaining the results. It is worth mentioning that the data entered in the matrices were normalized to avoid bias in processing the information, and the scales were reordered to find the best option with the same interpretation.

Scenario 1. Sensitivity analysis respects all-natural disasters happened in the country.
Scenario 2. Sensitivity analysis respect to natural disasters happened in the country that could affect the supply chain.
Scenario 3. Sensitivity analysis respect to the number of airports and their operations.

6.5 Results

One of the crucial issues in Disaster Relief Operations (DROs) has been the logistics and supply chain activities. Logistics in DRO context include such activities as assessing demand, procuring goods, determining priorities as well as receiving, sorting, storing, tracing, and tracking deliveries. It is possible to classify logistical activities to the distribution channel and the outbound logistics that deliver supplies from the distribution center to the affected areas (Sheu, 2007).

Saaty's methodology exposed in Chap. 3 of this book was the one used for the evaluation of the case study.

According to the criteria matrix solved, as shown in Table 6.4, the most important criterion to select the state to establish the humanitarian aid logistics cluster was the frequency that a Natural Disaster impacts considering the past ten years (2008–2018), according to the information provided by the National Center for Disaster Prevention (CENAPRED, 2018). Moreover, the second significant criterion was the security of goods transportation. The weight for each criterion does not change, it will maintain on each scenario, the unique data that were modified was the matrix information of the variables considering each criterion like the selection of natural disaster that affects the supply chain or the increase of the airport operations.

By the determination of the leading weight, the best location was determined. The first alternative was Coahuila in the three scenarios; the second option was Puebla, and the third considers Guanajuato and Michoacan as the best options for each scenario (Table 6.5).

For the first scenario of this analysis, Coahuila is the best option to establish the cluster, followed by Puebla, Guanajuato, and Colima. The natural disaster considered in this scenario was earthquakes, flood, hurricanes, severe rains, freeze, landslides, droughts, volcanos.

For the second scenario, the main options were consistent; The order of the first five options is maintained between scenarios 1 and 2 if we consider just the natural disasters that could stop the operations in the supply chain and eliminate the ones that have no incidence like freeze and droughts.

The order of the first five options for the third scenario changed. The schedule of airport operations must be as fast and efficient as possible so that a state with a more significant number of airports and many operations can afford the operations. Necessary in case of an emergency.

Table 6.4 Weighted index for the evaluation of criteria

Criteria	Weight
Security	0.26325369
Traffic	0.09878977
Natural disasters	0.49346672
Roads to location	0.11347965
Airport operations	0.03101017

Table 6.5 Scenario 1, 2 and 3. Total natural disasters evaluated

Scenario 1		Scenario 2		Scenario 3	
State	%	State	%	State	%
Coahuila	3.39	Coahuila	3.44	Coahuila	3.54
Puebla	3.37	Puebla	3.38	Puebla	3.36
Guanajuato	3.33	Guanajuato	3.32	Michoacan	3.31
Colima	3.29	Colima	3.29	Guanajuato	3.30
Michoacan	3.24	Michoacan	3.24	Colima	3.28
Oaxaca	3.24	Morelos	3.22	B California	3.25
Morelos	3.23	Oaxaca	3.21	Oaxaca	3.24
Guerrero	3.22	Mexico	3.21	Tamaulipas	3.20
Mexico	3.21	Guerrero	3.21	Mexico	3.19
Chihuahua	3.18	B California	3.18	Guerrero	3.19

This work shows the potential and safe states for the installation of the national logistic cluster to be able to attend the different disasters occurring in the North America area and probably once it can be consolidated as an industry in Mexico to global markets. With this research, the corresponding authorities in Mexico and the world can select options for a logistics cluster and integrate other criteria or define the national priorities for locating a better location based on geographical, meteorological, infrastructure conditions, among others. In Appendix 6.A is found the matrices that were used in each scenario.

6.6 Conclusions

The conditions that there are in Mexico, making an essential place in global logistics. National capacities are a solid basis to strengthen their participation in global chains, not only as an essential partner in the transformation activities but also as a provider of high added value logistics services. Its geographical position and the commercial agreements have established place it in proximity to the leading economies of all continents. Moreover, correcting the deficiencies that separate it from them is a relatively simple task for an economy that has demonstrated very advanced competencies in activities of high sophistication, the same that have placed it among the first places in high technology sectors.

It is of the utmost importance that the authorities take climate change into account and how it will affect in the future and continue with this and other studies, relying on groups of experts in the humanitarian field to solve all the questions that arise in this matter.

Appendix 6.A

Scenario 1. 32 × 5 matrix					
State	Security	Traffic	Natural disasters	Roads	Airport ops
Aguascalientes	0.032258065	0.027940691	0.032258065	0.029758316	0.031992911
Baja California	0.032258065	0.031494557	0.032143674	0.029922727	0.030798449
Baja California Sur	0.032175983	0.020298668	0.031800503	0.015591604	0.031035334
Campeche	0.032175983	0.032477316	0.032258065	0.021729599	0.032037199
Coahuila	0.032093901	0.035296332	0.032029284	0.049103962	0.032014037
Colima	0.032258065	0.026971224	0.032029284	0.043596208	0.032087952
Chiapas	0.031847657	0.040509099	0.030770991	0.026305694	0.031863415
Chihuahua	0.032258065	0.027033194	0.029741478	0.043102976	0.03145868
Ciudad De Mexico	0.029549372	0.016734572	0.032143674	0.022168028	0.021943527
Durango	0.032258065	0.031545514	0.031457332	0.024771195	0.032111844
Guanajuato	0.027743577	0.03224781	0.031800503	0.052693593	0.031650749
Guerrero	0.03201182	0.03270806	0.030885381	0.036581356	0.031840126
Hidalgo	0.03201182	0.031149415	0.032258065	0.025401436	0.032258065
Jalisco	0.029056883	0.027430755	0.031686113	0.032306681	0.028640023
Mexico	0.032093901	0.028610606	0.031571723	0.037238998	0.031978751
Michoacan	0.031683493	0.025258246	0.031686113	0.043459199	0.032034861
Morelos	0.032093901	0.032959616	0.032029284	0.032799912	0.032240108
Nayarit	0.032258065	0.032431998	0.032029284	0.018825012	0.032174718
Nuevo Leon	0.031929738	0.030373212	0.032258065	0.026744122	0.029582585
Oaxaca	0.031847657	0.041393242	0.030199039	0.033019126	0.031715962
Puebla	0.030944759	0.037063482	0.031800503	0.046308982	0.032107292
Queretaro	0.027169006	0.035387303	0.032258065	0.025127418	0.031843396
Quintana Roo	0.032175983	0.027427381	0.031457332	0.020003288	0.028359025
San Luis Potosi	0.030123943	0.029054422	0.031914894	0.025593248	0.031934136
Sinaloa	0.03151933	0.028379467	0.0306566	0.025045213	0.031430789
Sonora	0.031765575	0.034059458	0.029627088	0.020414315	0.031607063
Tabasco	0.031929738	0.036224022	0.031457332	0.026524908	0.031819881
Tamaulipas	0.031929738	0.032205708	0.0306566	0.028717049	0.031708367
Tlaxcala	0.028482311	0.030916894	0.032258065	0.027511372	0.032258065
Veracruz	0.027825659	0.036268631	0.021276596	0.052611388	0.031669711
Yucatan	0.032258065	0.031523271	0.032258065	0.028141612	0.031659652
Zacatecas	0.03201182	0.036625835	0.031342942	0.02888146	0.032143331

Scenario 2. 32 × 5 matrix					
State	Security	Traffic	Natural disasters	Roads	Airport ops
Aguascalientes	0.032258065	0.027940691	0.032258065	0.029758316	0.008134185
Baja California	0.032258065	0.031494557	0.032143674	0.029922727	0.044687071
Baja California Sur	0.032175983	0.020298668	0.031800503	0.015591604	0.038280362
Campeche	0.032175983	0.032477316	0.032258065	0.021729599	0.007101148
Coahuila	0.032093901	0.035296332	0.032029284	0.049103962	0.009834228
Colima	0.032258065	0.026971224	0.032029284	0.043596208	0.005325077
Chiapas	0.031847657	0.040509099	0.030770991	0.026305694	0.012499118
Chihuahua	0.032258065	0.027033194	0.029741478	0.043102976	0.026588549
Ciudad De Mexico	0.029549372	0.016734572	0.032143674	0.022168028	0.32030803
Durango	0.032258065	0.031545514	0.031457332	0.024771195	0.004561665
Guanajuato	0.027743577	0.03224781	0.031800503	0.052693593	0.017443273
Guerrero	0.03201182	0.03270806	0.030885381	0.036581356	0.01335737
Hidalgo	0.03201182	0.031149415	0.032258065	0.025401436	0
Jalisco	0.029056883	0.027430755	0.031686113	0.032306681	0.111263863
Mexico	0.032093901	0.028610606	0.031571723	0.037238998	0.00924325
Michoacan	0.031683493	0.025258246	0.031686113	0.043459199	0.007293961
Morelos	0.032093901	0.032959616	0.032029284	0.032799912	0.000583141
Nayarit	0.032258065	0.032431998	0.032029284	0.018825012	0.0025967
Nuevo Leon	0.031929738	0.030373212	0.032258065	0.026744122	0.080322922
Oaxaca	0.031847657	0.041393242	0.030199039	0.033019126	0.014889681
Puebla	0.030944759	0.037063482	0.031800503	0.046308982	0.005651918
Queretaro	0.027169006	0.035387303	0.032258065	0.025127418	0.010741075
Quintana Roo	0.032175983	0.027427381	0.031457332	0.020003288	0.120351922
San Luis Potosi	0.030123943	0.029054422	0.031914894	0.025593248	0.010184583
Sinaloa	0.03151933	0.028379467	0.0306566	0.025045213	0.029141357
Sonora	0.031765575	0.034059458	0.029627088	0.020414315	0.020805737
Tabasco	0.031929738	0.036224022	0.031457332	0.026524908	0.01291296
Tamaulipas	0.031929738	0.032205708	0.0306566	0.028717049	0.017279461
Tlaxcala	0.028482311	0.030916894	0.032258065	0.027511372	0
Veracruz	0.027825659	0.036268631	0.021276596	0.052611388	0.018093036
Yucatan	0.032258065	0.031523271	0.032258065	0.028141612	0.016972215
Zacatecas	0.03201182	0.036625835	0.031342942	0.02888146	0.003552142

Scenario 2. 32 × 5 matrix					
State	Security	Traffic	Natural disasters	Roads	Airport ops
Aguascalientes	0.032258065	0.027940691	0.032258065	0.029758316	0.013157895
Baja California	0.032258065	0.031494557	0.032143674	0.029922727	0.052631579
Baja California Sur	0.032175983	0.020298668	0.031800503	0.015591604	0.052631579
Campeche	0.032175983	0.032477316	0.032258065	0.021729599	0.026315789
Coahuila	0.032093901	0.035296332	0.032029284	0.049103962	0.065789474
Colima	0.032258065	0.026971224	0.032029284	0.043596208	0.026315789
Chiapas	0.031847657	0.040509099	0.030770991	0.026305694	0.039473684
Chihuahua	0.032258065	0.027033194	0.029741478	0.043102976	0.026315789
Ciudad De Mexico	0.029549372	0.016734572	0.032143674	0.022168028	0.013157895
Durango	0.032258065	0.031545514	0.031457332	0.024771195	0.013157895
Guanajuato	0.027743577	0.03224781	0.031800503	0.052693593	0.026315789
Guerrero	0.03201182	0.03270806	0.030885381	0.036581356	0.026315789
Hidalgo	0.03201182	0.031149415	0.032258065	0.025401436	0.013157895
Jalisco	0.029056883	0.027430755	0.031686113	0.032306681	0.026315789
Mexico	0.032093901	0.028610606	0.031571723	0.037238998	0.026315789
Michoacan	0.031683493	0.025258246	0.031686113	0.043459199	0.052631579
Morelos	0.032093901	0.032959616	0.032029284	0.032799912	0.013157895
Nayarit	0.032258065	0.032431998	0.032029284	0.018825012	0.013157895
Nuevo Leon	0.031929738	0.030373212	0.032258065	0.026744122	0.026315789
Oaxaca	0.031847657	0.041393242	0.030199039	0.033019126	0.039473684
Puebla	0.030944759	0.037063482	0.031800503	0.046308982	0.026315789
Queretaro	0.027169006	0.035387303	0.032258065	0.025127418	0.013157895
Quintana Roo	0.032175983	0.027427381	0.031457332	0.020003288	0.052631579
San Luis Potosi	0.030123943	0.029054422	0.031914894	0.025593248	0.026315789
Sinaloa	0.03151933	0.028379467	0.0306566	0.025045213	0.039473684
Sonora	0.031765575	0.034059458	0.029627088	0.020414315	0.065789474
Tabasco	0.031929738	0.036224022	0.031457332	0.026524908	0.013157895
Tamaulipas	0.031929738	0.032205708	0.0306566	0.028717049	0.065789474
Tlaxcala	0.028482311	0.030916894	0.032258065	0.027511372	0
Veracruz	0.027825659	0.036268631	0.021276596	0.052611388	0.065789474
Yucatan	0.032258065	0.031523271	0.032258065	0.028141612	0.026315789
Zacatecas	0.03201182	0.036625835	0.031342942	0.02888146	0.013157895

References

Annual statistical report. (2017). *United Nations Procurement Division (UNPD)*. New York, NY, USA. https://www.ungm.org/Public/ASRDataArchive.

Ballis, A. (2003). Airport sites election based on multicriteria analysis: The case study of the island of Samothraki. *Operational Research 3*(3), 261–279. http://d6.doi.org/10.1007/BF02936405.

Cambridge Systematics. (2008). *Estimated cost of freight involved in highway bottlenecks*. Washington, DC, USA: Federal Highway Administration.

CENAPRED, 2006–2018. Declaratorias sobre emergencia, desastre y contingencia climatológica a nivel municipal entre 2000 y 2018. National Center for Disaster Prevention. Retrieved February 13th, 2019, from https://datos.gob.mx/busca/dataset/declaratorias-sobre-emergencia-desastre-y-contingencia-climatologica/resource/41444ebe-6a35-4631-8f91-9237d5114488 [in Spanish].

Charles, A., Lauras, M., & Van Wassenhove, L. (2010). A model to define and assess the agility of supply chains: Building on humanitarian experience. *International Journal of Physical Distribution & Logistics Management, 40*(8/9), 722–741.

Disney, S., Naim, M., & Towill, D. (1997). Dynamic simulation modelling for lean logistics. *International Journal of Physical Distribution and Logistics Management, 20*(3–4), 194–196.

Ergun, O., Karakus, G., Keskinocak, P., Swann, J., & Villarreal, M. (2009). Humanitarian supply chain management—An overview. In *Models and Algorithms for Optimization in Logistics Dagstuhl Seminar Proceedings*, vol. 10.

Frazelle, E. (2018). *Supply chain strategy: Unleash the power of business integration to maximize financial, service, and operations performance*. New York: McGraw-Hill Education.

Gil, J. C., & Ángel, S. P. (2017). *El proceso analítico jerárquico*. Madrid: UNED—Universidad Nacional de Educación a Distancia.

Humanitarian supply management and logistics in the health sector. (2002). Washington, DC: WHO Regional Office for the Americas.

Inter-Agency Standing Committee (IASC), *The Centrality of Protection in Humanitarian Action*, December 17th, 2013. Retrieved July 26th, 2019, from https://www.refworld.org/docid/52d7915e4.html.

Jain, K., & Hailemariam, M. (2010). *Managing airport supply Chain: An Ethiopian case study*. India: Indian Institute of Technology Bombay.

Kovacs, G., & Spens, K. M. (2007). Humanitarian logistics in disaster relief operations. *International Journal of Physical Distribution and Logistics Management, 37*(2), 99–114.

Kunz, N., & Reiner, G. (2013). Government restrictions on relief supply chains. In *Proceedings of the 20th EurOMA Conference*, Dublin, pp. 1–10.

Liberatore, F., Pizarro, C., Simón de Blas, C., Ortuño, M.-T., & Vitoriano, B. (2013). Uncertainty in humanitarian logistics for disaster management: A review. In B. Vitoriano & J.-M. Da Ruan (Eds.), *Decision aid models for disaster management and emergencies* (pp. 45–74). Amsterdam: Atlantis Press.

Mason-Jones, R., Namim, M., & Towill, D. (1997). The impact of pipeline control on supply chain dynamics. *International Journal of Logistics Management, 8*(2), 47–61.

Miao, X., & Banister, D. (2012). Coping with natural disasters through resilience. Working Paper No. 1059, Transport Studies Unit, University of Oxford, Oxford.

Merminod, N., Nollet, J., & Pache, G. (2014). Streamlining humanitarian and peacekeeping supply chains. *Society and Business Review, 9*(1), 4–22.

Neufville, R. D. (2012). *Airport systems: Planning, design and management*. New York: Mcgraw-Hill.

Ortiz, E. G. (2006). Guia básica para elaboración de atlas estatales y municipales de peligros y riesgos: Conceptos básicos sobre peligros, riesgos y su representación geográfica. México: Secretaría de Gobernación. Retrieved May 19th, 2019 from http://mapserver.inegi.org.mx/eventos/cng2007/cng2007/oscarzepeda_metodologiasriesgo.pdf [In Spanish].

Pan American Health Organization (PAHO). (2001). Humanitarian supply management in logistics in the health sector Washington, D.C. Retrieved Febraury 10th, 2021 from https://www.paho.org/English/DD/PED/HumanitarianSupply.

Retrieved November 19th, 2018 from. (2017). *Diagnostico sectorial*. Retrieved November 19th, 2018 from https://promexico.mx/documentos/diagnosticos-sectoriales/dispositivos-medicos.pdf [In Spanish].

Secretaria de Comunicaciones y Transportes SCT. (2017). Aviación Mexicana en Cifras 2017. Distrito Federal, Mexico. Retrieved March 3th, 2019 from http://www.sct.gob.mx/fileadmin/DireccionesGrales/DGAC-archivo/modulo5/amc-2017-i.pdf [In Spanish].

Secretariado Ejecutivo del Sistema Nacional de Seguridad Publica. (2018). Estadistica de Incidencia delictiva del Fuero Comun. Retrieved May 27th, 2019, from http://secretariadoejecutivo.gob.mx/docs/pdfs/nueva-metodologia/CNSP-Delitos-2018.pdf [In Spanish].

Sergio Ricardo Argollo da Costaa, Vânia Barcellos Gouvêa Campos, Renata Albergaria de Mello Bandeira. (2012). Supply chains in humanitarian operations: Cases and analysis. *Procedia—Social and Behavioral Sciences, 54* (2012), 598–607.

Sennaroglu, B., & Varlik Celebi, G. (2018). A military airport location selection by AHP integrated PROMETHEE and VIKOR methods. *Transportation Research Part D: Transport and Environment, 59,* 160–173. https://doi.org/10.1016/j.trd.2017.12.022.

Sheffi, Y. (2013). logistics-intensive clusters: Global competitiveness and regional growth. In J. Bookbinder (Ed.), *Handbook of global logistics: Vol. 181. International series in operations research & management science*. New York, NY: Springer.

Sheu, J. B. (2007). Challenges of emergency logistics management. *Transportation Research Part E, 43*(6), 655–659.

Short, J., Trego, T., & White, R. (2010). Developing a methodology for deriving cost impacts to the trucking industry that generate from freight bottlenecks. *Transportation Research Record, 2168,* 89–103.

Syahrir, I., & Vanany, I. (2015). Healthcare and disaster supply chain: Literature review and future research. *Procedia Manufacturing, 4,* 2–9.

The Atlas of Economic Complexity. (2014). Cambridge: MIT Press Retrieved August 27th, 2018 from https://atlas.media.mit.edu/publications/.

Thomans, A. S., & Kopczak, L. R. (2005). *From logistics to supply chain management. The path forward in the humanitarian sector*. Fritz Institute. Retrieved October 15th, 2018 from http://www.fritzinstitute.org/PDFs/WhitePaper/FromLogisticsto.pdf.

Thomas, A. S. (2003). *Humanitarian logistics: Enabling disaster response*. Fritz Institute.

Thomas, A. (2010). *Supply chain security: International practices and innovations in moving goods safely and efficiently*. Department of Marketing. Paper 16. http://ideaexchange.uakron.edu/marketing_ideas/16.

United Nations Development Program (UNDP). (2014). *Reducing disaster risk: A challenge for development New York: UNDP bureau for crisis prevention and recovery*. Available at www.undp.org/bcpr.

United Nations Office for the Coordination of Humanitarian Affairs (OCHOA). (2019). *Visible body: OCHA*. New York, EU: OCHA Press. Retrieved October 28th, 2018, from https://www.unocha.org/our-work/coordination.

UNISDR. (2015). Informe de evaluación global sobre la reducción del riesgo de desastres. Disponible en: http://www.preventionweb.net/english/hyogo/gar/2015/en/gar-pdf/GAR2015_EN.pdf [October 28, 2018].

Van Wassenhove, L. N. (2006). Humanitarian aid logistics: Supply chain management in high gear'. *Journal of the Operational Research Society, 57*(5), 475–489.

Weisbrod, G., & Stephen, F. (2011). *Traffic congestion effects on supply chains: Accounting for behavioral elements in planning and economic impact models*. Supply Chain Management - New Perspectives: IntechOpen. https://doi.org/10.5772/23057.

Weisbrod, G., Vary, D., & Treyz, G. (2001). *Economic Implications of Congestion. NCHRP Report #463*. Transportation Research Board, Washington, DC, USA.

World Health Organization WHO. (2002). The World Health Report 2002. Paris, France. https://www.who.int/whr/2002/en/whr02_en.pdf?ua=1.

Part II
During the Disaster

Chapter 7
19S Earthquake in Puebla, Mexico: Intervention of the Different Actors in Humanitarian Aid

Meredith-Janeth Fon-Galvez, Diana Sánchez-Partida, Damián-Emilio Gibaja-Romero, and Santiago-Omar Caballero-Morales

Abstract Nowadays, humanitarian logistics (HL) has caught the attention of researchers and practitioners, given its role in natural disasters, which differentiates it from commercial logistics. However, research conducted in this field is not well-documented in different parts of the world. So, the objective of this chapter is to document how HL was applied in Mexico. Mainly, it is explained how the process was carried out from the occurrence of the 19S earthquake until one year later. This analysis summarizes the behavior of all involved agents in such catastrophe (victims, volunteers, government, and non-governmental organizations (NGOs)), who also integrate the humanitarian supply chain (HSC). For example, the emergence of a civic organization whose objective was to provide support to those affected by the earthquake was documented. Among the findings, it was observed that the allocation of human and material resources for reconstruction is one of the central problems that face such organizations. Besides, it was found that that Information and Communication Technologies (ICTs) played a predominant role in the coordination of logistics activities between volunteers and victims. Aside from the documentation of HL activities during the September 19th, it is also provided a comparison between the literature existing in the HSC management, and a theoretical framework as well to apply HL in a real context.

Keywords Disaster · Humanitarian aid · ICTs · Volunteers · Humanitarian logistics and supply chain (HLSC)

Electronic supplementary material The online version of this chapter (https://doi.org/10.1007/978-3-030-67295-9_7) contains supplementary material, which is available to authorized users.

M.-J. Fon-Galvez (✉) · D. Sánchez-Partida · D.-E. Gibaja-Romero · S.-O. Caballero-Morales
Department of Logistics and Supply Chain Management, Universidad Popular Autónoma del Estado de Puebla (UPAEP University), 17 Sur 901, Barrio de Santiago, CP 72410 Puebla, Puebla, Mexico
e-mail: meredithjaneth.fon@upaep.edu.mx

D. Sánchez-Partida et al. (eds.), *Disaster Risk Reduction in Mexico*,
https://doi.org/10.1007/978-3-030-67295-9_7

7.1 Introduction

There is an increasing interest in understanding Humanitarian Logistics (HL) mechanisms by a look of humanitarian challenges that society has faced in recent years, which are mainly the consequence of natural disasters like earthquakes (Dubey & Gunasekaran, 2015). Exists a theoretical framework that identifies a chain of actors and their associated responsibilities. Formally, HL is defined as the process of planning, implementation, and control of the efficient and profitable flow and storage of goods and materials, as well as related information, from the point of origin to the point of consumption in order to comply with the requirements of the final beneficiary (Oloruntoba & Gray, 2006; Thomas, 2005; Thomas & Kopczak, 2005). The difference with commercial logistics relies on the fact HL processes are activated to assist people affected by some disaster, such as earthquakes, tsunamis, hurricanes, epidemics, droughts, famines, terrorist attacks, and war situations or a combination of several disasters that happen simultaneously (Ludema, 2000).

The growing interest in this field has motivated the elaboration of manuals, the most important ones elaborated by the United Nations through the International Strategy for Disaster Reduction program (UN/ISDR, 2004). These manuals are freely available and pretend to generate a resilient community by explaining basic terminology, potential risks, and active measures to take.

Despite the efforts to define and specify the protocols that activate and manage HL processes, the uncertainty that surrounds natural disasters makes these processes a challenge by themselves. For example, the unexpected way in which the natural disaster impacts society and industry in a region are almost impossible to predict, making it difficult to measure the total impact of the catastrophe on demand and supply. However, it is possible to identify the risks that make a community more vulnerable to natural disasters; Paz (2018) points out the following risks

- The existence of multi-threat territories.
- Climate change or variety.
- The tremendous urban expansion and the weakness of regional planning.
- Increased exposure of people and their assets to threats.
- Territorial modifications due to the globalization of the economy.
- Socio-spatial segregation.
- The limited participation of the population in decision-making
- Risk management is weakly considered in regional planning processes.

Hence, HL designing deals with increasing delivery times and the lack of resources (Balcik & Beamon, 2008). Among the primary deficiencies related to HL processes, Ilahn (2011) lists the following

- Inadequate IT infrastructure that difficult the gathering and storage of necessary information.
- The unpredictable, chaotic, and unknown environment that remains after the disaster happens.
- The lack of encouragement for organizations to measure performance.

- The potential focus on the media that may generate negative externalities such as information distortion.
- The lack of human resources with adequate training
- Poor organizational culture in the region impacted by the natural disaster.
- No identification of short-term and long-term objectives that results in no HSC planning.

It is necessary to understand what a disaster means in order to deal with the previous deficiencies. A disaster can be defined as a severe alteration of the functioning of the community, which represents a significant and widespread threat to human life, health, as well as to real estate assets or/and the environment; whether by accident, nature or human activity and develops suddenly or because of long-term processes (UN/ISDR, 2004). The magnitude of this disaster may vary depending on the vulnerability of the community.

The previous definition allows HL decision-makers to clear the necessary activities to deal with the disaster. Even more, it establishes HS echelons based on the different phases of a disaster. Such phases are shown in Fig. 7.1.

The phases of a natural disaster can be divided by taking into account temporalities. The activities were identified that are carried out immediately before and after a disaster happens to mitigate the suffering of the victims (such as prevention or reduction of disasters, mitigation of the consequences of disasters and emergency preparedness). Also, it was observed that long-term rehabilitation activities oriented to rebuild and generate a sustainable process for the community that suffers the disaster (Ilhan, 2011). By analyzing the phases of disasters and their associated activities, it is essential to bear in mind that, in the case of seasonal disasters, it is possible to plan the location of critical assets and supplies, as well as planning the evacuation of the affected community using information from previous disasters (Apte, 2010). In other words, seasonal disasters induce a continuous process to address slow-onset phenomena with a need for long-term supplies.

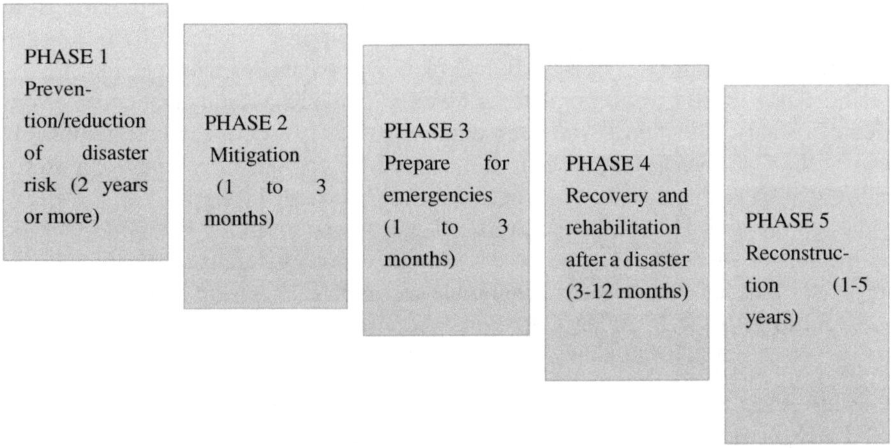

Fig. 7.1 Phases of a disaster, own drafting with information by (Ilhan, 2011)

The phases of a disaster make HSC have a specific structure where different decision-makers interact. Hence, in humanitarian logistics, it was observed that the government, commercial organizations, non-governmental organizations, and society actively participate in attending the community/region needs at each disaster's phase. An exciting feature is a fact that each of these actors performs different functions within the chain (Abidi et al., 2016), and typically, no individual actor has enough resources to respond to a significant disaster effectively or to cope with all the needs at each phase (Bui et al., 2000). However, each of the involved actors, in response to disasters, has the same general objective, which is to help people and relieve suffering. To reach this end, HSC agents face two core problems: the coordination of activities and efforts, and operational limitations caused by the disaster (Van Wassenhove, 2006).

Concerning Mexico, it has a high prevalence of natural disasters since, throughout the year, it is prone to threats of different types. The country is exposed to hydrometeorological, seismic, volcanic phenomena, among others, due to its geographical localization in the "ring of fire." Economically, Mexico is the seventh country with the most economic losses derived from natural catastrophes, with a total loss of 2 thousand 957 million dollars; moreover, on average, it has a loss of 142 annual lives due to these catastrophes (Eckstein & Schäfer, 2017).

That is the reason for this research is documented how humanitarian logistics were applied before and after the earthquake of September 19th, 2017, in Puebla, Mexico. In this chapter, are investigated the actors that intervened in the disaster's phases and the activities that they performed. Moreover, analyze to them to identify the main challenges that they faced during this phenomenon.

7.2 Antecedents of Seismicity in Central Mexico

Mexico is a seismic area since it is in the Circum Pacific Belt, a region characterized by a high occurrence of earthquakes since it concentrates a large part of the seismic activity in the world. Specifically, the Mexican Geological Service (2018) indicates that the high seismicity is due to the interaction between North American, Cocos, Pacific, Rivera, and Caribbean plates, as well as some local faults that cross several Mexican states. The North American plate separates from the Pacific one but rubs against the Caribbean plate; also, it collides the Rivera and Cocos plates, an interaction that causes the majority of earthquakes that occur in the Mexican territory. Although the areas where the earthquakes' epicenters are in the Pacific Ocean, the central area of the country has become a seismic receiver given its proximity with the epicenters but mainly due to its orographic features, such as soft soil.

According to the Mexican Geological Service 2018, the Mexican Republic is divided into four seismic zones:

- Zone A: In this zone, there are no records of earthquakes in the last 80 years. Hence, no accelerations of the earth are expected above 10% of the acceleration of gravity because of tremors.

- Zone B and C: these zones are also named as intermediate earthquakes zones since, despite to register earthquakes in the last 80 years, the frequency of seismic phenomena is low. Moreover, high ground accelerations do not exceed 70%, which induce small damages in those areas affected by such phenomena.
- Zone D: this area characterizes by registering a large number of earthquakes, per annum, and reports a large number of historical earthquakes (those that generate significant damage). Also, this zone identifies a place of occurrence (a location where its earthquakes are originated), and the acceleration of the soil exceeds 70% during a seismic phenomenon.

Figure 7.2 presents a map of Mexico that divides it into the four zones mentioned in the paragraph above. Note that Puebla is located in Zone B and intermediate-risk zone where earthquakes with significant magnitude are not frequent but may happen (as it is the case of the 19S) is possible to see each of these zones throughout the country place.

Seismographs and accelerographs have studied the frequency of seismic activity in Mexico. In this sense, it is essential to mention that it is not possible to predict when or where or how significant an earthquake can be. However, although research studies are inconclusive, a high magnitude earthquake is expected from the coast of Guerrero since this zone reports a high accumulation of energy and no significant seismic activity. Also, it is known the existence of a gap that goes from the southeast of Petatlán, in the state of Guerrero, almost up to Pinotepa Nacional, in the state of Oaxaca, that may generate an earthquake could hit a magnitude above eight on the Richter scale, if this gap fractures in a single movement. Given the

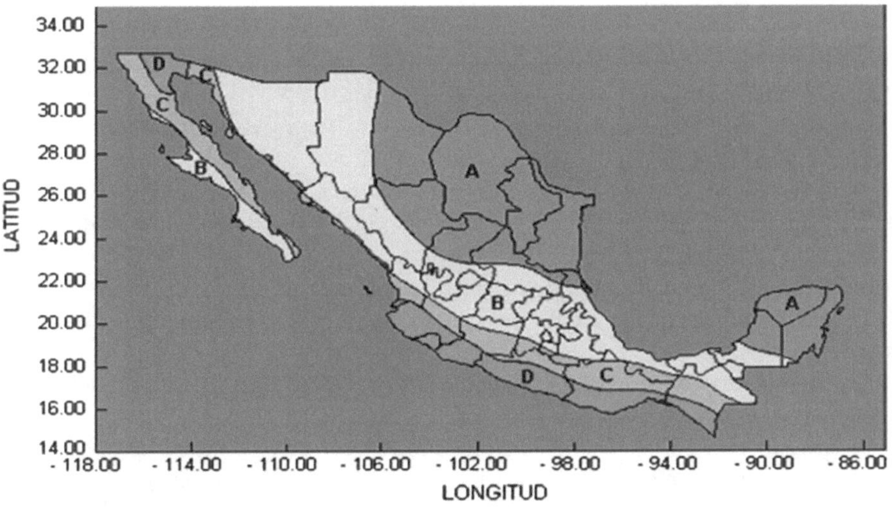

Fig. 7.2 Seismic areas of Mexico (Servicio Geológico Mexicano, 2018)

proximity of Puebla with Oaxaca and Guerrero, both possible earthquakes make necessary the elaboration of a prevention plan in Puebla, despite there is no certainty about date and magnitude.

The devastating effect of an earthquake depends on a set of factors such as its magnitude, depth, distance to population centers, type of soil, type of construction, and duration (Trujillo et al., 2010). Among the most destructive earthquakes that have been recorded in the Mexican territory, the one that happened on September 19th in 1985 is one of the most devastating phenomena recorded in Mexico, mainly because of the damages that Mexico City registered. This earthquake had a magnitude of 7.6 with epicenter in Zihuatanejo, Guerrero, and had a replica that reached a magnitude of 8.1 with an epicenter in the Pacific Ocean, near the mouth of the Balsas River, on the coast of the state of Michoacán.

On September 7th, 2017 occurred an earthquake of magnitude 8.2, with an epicenter in Pijijiapan, Chiapas. This earthquake affected the southern region of Mexico that comprises the states of Chiapas, Tabasco, and Oaxaca. Almost two weeks later, another earthquake occurred on September 19th, with a magnitude of 7.1, and with epicenter in Axochiapan, Morelos, it generated significant affectations in the states of Puebla, State of Mexico, Morelos, Guerrero, and Oaxaca. Although Puebla is located in zone B, which is characterized by not being the origin of severe seismic activity, its closeness with Axochiapan (79 km), located in zone D, makes necessary the elaboration of an HL plan that manages seismic contingences in Puebla (National Seismological Service, 2017).

7.3 Case Study: Earthquake 19S in Mexico

The 2017 19 S earthquake in Mexico struck at 13:14:40 h, with an epicenter located 12 km southeast of Axochiapan, Morelos (Latitude 18.4° Longitude 98.72°), with a depth of 57 km. From this date until January 2018, more than 100 aftershocks occurred (National Seismological Service, 2017).

The earthquake had a meaningful impact on the central region of Mexico. The federal entities of Morelos, Puebla, and Mexico City registered material damages, more than 200 buildings collapsed, and human casualties, around 370 dead and more than 6000 injured. Although there is a distance of 120 km between Mexico City and Axochiapan, the capital registered significant affectation due to the soil where it is located.

In Puebla, approximately 112 localities were affected by the earthquake with meaningful affectations as the fall of the church of the pyramid in Cholula (Protección Civil, 2017). The affected locations received an emergency declaration, which are reports made during extraordinary events. Such reports record the activities that the government and its institutions made to provide support to the communities in an emergency.

Unfortunately, it is not possible to say that affected cities in Puebla had an effective warning system for such a phenomenon. In general, the predictability of a disaster plays a significant role in determining the possibilities of response since the occurrence and affectations of specified disasters are linked to specific geographic areas and features; muddy soil, for example (Chang et al., 2007). In this sense, the criticism on the authority's response to the September 19th, 2017 relies on its capacity to warn the population until the structure of an earthquake. Also, critics on the authority's response rely on the prevention activities that, during the quake, appeared to be insufficient. This threat reached vulnerable localities and, consequently, it ended in a disaster of a larger magnitude. At the time of the earthquake, communication was lost, and it was not known how substantial the impact had been on the infrastructure and social. Once the internet and telephone communications were re-established, it was possible to observe how significant the impact of this event had been. Part of the consequence of this phenomenon were 369 fatalities, thousands of victims, and substantial material damage to public buildings, as well as commercial and private property (Ureste & Aroche, 2017).

Van Wassenhove (2006) points out that nobody has the necessary preparation to deal with a disaster, which is understood as "an interruption that physically affects a system as a whole and threatens its priorities and objectives." In this sense, the enhancement and promotion of humanitarian logistics arise as a possible solution to deal with an unpredictable context, just as it happened in the 19S, 2017.

Within this research, interviews were carried out with different actors that participated in the earthquake of September 19th, 2017. It was done to recognize the action protocols of each organization and to know more about the emergence of the 19S Organization, an organization from the civil society that was a pillar in implementing strategies to mitigate the suffering within the first 72 h after the earthquake of September 19th. In the same way, it was performed interviews with volunteers and victims. The latter provides essential information on the impact that the intervention of each involved organization had during and after the disaster.

7.4 Intervention of the Actors

7.4.1 Civil Protection—Government Actor

Civil Protection is an organization that has governmental funds, whose purpose is to provide support and attention to the population affected by a disaster, whether natural or social. This organization previously had a reactive intervention, but today has a defensive philosophy since it has the objective of strengthening vulnerable populations in a way that the impact of a threat can be mitigated, or avoiding a disaster if such a scenario is possible.

In an interview with this organization (Gower, 2018), its stakeholders declare that the disaster's management in Mexico is organized into three levels: municipalities, political entities (states), and the federal government. Each level has a specific function that depends on the disaster features such as magnitude, nature, and its capacity to attend the affectations.

Regardless of the phenomenon, the municipal civil protection is the one that responds first. If this level cannot contain the disaster and requires more support, the state level intervenes. The federal-level acts when, in an emergency, none of the above can cope with the population needs. Hence, the three levels can manage in coordination, or not, the activities to overcome the affectations of a natural disaster. The case of September 19th, 2017, is an example of an emergency that required the intervention of the three levels in coordination. The law of the state Civil Protection system registers the general framework to activate emergency protocols as well as to determine which authority level must intervene in a disaster (Ley del Sistema Estatal de Protección Civil, 2017).

Summarizing, the Civil Protection Organization oversees coordinating the three authority levels in case of disasters, when it finishes its activities within the first days of the mitigation phase, it gives way to other instances of government, such as the national army. The last intervenes through relief plans to support the population in some basic needs. Moreover, Civil Protection intervenes reviewing properties that were affected to determine with a quick review of the property represents a risk or not.

Concerning earthquakes, at the municipal level, the Civil Protection System (CPS) has the function of responding to the disaster at the local level. At the state level, the authority has the task of inspecting if buildings have damages, that report to the federal government. The last evaluates if the previous levels require additional resources to deal with the affectations caused by the earthquake. Moreover, those are the three-step process to get resources from the federal government; those are illustrated in Fig. 7.3. The procedure is as follows, first, the state civil protection makes a quick inspection to inform the damages to the Federal Department of Infrastructure, which is independent of CPS, and it determines if a building is habitable or not. Later, this report is channeled to the Natural Disasters Fund (FONDEN), which is a public agency that evaluates and calculates the affected population that may require support, either in-kind or economic, depending on what

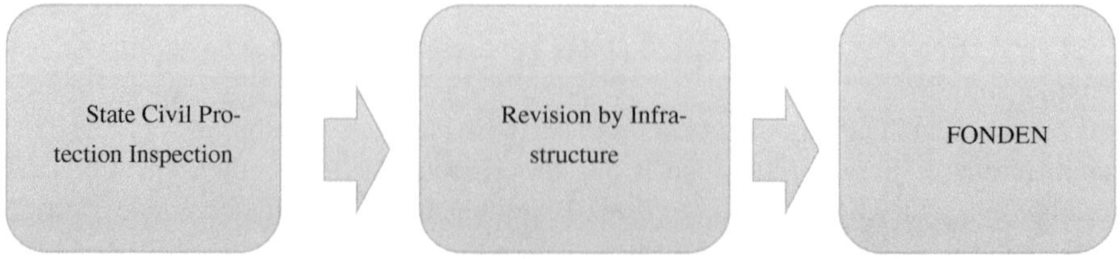

Fig. 7.3 The protocol of the state civil protection system in the revision of immovable

they need. Later, resources are sent to the place where the disaster happens, and local authorities, in coordination with the other levels, distribute them.

7.4.2 Movement 19S—Civil Society Actor

Now, the role played by the civil society in the 19S earthquake of 2017 will be discussed. Before describing the participation of the 19S Movement, it is essential to remark that civil society has no primary role or mandatory participation within the structure of the traditional Humanitarian Supply Chain (HSC). Typically, HSC literature considers the civil society as a secondary actor mainly because it is overshadowed by proper protocols, the government, and international organizations. However, in the case of analysis, civil society played an essential role during the response phase, i.e., in the activities performed after the disaster happened, given the difficulties to establish communication between affected communities and official organizations. In this sense, Howden (2009) mentions that information and communication play a central role in humanitarian logistics since there is a decoupling between the final recipient of the aid and the providers; the providers do not influence what the beneficiary receives or not. So, disinformation is a considerable disadvantage for the successful implementation of humanitarian logistics. Specifically, it does not allow us to know with certainty the real needs of the affected population, and the place where support is needed precisely. In the case of 19S, ICTs were the bridge for the delivery of aid to be carried out more efficiently.

The 19S Movement is an example of the emergence of an organization formed by civilians, who must deal with the inefficiencies of the formal institutions. This organization was structured almost immediately, in the first 48 h after the phenomenon, with the support of volunteers belonging to different disciplines. During the first hours after the disaster, the government did act immediately since all levels have to fulfill specific protocols of action before to develop some strategy. Such protocols can be found in the Official Gazette of the Federation (DOF) (Agreement by which the General Rules of the Naturals Disaster Fund are issued, 2010) and AGREEMENT that establishes the FONDEN Emergency Care Fund Guidelines (DOF, 2010). Hence, the movement acts as a link between several communities in Puebla, Morelos and Oaxaca, and official organizations by using Information and Communication Technologies (ICTs) since Civil Protection could not establish efficient communication with such communities.

During the elaboration of this research, we had the opportunity to interview one of the leaders of the 19S movement. Reyes (2017) points out that this movement arose from a desire to provide support and attention to affected communities in the best possible way. Initially, the organization started with four members and focused their activities on identifying and managing problems that nobody else was attending. For example, pollution in food and water because such resources were sent in PET containers; vehicular traffic in the access roads of the affected localities; the neglect of the affected localities in the reception, administration and distribution of humanitarian aid; the lack of coherent and verified information; the absence of control in the distribution of help, and the flow of food that was not required by locals since consumption habits and food culture change from one place to other.

Later, the 19S Movement launched a call to incorporate more people in its activities; funders of this civic organization used social networks to coordinate their activities, and to create links with other civic organizations, businesspeople, government and civil society that would like to help. Subsequently, all the volunteers met and set up work plans. During the meeting, many of these people had already been in the affected localities and could provide valuable information, based on their experience, to generate solutions for the problems raised above.

This synergy allowed the movement to have a useful organizational chart after a few days. By having volunteers and work plans, the organization decided to form committees; Fig. 7.4 presents in detail the structure of the organization. Note that it was divided into the committees of housing, food, transport and logistics, network, and health. Committees were carried out based on resources, civic organizations and personal abilities; later, each committee started to work in different areas, and the communication among them contributes to identify specific needs and implement efficient strategies to deal with them.

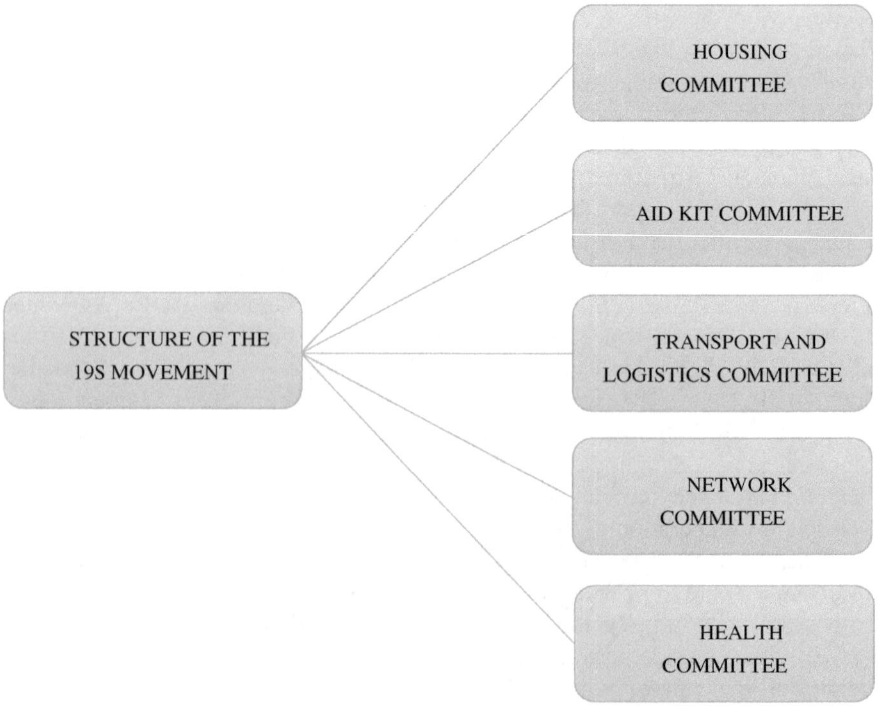

Fig. 7.4 Structure of the 19S movement committees

Each committee met separately to identify the needs of their area of work. Each area had its degree of complexity; for example, there were no platforms to monitor and inventory; they had not there the latest technology to carry out each activity. Everything was handled through necessary and available instruments such as mobile phones, a notebook, and an online database that was filled with information gathered from social networks and volunteers. Similar efforts were found to achieve better coordination and break down the disinformation barriers in disaster management. Yang et al. (2014) use data development analysis to build a network of emergency response in case of earthquakes, while Anparasan and Lejeune (2017) emphasize the importance of verified information to attend epidemics. The last authors propose a model of emergency responses to epidemics that can be used in countries with limited resources, and Sushil (2017) proposes the use of a qualitative and interpretative framework called SAP-LAP in the context of disaster management. However, it would be worthwhile to develop an information system to have a functional integration, coordination, and decision making of all the actors before the appearance of a disaster.

One of the reasons why this movement was strengthened was the active participation of volunteers and their desire to help at first hand. They created exclusive pages for the movement in social networks to deal with information asymmetries. Among their main results, it is worth mentioning that they generate accurate reports of affected communities' stage; in other words, the organization provided information to civil society and governmental authorities about the conditions of the localities on which they had been able to help. Through these means, the volunteers began to join, and together with them, private companies offered transportation.

Volunteers had the most information since they were the ones who went directly to deliver aid to the affected localities, and therefore, they were coordinating the chain. They began to receive calls from people, institutions, companies inside and outside the state that offered shipments of food, construction materials, and collection centers. Finally, the government recognized the activities of the 19S Movement and worked with its members.

An important feature to recall in this movement is the way that it positions civil society as a driver of an HSC. As Howden (2009) says, civil society can provide/generate the information that the humanitarian supply chain needs, which is necessary to satisfy some aspects in the relief process; the effectiveness of such intervention depends on information veracity and velocity (Reyes, 2017), such a problem was successfully overcome by the Network Committee of the 19S movement, whose central task was to check and corroborate. In other words, the 19S movement established the Verified area to check the needs of each locality, make a map of contact information in each point that they were volunteers in their locality and to analyze the information that the brigades brought to their return. All this information was introduced to the online database, which was updated continuously and publicly available through social networks. The Verified Area was also the one that coordinated and established the activities of each committee and internal department in the movement by transforming the gathered information into knowledge. Figure 7.5 presents a diagram in which can be observed how the

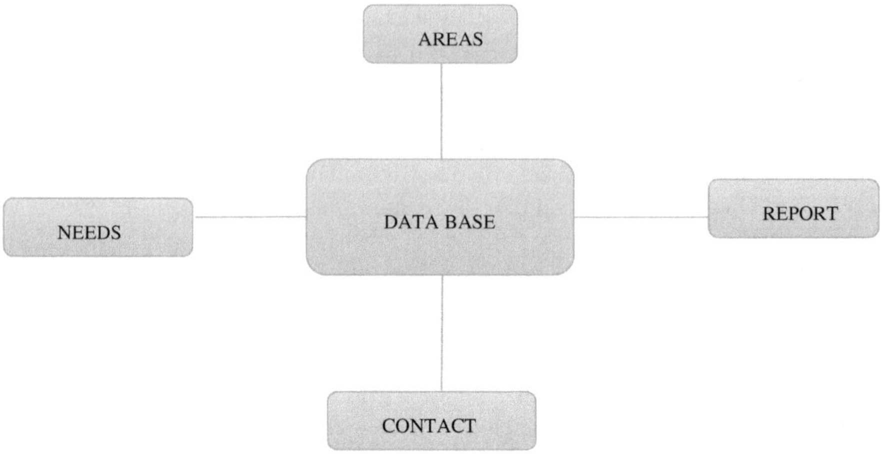

Fig. 7.5 Interaction of the information in the database in the case of 19S

database was essential during the emergency response phase. Such a database served as a reference to estimate the demand for aid in real-time, that allowed to design and implement the corresponding actions and determine the areas that needed to be involved. Also, civil society could use this database to ask for specific support and get an efficient response from authorities or other humanitarian organizations.

As in the case of the 19S movement, civil society movements started throughout Mexico to deal with the slow responsiveness of official authorities, in the first hours after a disaster. For example, Social TIC (n.a.) is a website that documented each of the web platforms and movements in social networks that were useful during the 19S phenomenon. Both movements recognize and implement ICTs to diminish information asymmetries, facilitate coordination, and share information via the Internet. Therefore, ICTs are the necessary technologies for the management and transformation of information. Notably, the use of computers and programs allow the creation, modification, storage, protection, and recovery of valuable information to prevent a disaster (Duarte, 2008).

Moreover, the civil society that could not act on the streets offer support and help from their homes by using digital media. There were many ways to support the networks (software, databases, collaborative maps, free Wi-Fi, calls, verification of messages, and dissemination). Many Twitter accounts were created (@JuntosSismoCDMX, @comoayudarmx, @Brigada19S, @horizontalmx, @Verificado19S, and @Verificado19SEstados). Hence, ICTs were the channel that provided information in real-time to know how to act and how to help.

In the same way, different platforms reactivated tools that were very useful since they served to indicate the status of security personnel and communicate the status of well-being during the first hours after the earthquake. The most important ones were:

- Safety check on Facebook.
- Google People Locator.

Concerning digital media that were used to collect and organize information, we can mention the following online platforms:

- How to help (different ways to help in the earthquake).
- Blooders (list of blood hospitals).
- InfoSismoMx (information related to the earthquake).
- Rebuildingmexico.com (concentrates a list with the organizations that receive donations for the rebuilding phase).

As we mentioned before, the location of the affected communities and their needs is necessary for the successful functioning of the HSC. Below are mentioned the most important ones.

- Collaborative map (civil society and Google).
- Open data via datos.gob.mx (collaborative mapping on damages, shelters, collection centers, hospitals, and wi-fi connection points.
- Map of seismic zones and damages (brigades UNAM, CENAPRED).
- Map of potentially affected areas (European Space Agency and NASA).
- Map with Ushahidi (mapping of humanitarian and political impact.
- Map against oblivion (information about victims of the earthquake.

Journalism is necessary to provide the population with the necessary information, and it is crucial to monitor the efforts made by civil society organizations to regulate their activities. In this sense, the Mexican population could check:

- Mexicans against corruption.
- Here the initiative called datera.
- Epicenter (monitor investment for reconstruction).
- 19S Citizenship.
- Budget transparency.

Finally, specialized software was necessary to solve problems in affected communities. For example:

- Fuerza México (geolocalized disaster capture).
- RescueCDMX (search for rescued and disappeared).
- API of collection centers (integrate into other applications).
- Ready-mx (list of missing persons and damaged buildings).
- SuperCívicos (integrates damage and complaints report).
- Human chain (digital money to support victims).

All the previous tools were of vital importance for the involved agents in the HSC. Mainly, they were used by the volunteers in the response phase of the phenomenon of study.

In addition to the participation of the 19S Movement, different groups emerged during the mitigation stage of this disaster. Each of these had as support each of the volunteers who complied with specific activities; however, all it was done without prior preparation that reflects a lack of planning from the Mexican government and society to face a natural disaster. Thus, it is crucial to have a planning process, and that integrates and organizes all possible involved actors; such planning must be publicly available by official media to be in harmony with the protocols and actions of the government.

7.4.3 OXFAM—Non-governmental Organizations Actor

Another important actor in the HSC is the Non-Governmental Organizations (NGOs) that act independently from the governmental organizations. They should take care of no interfering in the protocols carried out by the government. Many of these NGOs are international, as the Oxford Committee for Famine Relief (OXFAM). In an interview in 2019, d'Hyver, Coordinator of the Humanitarian Action department within OXFAM, mentions that the disaster caused by the earthquake of September 19th, 2017, was different from other disasters that OXFAM has attended in their record. The difference relied on the affectations that OXFAM Mexican offices suffered, which delayed a few hours the deployment of aid. However, the organization maintained his purpose of supporting the affected localities. First, OXFAM was observer because the civil society actively intervened in the affected localities, while the government applied its protocols. Then, OXFAM acts in the localities that cannot be assisted yet supporting with the proper help. They carry out long-term projects to restore the quality of life in each of the localities in the post-disaster phases. Likewise, they detected that in cases of disasters, there was a lack of psychosocial attention in the localities, but Oxfam does not have the experience or mandate to address it.

Some of the actions they carried out were: the implementation of water filters, sanitation, the construction of bathrooms in schools, and training committees in Tochimilco, Puebla, for risk reduction. Some of the challenges they faced as an organization, and which they mention (d'Hyver, 2019), was the lack of coordination between organizations. Nowadays, OXFAM continued to carry out projects for the restoration of localities for 18 months after the date of the disaster.

It is essential to mention that many of the improvements for the recovery of this type of disaster come from the NGOs. They are actors of utmost importance within the chain since they seek to correct the aid in the unassisted localities; this reduces the suffering of those affected.

7.4.4 Volunteers

People who collaborated as a volunteer, being part of an organization or not, tell in a different way each of the difficulties they faced in their work. The interview was conducted with six volunteers; five of the interviewees belong to the municipality of Puebla in the state of Puebla and one more from the municipality of Zacapoaxtla, which is also located in the state of Puebla. It is worth mentioning that two of the volunteers were part of the 19S Movement.

From these interviews, we observe that difficulties are the same for all volunteers, but they vary across contexts and are based on the activities that they performed. The time of collaboration was very different, given their rhythm of life; their activities impose constraints to the time allowed to carry out their volunteering. So, we observe that volunteering time ranged from 24 h to two months. Among the answers obtained regarding the activities carried out, the majority oversaw the collection and delivery of provisions; in the last one, they report almost insignificant participation.

Among the challenges that they face, all volunteers agree in the lack of adequate information to carry out related activities. Also, finding information processes lead delays in delivering humanitarian aid; i.e., satisfying the immediate demand of a locality was not known. The lack of organization of independent volunteers, who did not have the necessary preparation, also complicated each of the tasks to be performed. The transport part also became a difficult task, since different localities were challenging to access because some stretches of road were affected by the earthquake and suffered a breakdown. None of the volunteers interviewed were aware of the different platforms that were created to facilitate these tasks; many of them only had access to the database created by the 19S Movement.

Among the answers obtained by the volunteers, 50% learned of volunteering through the calls that were made on social networks, the other 50% learned through family and friends. Social networks also were a vital communication means between the volunteers. Much of the information about the affected localities and the needs of each one were published on social networks. However, most of the people who volunteered did not know any of the platforms created by other volunteers from different parts of the country to handle the information.

The questions applied for the interview were the following:

- Question 1: How did you find out about volunteering?
- Question 2: How long did you serve as a volunteer?
- Question 3: In what activities did you provide support?
- Question 4: What do you consider the most significant challenge?
- Question 5: What would you improve in the processes that were carried out?
- Question 6: What medium did you use to communicate?
- Question 7: Did you know any of the platforms that were created for 19S?

Figure 7.6 presents the percentage of responses obtained in the interviews. This information is essential because volunteers also deal with problems that belong to the HSC, but that are ignored due to the disaster's features. As before, the role of volunteers emphasizes the importance of civil society in the solution of inefficiencies of HL, as well as to improve the effectiveness of the HSC. In other words, in the future, we can improve HSC processes to perform a better intervention in the

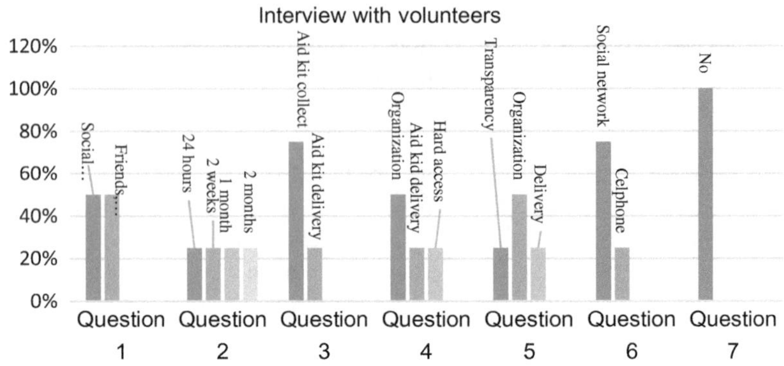

Fig. 7.6 Percentage of responses obtained in volunteer interviews

treatment of disasters. From the in-depth interviews, we conclude that HSC, in opposition with commercial ones, needs the involvement of all agents; i.e., it is necessary that all involved actors in a disaster be aware of the current situation and, if it is possible, such information must be delivered in real-time. Concerning the planning process, it must include alternate routes in case some section is destroyed and notify the actors of this situation.

7.4.5 *Academic Institutions*

An interview was conducted with Dr. Soto (2018), an active member of the JUNTOS 19S collaboration network. This network is made up of academic institutions and organizations dedicated to humanitarian work, such as OXFAM and the United Nations (UN), through the United Nations Development Program (UNDP), among others. They intervened from the first days of the disaster through the establishment of collection centers: later, academic institutions worked together with other organizations such as OXFAM and the 19S Movement. There were no strictly structured protocols to activating the participation of academic institutions, but there were previous experiences and work carried out under that line. It is essential to mention that academic collaboration was not optimal since unforeseen issues arose; for example, the lack of coordination in the participation with the civil society, the garbage that was accumulated in the communities by the PET bottles, and the food that was received for the foodstuffs were not part of the diet of the region. These unforeseen events are part of the external factors that can affect a protocol, and that can be anticipated with this accumulated experience for future disasters.

Soto (2018) mentions that the needs should be seen from the affected, not from a generalized point of view. In the 19S case, the effects occurred in rural areas where the needs are different, and everything has a protocol at the administrative level by the state. In other words, the necessary permits must be requested for the deployment of aid in the region.

The phase where academic institutions have been most involved is in the Reconstruction Phase. They have restoration plans for some communities, preferably with families that did not get support from FONDEN. The rebuilding of houses considers a design that incorporates the characteristics of the area to construct safer new buildings. All this work is currently in process and will continue in the following years; its time of completion is not known because the phase is long-lasting.

7.4.6 Victims

In order to have a broad view of the context, the victims of the 19S case were also surveyed. In this case, we proceeded to survey victims belonging to the State of Puebla. The construction of the surveys is shown in the Appendix 7.1.

To determine the locations where the surveys were to be conducted, the geographical location factors and the FONDEN emergency declaration records were taken into account.

Table 7.1 shows the four most affected localities belonging to the State of Puebla, from a total of 112 locations that suffered damages, this based on statistics elaborated with information of the declaration of extraordinary emergency issued by Civil Protection, whose access is authorized by FONDEN (FONDEN, 2018).

Figure 7.7 presents the five localities in Puebla that are close to the epicenter of the earthquake on September 19th, 2017.

After selecting the locations where the surveys were carried out, we proceeded to calculate a sample to obtain the representative number of surveys that we need to apply. Hence, we obtain a reliable result with a low margin of error, 20% in our case.

Triola (2009) mentions that when the proportion of the population (*p*) is known, the Eq. 7.1 should be used to find the sample size.

In this case, the use of this formula is appropriate, given that the affected population is known based on the data provided by FONDEN, but we do not know with certainty the exact number. This formula is independent of that of the total population affected. However, this can show us an estimated number of survey applications to obtain a confidence level of 80%.

Table 7.1 Most affected localities in the State of Puebla (FONDEN, 2018)

Localities	Affected population
Izúcar de Matamoros, Puebla, Me7.	77,601
Chiautla, Puebla, Me7.	20,155
Tochimilco, Puebla, Me7.	17,956
Atlixco, Puebla, Me7.	134,364

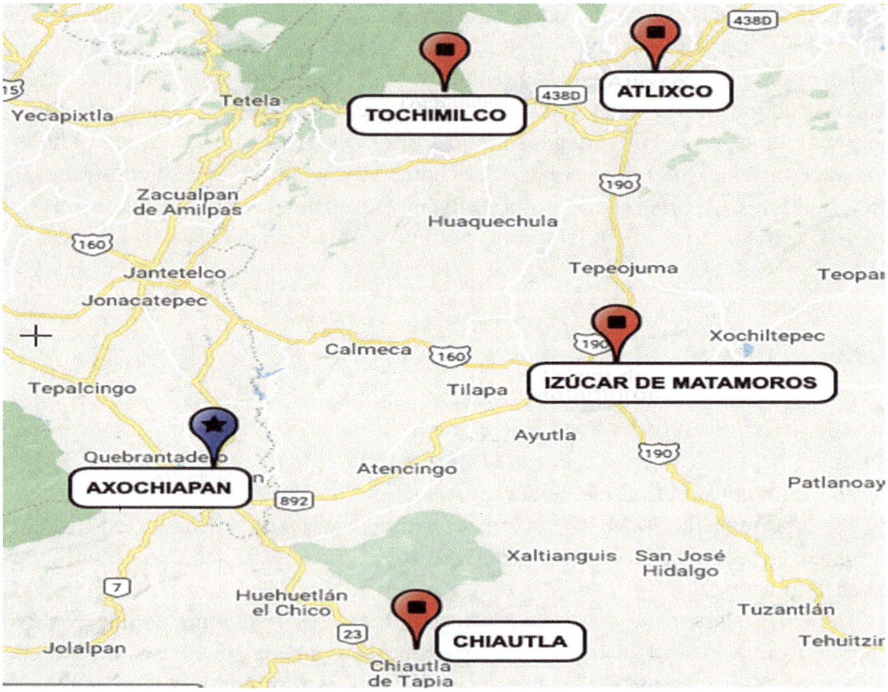

Fig. 7.7 Map showing the locations of Atlixco, Tochimilco, Izúcar de Matamoros, and Chiautla and Axochiapan as the epicenter

We opted for a level of confidence of 80%, given the nature of the phenomenon, which is natural disasters. This confidence level is lower than that usually established, but in this case, this percentage was used for the number of the total population affected and the number of poorly answered surveys per 100.

$$n = \frac{\left[Z_{\alpha/2}\right]^2 pq}{E^2} \quad (7.1)$$

where:

n Sample size
p Population proportion
α Confidence level
q The sample proportion of failures in n
E Error margin
$Z_{\alpha/2}$ Critical value

The critical value (was used for obtaining the confidence level; it was obtained with the two-tailed Z-table).

This considering that we look for a confidence level of 80%. When looking for that value in the two-tailed Z table, it gives us a critical value result of $Z_{\alpha/2} = 1,285$.

The clear of Eq. 7.2 is performed to determine the margin of error (E), resulting in the following expression:

$$E = Z_{\alpha/2}\sqrt{\frac{pq}{n}} \tag{7.2}$$

Within this clearance, the following values are assigned:

n	100
p	0.96
$Z_{\alpha/2}$	1.285 (Confidence level 80%)
q	0.04
E	¿?

These numbers were given from pilot surveys that had been conducted previously.

The values of n, p, and q are determined by the total of answered surveys per 100. Taking as reference q = 4 poorly answered surveys for each n = 100 applied, resulting in p = 96 useful surveys.

$$E = 1.285\sqrt{\frac{(0.96)(0.04)}{100}} \tag{7.3}$$

$$E = 0.02518075p \tag{7.4}$$

Thus, we have an error margin of 0. 02518075. Since we have the required values for Eq. 7.1, we can now compute the sample size.

$$n = \frac{\left[Z_{\alpha/2}\right]^2 pq}{E^2} \tag{7.5}$$

In step 2, we perform the substitution; these values are those that were determined after testing with the data found in Table 7.2 and decide a level of confidence of 80% with a proportion of the population of 0.95 and a sample proportion of 0.05 failures.

$$n = \frac{[1.285]^2(0.95)(0.05)}{(0.02518075)^2} \tag{7.6}$$

$$n = 123 \tag{7.7}$$

The values of n, p, and q are determined by the total number of poorly answered surveys. It is considered a low value because the surveyor will assist the respondent

Table 7.2 Test percentages to decide the level of confidence

Confidence level	Critical value	The proportion of the population	The sample proportion of failures	Error range	n
80%	1.285	0.95	0.05	0.02518075	123

in trying to understand the scope of each of the questions. Resulting in $E = 0.02518075$, that is, approximately 2% error.

Finally, the sample size was calculated with a percentage of 80%, which gave us a result of 123 surveys, which were applied among the four localities. Due to the characteristics of the phenomenon, no more surveys could be conducted.

In Table 7.3, the distribution of the survey application is shown.

The survey was applied in two months, which initialized in October of 2018 and finishes in November of 2018. Each group of surveys, separated by location, yielded very different answers, despite the geographical proximity. The difference relies on the question that we use to locate victims in the context that they lived. Within the context variable, we include the type of affectations that they suffered, the different city councils, and protocols followed by them. We include such items since Atlixco and Tochimilco have the Magical Town certification granted by the National Institute of Anthropology and History (INAH); such status provides them with protocols that must be activated during specific phenomena.

Concerning descriptive statistics, seventy-seven percent of people surveyed in Tochimilco have not yet covered the reconstruction of their homes. Under the perception of the victims of Izúcar de Matamoros, the recovery has been too slow, they have not had much support from the corresponding government, and more than a year after the earthquake, they still cannot recover. More than 84% of the population responded that the reconstruction process has been slow. In Chiautla, the government supported them with cards loaded with money that they could use to buy clothes, food, and products they needed. We can say that Chiautla is the place that implemented the most efficient and beneficial strategies for the respondents. Concerning the reconstruction process, 64% of respondents agree that their needs have been met.

In Tochimilco, families had significant losses. From these families, some of them received financial support from FONDEN that supports the reconstruction of houses, but the resource arrived incomplete according to the people surveyed.

Table 7.3 Distribution of the application of the surveys in the localities

Localities	Affected population	Applied surveys
Izúcar de Matamoros, Puebla, Me7.	77,601	16
Chiautla, Puebla, Me7.	20,155	51
Tochimilco, Puebla, Me7.	17,956	13
Atlixco, Puebla, Me7.	134,364	43

Several buildings remain unsettled, given that they are protected by the INAH and are waiting for it to grant the necessary permits to begin the reconstruction. In Atlixco, which also is considered as a Magical Town, the situation is similar to the one reported in Tochimilco. Many downtown buildings cannot be arranged, including some churches due to INAH's restrictions. Specifically, any intervention to modify such buildings could lead to the imposition of very high-cost fines. However, families whose houses are not watched out by INAH benefited from the economic resources granted by FONDEN. Also, they were subject to the local government administration, that assigned economic support, as well as labor and material for reconstruction of the total number of people surveyed in the localities, the majority did not receive support for the moment of the earthquake or subsequent days. 53% of the affected population in Atlixco did not receive any help at the time.

Some people surveyed mentioned that the government offered them support for the revision of their homes to get an opinion on whether the property was habitable or not. In Atlixco, 53% of the respondents received that support, while 52% of respondents mention not having received the support for the reconstruction because they were not at home, although their houses were inhabitable. Others mention that their needs have been completely satisfied in terms of reconstruction.

On the other hand, an average of more than 50% of respondents in the localities agrees that, within the first days of the emergency, it was the civil society that came to support the community. The government made its intervention days later, as marked by the civil protection protocol.

7.5 Discussion

In the different interviews that were conducted, we observe points in common to highlight some of the problems that arise within the HSC. In the interview with Reyes (2017), he mentions that, at the beginning of the 19S, the HSC organization deal with problems related to pollution by using non-recyclable containers. Also, the access roads to the localities were saturated since there was no organization for the delivery of supplies; such a problem as a consequence of the lack of information. We can also observe this conclusion in the answers given by the volunteers interviewed, that, being part, or not, of an organization started on September 19, they had similar problems at the time of volunteering.

A problem that can be identified in the interviews with Reyes (2017) and Soto (2018) was the lack of knowledge at the cultural level of the affected communities. Consequently, the provision of support and help does not satisfy specific needs and social features since HL mechanisms considered general features. So, the supplies that were distributed do not entirely meet the needs of the community because society's needs and preferences were partially known or unknown, in some cases.

The respondents pointed out that humanitarian aid includes unnecessary good for some communities, i.e., the supplies that were delivered did not match with the consumption habits and social features like age, gender.

The use of ICTs is another critical point that was identified in the study; the ICTs were the basis in the intervention of the civil society. Moreover, these served as a tool to obtain updated information concerning the needs of the affected communities. Social networks and specialized software provide verified data that SCH agents could transform into information that they used as guidelines for more accurate decision making. It is worth to mention that online platforms played a central role during the mitigation phase given its capacity of fast response; agents of any type in the HSC agreed that online platforms fulfilled their objective immediately.

The surveys that we carried out also provide information about the reconstruction period, which is still ongoing. Results show that some of the affected people have fully met their needs regarding reconstruction. However, this status does not apply to all respondents since the building is slowed by location factors, protocols, and permits that INAH must provide. Therefore, the reconstruction process is stagnant in those localities with the Magical Town certification, which has originated economic problems due to tourism diminished in the months after the disaster.

Concerning other localities, the majority of people affected remained in the same condition even after a year of the earthquake, due to the mismanagement of the resources that were granted to local governments. They cannot take a proper reconstruction or at the time they should. Remarkably, few people received pantries and other types of support in the mitigation phase because they were taking care of their belongings in their homes. Moreover, the insecurity forced affected people to monitor their properties, which limited their possibilities to get support and humanitarian from civil society organizations. At the moment we applied the interviews, we find that respondents agree in a feeble response from the government; specifically, the distribution of humanitarian aid was chaotic and did not meet the needs of the affected populations.

It is very complicated to be able to carry out a chain of humanitarian logistics in a practical way when they find themselves through such severe problems as mismanagement from the delivery of foodstuffs to the actions of the local government. The actors have different objectives, procedures, and response times during their participation in an HSC. However, withal of them have the same purpose, which is to support the affected people. Figure 7.8 illustrates the phase in which the HSC's actors intervened in our case of analysis.

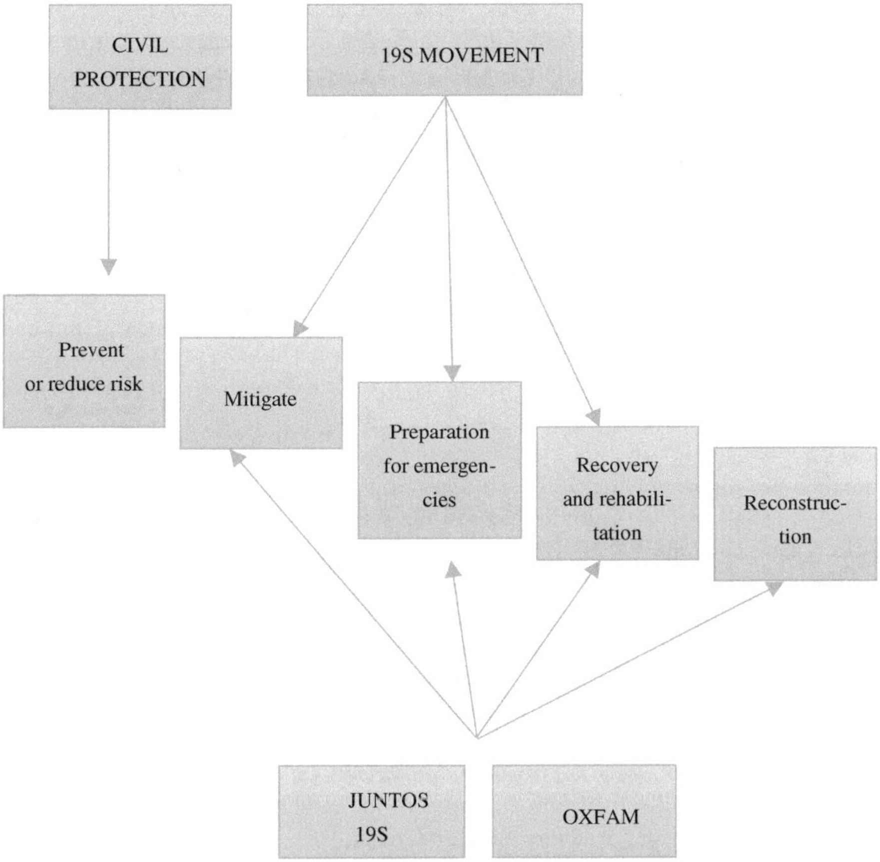

Fig. 7.8 Participation of the actors in the HSC in the case of the earthquake of September 19th, 2017

7.6 Conclusion

By analyzing the humanitarian logistics before and after the 19S earthquake, we observe that civil society had a high incidence within the HCS during the mitigation phase. Given the protocols established by the government, there is an unattended period between the time the disaster happened and the moment when they intervene in the community. Given that victims have to need to be satisfied, the civil society emerges as an efficient option to cope with the affected localities needs.

The 19S Movement is an example of how civil society organizes. So, our main contribution is to provide a record of how it originated and evolved in the different phases after the phenomenon; adding to this result, the interviews that w performed point out the challenges that HSC entails in a country like Mexico. The geographical location of Mexico makes necessary the designing of humanitarian logistics that can prevent, mitigate, and solve the affectation caused by any disaster.

Note that not all problems depend on the organization that civil society must generate, the HSC also faces other difficulties given the features of each involved agent and the disaster affectations magnitude. This research indicates that goodwill is necessary to provide adequate aid to victims, but it is not enough; the generation of useful information, communication, and coordination are also crucial factors that decision-makers need to watch out in all echelons of the HSC. Moreover, it is vital to have in mind the high level of uncertainty before and after a natural disaster, which leads to the development of contingency plans for those situations out of our control.

Finally, it is essential to emphasize how ICTs contributed to diminishing information asymmetries in the HSC. In other words, the emergence of the civil society as a protagonist in the 19S is the result of the fast responsiveness that ICTs provide and their capacity to identify the source of specific requisitions, the identification of special needs, and the establishment of differences among affected locations. Data, information, and knowledge increase their effectiveness in humanitarian logistics when ICTs are appropriately used by all government levels and other involved agents in the HSC. In the future, it will be essential to develop institutional platforms, databases, and tools for an adequate flow of information that will reduce the uncertainty that initially exists.

Appendix:

The survey[1] seeks to assess the impact of the effects on the population that belongs to some of the localities near the epicenter of the earthquake, within the stages of the disaster. This survey was conducted based on the Population Survey on flood risk due to heavy rains (Godefoy et al., 2017), which also seeks to measure the impact of the disaster, including the government, infrastructure and society dimension.

For the measurement of responses, the Likert Scale was used, which is intended to measure attitudes; Individual predispositions to act in a certain way in specific social contexts (García et al., 2011).

In this case, the questions used were of open modality. The application of the Likert Scale is applied in the repetition of the sentences used by the respondents, having as a scale of 1 the negative answers, 3 for the answers in which the knowledge was not known, 5 for positive answers.

The blocks of questions and their scores are listed in Appendix 7.A of the spreadsheet.

Reliability refers to the consistency or stability of a measure (Quero, 2010). To measure the reliability of the instrument in this case, Cronbach's Alpha was used. The formula is the following:

[1]To know the survey in detail you can contact the author.

$$\propto = \frac{K}{K-1}\left[1 - \frac{\sum Vi}{Vt}\right] \tag{7.8}$$

where:

K Ítems
$\sum Vi$ Sum of individual variances
Vt Total variance

The variances of each item are obtained within the table, resulting in:

Q1	Q2	Q3	Q4	Q5	Q6
1.442356392	2.79288285	2.46941224	1.38198054	1.94842063	3.8118086

Q7	Q8	Q9	Q10	Q11	Q12
1.64787418	2.43196055	2.54631481	4.01119552	4.01119552	2.270825

The sum of the individual variances is equal to 30.7662268

The total variance is obtained from the sum of the scores of each survey. Subsequently, the variance is calculated in the sum of all the scores.

Total variance: 102.77609

Value assigned:

K 12
$\sum Vi$ 30.7662268
Vt 102.77609

$$\propto = \frac{12}{12-1}\left[1 - \frac{30.7662268}{102.77609}\right]$$

$$\propto = 0.76434328$$

The reliability is 0.76434328, in a margin between 0 and 1. Taking one as the total confidence.

References

Abidi, H., Jensen, L. M., & Klumpp, M. (2016). Roles of humanitarian actors in the humanitarian logistics. In *POMS 27th Annual Conference*. Orlando, FL: FOM Hochschule.

Acuerdo por el cual se emiten las Reglas Generales del Fondo de Desastres Naturales. (2010). Diario Oficial. Retrieved September 10th, 2018, from: https://www.dof.gob.mx/nota_detalle.php?codigo=5169686&fecha=03/12/2010 [in Spanish].

Anparasan, A., & Lejeune, M. (2017). Resource deployment and donation allocation for epidemic outbreaks. *Annuals of Operations Research*. Online ISSN 1572-9338. https://doi.org/10.1007/s10479-016-2392-0.

Apte, A. (2010). Humanitarian logistics: A new field of research and action. *Foundations and Trends in Technology Information and Operations Management, 3*(1), 1–100. Online ISBN: 978-1-60198-337-4. http://d7.doi.org/10.1561/0200000014.

Balcik, B., & Beamon, B. (2008). Facility location in humanitarian relief. *International Journal of Logistics: Research and Applications, 11*(2), 101–121. https://doi.org/10.1080/13675560701561789.

Bui, T., Cho, S., Sankaran, S., & Sovereign, M. (2000). A framework for designing a global information network for multinational humanitarian assistance/disaster relief. *Information Systems Frontiers, 1*(4), 427–442. https://doi.org/10.1023/A:1010074210709.

Chang, M., Tseng, Y., & Chen, J. (2007). A scenario planning approach for the flood emergency logistics preparation problem under uncertainty. *Transportation Research Part E, 43*(6), 93–138. https://doi.org/10.1016/j.tre.2006.10.013.

D'Hyver, A. (2019, Febrero). OXFAM. (MJ Galvez, Interviewer).

DOF Report. (2010). DOF. Acuerdo por el cual se emiten las Reglas Generales del Fondo de Desastres Naturales. Retrieved February 20th, 2019, from http://dof.gob.mx/nota_detalle.php?codigo=5169686&fecha=03/12/2010.

Duarte, E. S. (2008). Las tecnologías de información y comunicación (tic) desde una perspectiva social. Educare, pp. 155–162. ISSN: 1409-42-58.

Dubey, R., & Gunasekaran, A. (2015). The sustainable humanitarian supply chain design: Agility, adaptability, and alignment. *International Journal of Logistics Research and Applications, 19* (1), 62–82. https://doi.org/10.1080/13675567.2015.1015511.

Eckstein, D., & Schäfer, L. (2017). Global Climate Risk Index 2018. ISBN: 978-3-943704-70-9. www.germanwatch.org/en/cri.

FONDEN Report. (2018). FONDEN Emergencias. Retrieved August 10th, 2018, from Protección Civil: http://www.proteccioncivil.gob.mx/es/ProteccionCivil/PUEBLA2017. [in Spanish].

García, J., Aguilera, J. R., & Castillo, A. (2011). Guía técnica para la construcción de escalas de actitud. Odiseo, revista electrónica de pedagogía, 8(16). ISSN 1870-1477. [in Spanish].

Godefoy, E., Acosta, N., Pedroso, I., Núñez, L., Bayón, P., López, C., Huergo, V., Rodríguez, I. (2017). Encuesta a la población sobre el riesgo de inundación por intensas lluvias. Fortalecimiento de Alerta Temprana Hidrometeorológico. Retrieved June 17th, 2019, from PNUD: https://www.undp.org/content/dam/cuba/docs/Desastres%20Naturles%20y%20Riesgo/Encuesta%20FORSAT_multimedia.pdf. [in Spanish].

Gower, J. (2018). Interview to Civil Protection, September, 19th. (M. J. Galvez, Interviewer) [in Spanish].

Howden, M. (2009). How humanitarian logistics information systems can improve humanitarian supply chains: A view from the field. In *6th International ISCRAM Conference*. Gothenburg, Sweden.

Ilhan, A. M. (2011). The humanitarian relief chain. *South East European Journal of Economics and Business*. https://doi.org/10.2478/v10033-011-0015-x.

Ley del Sistema Estatal de Protección Civil. (2017). Ley de Puebla. Retrieved September 10, 2018, from Congreso de Puebla: congresopuebla.gob.mx [in Spanish].

Ludema, M. (2000). Military and civil logistic support of humanitarian relief operations. Decade of progress—A new century of opportunity. In *Proceedings of the 10th Annual International Symposium of the International Council on Systems Engineering*, pp, 143–50. Minneapolis, MN: INCOSA.

Oloruntoba, R., & Gray, R. (2006). Humanitarian aid: an agile supply chain? *Supply Chain Management, 11*(2), 115–120. https://doi.org/10.2478/v10033-011-0015-x.

Paz, C. C. (2018). Conceptos básicos de Vulnerabilidades ante desastres socionaturales. Universidad de Chile. [in Spanish].

Protección Civil-DATA. (2017). Declaratorias de emergencia. Retrieved August 8, 2018, from: http://www.proteccioncivil.gob.mx/work/models/ProteccionCivil/Resource/2405/1/images/Insumos_Autorizados_PUE_313-17.pdf [in Spanish].

Quero, M. (2010). Alpha de Cronbach. *Telos, 12*(2), 248–252. ISSN 1317-0570 [in Spanish].

Reyes Garcia, A. (2017, Diciember). Interview to Organización 19S. (MJ Galvez, Interviewer).

Servicio Geológico Mexicano. (2018). Sismología de México. Retrieved August 8, 2018, from Servicio Geológico Mexicano: https://www.sgm.gob.mx/Web/MuseoVirtual/Riesgos-geologicos/Sismologia-de-Mexico.html [in Spanish].

Servicio Sismológico Nacional. (2017). Catálogo de sismos. Retrieved March 10th, 2018, from Servicio Sismológico Nacional: http://www2.ssn.unam.mx:8080/catalogo/ [in Spanish].

Social TIC. (n.d.). Tecnología, datos y participación durante sismo en México. Retrieved August 10, 2018, from Social TIC: https://socialtic.org/blog/tecnologia-datos-y-participacion-ante-el-sismo-en-mexico/ [in Spanish].

Soto, O. (2018, May). Interview to Juntos 19S. (MJ Galvez, Interviewer) [in Spanish].

Sushil. (2017). Theory building using SAP–LAP linkages: An application in the context of disaster management. *Annals of Operations Research.* Online ISSN: 1572-9338. https://doi.org/10.1007/s10479-017-2425-3.

Thomas, A. (2005). Logistics training: Necessity or luxury. *Forced Migration Review, 22*, 60–1. Online ISSN: 1460-9819.

Thomas, A., & Kopczak, R. (2005). *From logistics to supply chain management: the path forward in the humanitarian sector.* Fritz Institute.

Triola, M. (2009). *Estadística* (10th ed.). México: Pearson Educación. [in Spanish].

Trujillo, P., Carlos, H., Ospina Lopez, R., Parra Lara, H. (2010). Los terremotos: Una amenaza natural latente. *Scientia Et Technica, 16*(45), 303–308. ISSN: 0122-1701.

UN/ISDR. (2004). Living with risk: A global review of disaster reduction initiatives. Retrieved May 3, 2018, from United Nation Office for Disaster Risk Reduction: http://www.unisdr.org/eng/about_isdr/bd-lwr-2004-eng.htm.

Ureste, M., & Aroche, E. (2017, octubre, 19). Lo que el 19S nos dejó: las víctimas, daños y daminificados en México. Animal político. Retrieved August 8th, 2018, from https://www.animalpolitico.com/2017/10/cifras-oficiales-sismo-19s/.

Van Wassenhove, L. N. (2006). Humanitarian aid logistics: Supply chain management in high gear. *Journal of Operational Research Society, 57*(5), 475–489. http://d7.doi.org/10.1057/palgrave.jors.2602125.

Yang, F., Du, S., Liang, L., & Yuan, Q. (2014). Reserving relief supplies for earthquake: A multi-attribute decision making of China Red Cross. *Annals of Operations Research, 247*(2), 759–785.

Chapter 8
Optimization Model to Locate Pre-positioned Warehouses and Establish Humanitarian Aid Inventory Levels

Erika Barojas-Payán, Diana Sánchez-Partida, Damián Emilio Gibaja-Romero, José Luis Martínez-Flores, and Mauricio Cabrera-Rios

Abstract The focus of this work is to enable the dignified survival of the people affected in a natural disaster through the optimal location of pre-positioned warehouses and the determination of their optimal inventory levels aiming to provide emergency supplies efficiently. To this end, a *mixed-integer nonlinear mathematical* model is presented to be applied during the disaster preparation phase. The mathematical model integrates two elements immersed in *humanitarian logistics;* one that is based on determining the *optimal locations* for the installation of pre-positioned warehouses, and the other that approaches the establishment of the *inventory levels* in these facilities to supply food, equipment, water, and vaccines to *different kinds of groups of survivors* classified according to the *human life cycle*. The mathematical model is validated through the analysis of two regions of the State of Veracruz, the Mexican state with the highest number of declarations of emergency related to hydrometeorological phenomena. The installation of instances with two and three pre-positioned warehouses, each with a total of 29 municipalities, 14 of which have been subject to the issuance of declarations of emergency, are presented. The results show the optimal location for the pre-positioned warehouses and the number of kits allocated to each one depending on the type of

Electronic supplementary material The online version of this chapter (https://doi.org/10.1007/978-3-030-67295-9_8) contains supplementary material, which is available to authorized users.

E. Barojas-Payán (✉) · D. Sánchez-Partida · D.-E. Gibaja-Romero · J. L. Martínez-Flores
Logistics and Supply Chain Management, Universidad Popular Autónoma del Estado de Puebla, (UPAEP University), 17 South 711, Barrio de Santiago Puebla, Puebla 72410, Mexico
e-mail: erika.barojas@upaep.edu.mx

M. Cabrera-Rios
Industrial Engineering Department, University of Puerto Rico at Mayagüez, PR 00681, USA

D. Sánchez-Partida et al. (eds.), *Disaster Risk Reduction in Mexico*,
https://doi.org/10.1007/978-3-030-67295-9_8

demand. The model is a decision support instrument designed for the use of planners for the timely and efficient response to disasters to aid the people affected by a natural phenomenon, especially one of the hydrometeorological type.

Keywords Hydrometeorological phenomena · Location of facilities · Pre-positioned warehouse · Human life cycle · Mathematical programming

8.1 Introduction

Article 2, paragraph XLVIII, of the Mexican General Law of Civil Protection (2012) defines risk as to the probable damage or loss of an affected agent, resulting from the interaction between its vulnerability and the presence of a disturbing agent (Chamber of Deputies of the Congress of the Union, 2012a, 2012b). In paragraph XVI, it conceptualizes disaster as the result of the occurrence of one or more severe and/or extreme disturbing agents, concatenated or not, of natural or human origin. Occurring in a given time and in a particular area that causes damage and exceeds the response capacity of the affected community due to their magnitude (Chamber of Deputies of the Congress of the Union, 2012a, 2012b). Thus, the risk for disaster is determined by three key factors: (a) the risk; (b) the degree of exposure; and (c) the vulnerability of exposed assets (Government of the State of Veracruz; Secretariat of Civil Protection, 2018).

Regarding risk, this is found in the disruptive phenomenon itself, irrespective of its type: geological, hydrometeorological, chemical, sanitary, or socio-organizational. However, the probability that a hazard will materialize in some damage to the community depends on the degree of exposure, that is, the proximity or distance of these phenomena with the community, the population, or the infrastructure. Most important, however, is the propensity of individuals or material goods to be affected, that is, the fragility of society and its goods and the inability to self-restore (lack of resilience) (Government of the State of Veracruz, Secretariat of Civil Protection, 2018; Government of the State of Veracruz, Secretariat of Civil Protection, 2011).

When an unexpected event occurs, i.e., the presence of a disruptive phenomenon, a large number of relief supplies are needed in haste to cope with the emergency. The most notable characteristic of emergency problems is the urgency of time, so its reduction becomes the primary objective of the optimization. In other words, the delivery of necessary emergency and disaster relief and assistance supplies from the points closest to the disaster area should take place as soon as possible (PAHO, 2017).

In emergency operations, logistics is required to support the organization and implementation of response actions so that they are not only quick but also agile and productive. Therefore, humanitarian logistics is defined as the process of efficiently planning, implementing, and controlling the flow and storage of goods and materials, as well as related information; from the point of origin to the point of consumption in order to alleviate the suffering of people who are in situations of fragility and low resilience (OPS, 2001; Overstreet et al., 2011; Vela, 2014).

Tomasini and Van-Wassenhove (2009) identify four primary stages in the emergency humanitarian aid management cycle: (1) *Mitigation,* which aims to develop preventive actions that at least soften the effects of a possible future disaster, an example of which is to avoid building weak constructions in the course of a basin subject to high floods; (2) *Preparation* entails planning the actions to take in the event of a humanitarian disaster; the Early Warning Systems (EWS) and the training of vulnerable population groups are clear examples of this; (3) *Response* takes place during the disaster and its immediate aftermath and includes the initial assessment of the impact of the crisis, as well as the rescue and immediate care of the victims; (4) *Rehabilitation* involves a strong interaction with development cooperation projects. The objectives of the reconstruction phase should be to avoid a return to the pre-crisis status quo and to try to improve the conditions of the affected community to reduce both the likelihood of suffering a disaster in the future and its degree of vulnerability in the event of its occurrence.

Natural disasters constitute a significant source of fiscal risk in countries that are highly exposed to natural disasters, thus presenting contingent liabilities of considerable magnitude for the governments of such nations. The absence of efficient emergency preparation and response mechanisms, as well as adequate financial planning to cope with disasters, can create difficulties and delays in response, which could aggravate the consequences regarding human and economic losses (FONDEN, 2012).

For the reasons offered previously, this work presents a mixed-integer nonlinear optimization model integrating the location of facilities and the establishment of inventory levels, which is resolved to global optimality. This optimization model will allow identifying the optimal location of several pre-positioned warehouses during the preparation phase, as well as their inventory levels to efficiently provide relief supplies and thus avoiding human losses.

8.2 State of the Art

In recent years, different institutions, researchers, associations, and organizations have developed methods, techniques and logistical tools to meet the basic and non-core needs of people affected by a natural event, focusing on logistical aspects such as location, storage, routing, and collection, to name a few. The research project presented here is based on the location of facilities and their inventory levels. The research carried out in recent years, which forms an essential basis for this document is reviewed next.

Rawls and Turnquinst (2010) developed an emergency response optimization model using a tool to determine the location and quantities of various types of supplies. The article presents a two-stage stochastic mixed index program (SMIP). In stage one, unanticipated decisions at this stage are made in the presence of uncertainty about future performance (effects of the hurricane, changes in demand, and damage to the network) considering location, storage, and diversity of supplies. In stage two, the decisions of the phase (appeal) are made after the realizations of the elements of the previous problem are known and considered. This work uses a heuristic algorithm called Lagrangian, developed for the resolution of large-scale cases of the problem and focuses on the threat of a hurricane off the U.S. Gulf Coast.

The document by Campbell and Jones (2011), which examines the decision to pre-position supplies as a preventive measure in the event of a disaster, sets out the hypothesis of independent destruction probability for each supply point. The authors address the risks of proximity to the potential impact zone through the derivation of equations and determine the optimal amount of stock, as well as the costs associated with delivery from the point of supply to that of demand. Computer experiments show a variety of distance-to-hazard relationships and show how they affect location decisions and inventory levels.

The document by Bai and Liu (2011) establishes a decision-making model with the aim of minimizing emergency response time, saving points, higher reliability of transport routes, and lower response costs. It uses the ideal diffuse point algorithm to transform multiple decision-making into a single target problem. The authors provide a method for limiting the duration of disasters, as demonstrated by the results of a simulation.

In the same line of work, the document presented by Deqiang and Xiaoming (2011) develops an emergency logistic location method that allows increasing the reliability of the location and reducing the complexity of the decision-making process through the combination of the Analytical Hierarchy Process (AHP) and Data Envelopment Analysis (DEA) methods.

The document by Timperio et al. (2016) proposes the design of a supply chain framework for multi-stage network design in the domain of disaster relief supply chains to improve the efficiency of operations by facilitating the safeguarding of commodities. The authors work through the Geographic Information System (GIS) technology, where they integrate essential information that leads them to calculate the optimal network design of emergency response services determined by a 72-hour time window. Moreover, they use a Fuzzy Analytical Hierarchy Process (FAHP), in which, unlike AHP, the criteria are compared using linguistic terms, which were used to make the comparison by pairs of matrices. The assessment of the location criteria for site selection was calculated from these synthetic matrices using the existing value method; the results were some possible locations (nodes) of the emergency response network.

The model proposed in this document is applicable in the disaster preparation phase and has the particularity of assigning product inventory levels optimal for each type of demand. Based on each phase of the human life cycle, it is possible to divide demand into food products, equipment (diapers, glasses, underwear, shoes, raincoats, cleaning supplies, and so forth), water, and medicines (protection and application materials). This capability can be afforded to other types of disruptive phenomena beyond the initial hydrometeorological application.

The following sections show the development of the research carried out. First, the problem statement is presented with a description of the State of Veracruz, Mexico, along with a quantification of the emergency declarations that have been deployed, and percentage analysis of these declarations. Afterward, the methodology used to solve the problem is described. This includes the description of the mathematical model, the location data, the inventory level data, the description of programming software, and the evaluation of the scenarios. Finally, the results, conclusions, and future works are presented.

8.3 Problem Statement

"Mexico is a country of significant differences and climatic versatility; this is derived from its geographical location, oceanic contiguity, complex topography, valleys, mountains, and plateaus among orographic shadows" (Sánchez & Cavazos, 2015). That is why the country is highly exposed to a wide variety of geological and hydrometeorological phenomena. In fact, "approximately forty percent of the Mexican territory and more than a quarter of its population are exposed to storms, hurricanes, and floods" (FONDEN, 2012). Within the national territory, "the highest amount of precipitation is concentrated in the southern and southeastern states, with average annual rainfall more than 1000 mm. It exposes the nation to extreme water events with effects mainly on coastal areas, flood areas, and mountain slopes" (City Council of Mexicali, B.C., 2011; Civil Protection, Guanajuato, 2015).

The National Center for Disaster Prevention (CENAPRED for its acronym in Spanish) in its infographics *Disasters in Mexico* within *social and economic impact* mentions that disasters are measured by the economic impacts that generate damage and losses. It also importantly considers the effects that occur in the social sphere, such as injured people and deaths, damaged houses, schools, and hospitals, among others. The document shows a list of disasters with the most significant impact from 1980 to 2014; the states affected, the cost in millions of dollars, and deaths of people. It also alludes to the fact that 90% of the damages and losses since 1991 are of hydrometeorological origin. Figure 8.1 graphically concentrates the data provided, which shows a more significant presence of this type of event in the State of Veracruz, at 15% (CENAPRED, 2015).

It is worth mentioning that the hydrometeorological phenomena that frequently occur in the State of Veracruz are cyclones, frosts, droughts, electric storms, hail, and extreme temperatures. Some examples of these are tropical cyclones, Karl and Matthew, in 2010, for 1,972.8 million dollars (CENAPRED, 2015; Mora et al., 2015; Government of the State of Veracruz, Ministry of Civil Protection, 2011; Ministry of the Interior, National Civil Protection System, National Center for Disaster Prevention, 2001).

"The State of Veracruz has an extension of 71,820 Km^2 equivalents to 3.7% of the national territory"; until 2010 it had a population of 7,643,194 inhabitants,

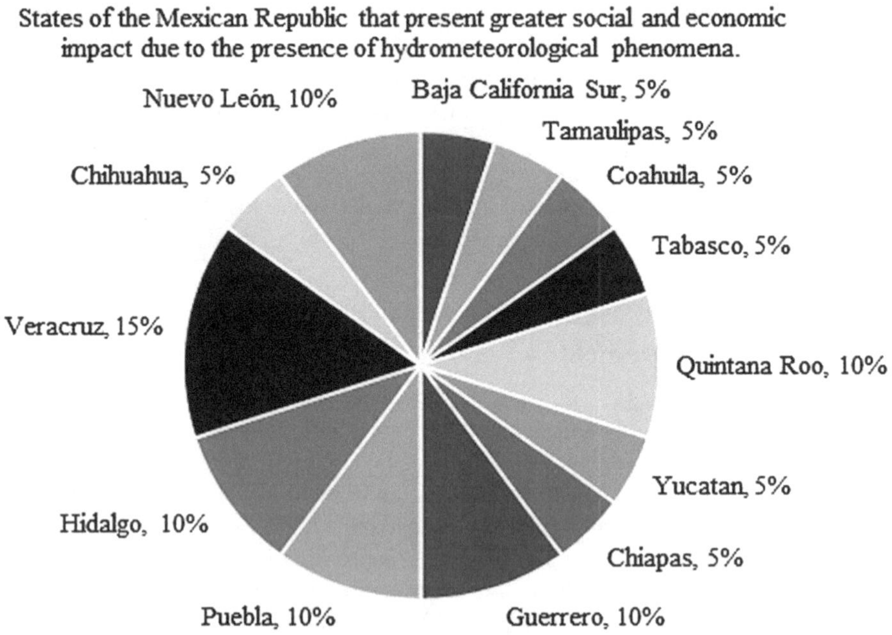

Fig. 8.1 Percentage of impacts in Mexico. *Source* CENAPRED 2015

showing an annual growth rate of 1%, which represents 6.9% of the total of the country, with 61% of the population being of urban-type and 39% of rural type. "Its capital is the city of Xalapa de Enríquez; the state is divided into 212 municipalities and is adjacent to the states of Tabasco, Chiapas, Oaxaca, Puebla, Hidalgo, San Luis Potosí, and Tamaulipas" (INEGI, 2016; Mora et al., 2015).

The Planning Committee for the Development of Veracruz SEFIPLAN divides the municipalities belonging to the State into ten regions: (1) *Capital* Region with 33 municipalities; (2) *Huasteca Alta* Region with 15 municipalities; (3) *Huasteca Baja* Region with 18 municipalities; (4) *Altas Montañas* Region with 57 municipalities; (5) *Nautla* region with 11 municipalities; (6) *Olmeca* Region with 25 municipalities; (7) *Papaloapan* region with 22 municipalities; (8) *Sotavento* Region with 12 municipalities; (9) *Totonaca* Region with 15 municipalities; and (10) *Los Tuxtlas* Region with four municipalities (Secretariat of Education, 2013).

In the period from 2000 to 2015, among the regions of the State with the highest number of declarations of emergency derived from any hydrometeorological phenomenon are the *Altas Montañas* Region, the *Olmeca* Region, and the *Capital* Region. They are home to the most significant number of municipalities within the State. A quotient analysis was carried out, based on the variables of some declarations and number of municipalities contained in the region (declarations/municipalities), in order to determine that the regions with the most significant impact are the *Nautla* Region and *Los Tuxtlas* Region, derived from the number of municipalities that they have. This relation can be seen in Table 8.1.

Table 8.1 Regions with their respective indicators *Source* Authors

# Municipalities	Region	% Municipalities (%)	% Declarations (%)	# Declarations/ # municipalities
33	Capital	15.57	13.48	6.85
15	Huasteca Alta	7.08	7.45	8.33
18	Huasteca Baja	8.49	10.08	9.39
4	Los Tuxtlas	1.89	2.86	12.00
57	Montañas	26.89	20.21	5.95
11	Nautla	5.19	7.45	11.36
25	Olmeca	11.79	16.28	10.92
22	Papaloapan	10.38	8.77	6.68
12	Sotavento	5.66	4.77	6.67
15	Totonaca	7.08	8.65	9.67
212		100.00	100.00	

These regions are further composed of the most significant number of municipalities, so for this document, the regions are analyzed from the resulting quotient. It results in the *Nautla* Region and *Los Tuxtlas* Region displacing the *Altas Montañas* and *Capital* Regions. Likewise, it is highlighted that the following are among the five municipalities with the highest number of declarations of emergency: (a) Minatitlán, and (b) Nanchital de Lázaro Cárdenas, which belong to the *Olmeca* Region. This region had the third-highest quotient, in addition to being the region that borders *Los Tuxtlas*; thus, based on this and the above, the evaluation of the model is determined from the *Olmeca* and *Los Tuxtlas* Regions.

8.4 Methodology

The present model introduces a problem of the location of facilities, which allows the location of several *pre-positioned warehouses* to satisfy the *highest number of people affected by a natural phenomenon*. These locations must comply with two conditions: to not have been impacted by the natural phenomenon and to have a minimum covered distance. Considering that it is not possible to predict the impacted demand accurately, the program includes terms and restrictions *to minimize costs and establish inventory levels* through *an extension to the model (q, R) of continuous review with uncertain demand*, based on the *level of service*.

The methodology developed for the solution of the problem described is shown in Fig. 8.2. It begins with the description of the mathematical model, which includes the assumptions, notation used, and model with the description of each of its equations. Subsequently, we have the supporting data for the location of a place

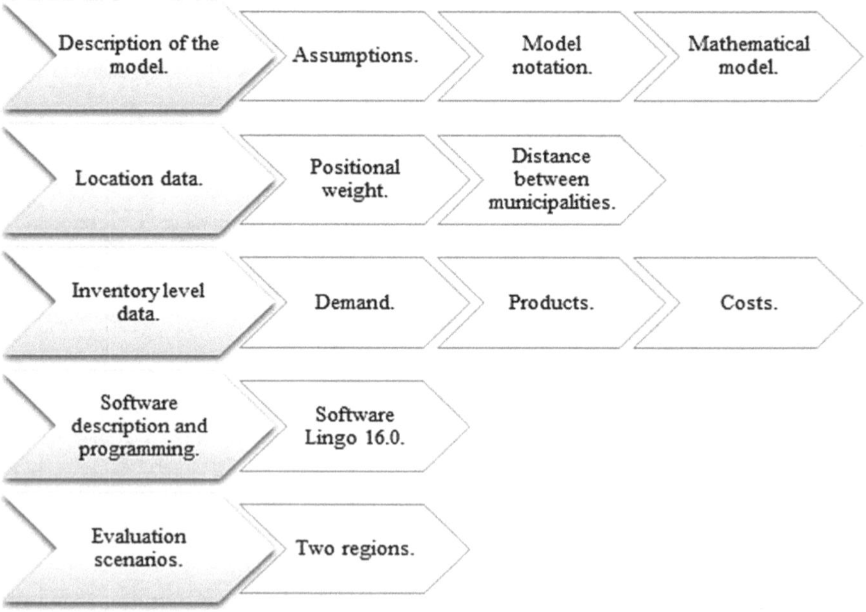

Fig. 8.2 Applied methodology *Source* Authors

that has the attributes determined by the National Civil Protection System and the National System for the Integral Development of the Family for the installation of a pre-positioned warehouse, and the distances between municipalities with the possibility of impact and non-impact of the phenomenon. Data on demand for each of the phases of the human life cycle are collected, the necessary products for the survival of those affected and their costs are the next step. Then, the model is coded in the optimization software Lingo, evaluating its behavior through two scenarios in two regions with the location of two and three optimal locations and inventory levels for each of them.

8.4.1 Description of the Mathematical Model

El siguiente modelo logístico, parte de una extensión del modelo publicado por Barojas-Payán et al. (2019), en el cual, se evalua un modelo que permite obtener una sola ubicación factible para la instalación de un almacén pre-posicionado y sus niveles de inventario en las distintas etapas de vida del ser humano, en el modelo que se presenta en los siguientes párrafos, se busca la localización de dos o más ubicaciones para la instalación de almacenes pre-posicionados con sus diferentes niveles de inventario dependiendo de la demanda a abastecer por cada uno de ellos, así mismo cabe resaltar una caracterización diferente en cuanto a la variable de peso posicional de los municipios que podrían ser sedes del almacén y el espacio geografico en el cual se evalua el modelo con un mayor número de municipios impactados por un fenómeno natural de tipo hidrometeorológico y de municipios no impactados con posibilidad de albergar un almacén de pre-posicionamiento.

8.4.1.1 Assumptions

For the planning purposes of the logistics model to be evaluated, the following assumptions are considered:

1. Different consumer and human products can be stored in the same pre-positioned warehouse;
2. Each municipality is stocked by only one pre-positioned warehouse;
3. The demand is a normally distributed random variable with mean μ mean and variance δ^2, i.e., $D \sim N(\mu, \delta^2)$ (Nahmias, 2014). For this purpose, the amounts were extracted from a database developed by the Natural Disasters Fund, which considered a normal distribution for the data.
4. For the efficient supply of food and equipment to the target population, the nutritional needs of people throughout the different phases of their life cycle are taken into account: (a) early childhood; (b) childhood; (c) adolescence; (d) adulthood; and (e) old age[1]; as established by the World Health Organization (WHO).
5. Product groups are considered for the different demands involved in the five stages of the human life cycle. These goods are those that belong to the food industry, those classified as equipment (baby diapers, adult diapers, clothes, towels, shoes, and similar), water, and medicines. A generic product is indicated as $p \in \{1, 2, \ldots, P\}$.
6. The Federal Entity requests inputs on a *4-day* basis for the urgent needs of the affected population until the emergency is over.

8.4.1.2 Model Notation

Sets: We indicate I as the set of all the municipalities that were not impacted by a hydrometeorological phenomenon, and that can be sites for a pre-positioned warehouse. J indicates the set of municipalities with the possibility of being impacted by a natural hydrometeorological phenomenon and P as the set of products to be supplied depending on the type of demand D_j.

Parameters: Parameters belonging to the following are proposed: (1) A number of pre-positioned warehouses to be installed (A); (2) Point of re-order or quantity of

[1]The WHO mentions that pregnancy and puerperium are particular stages of life in which there are particular nutritional needs, these stages will not be considered in the model (WHO, 2017).

products p which will serve as the basis for the supply order to meet the different types of demand of the municipality D_j, $\left(R^p_{D_j}\right)$; and *3*. The demand of each sector of the population of products p to be sent to municipality j, $\left(D^p_j\right)$.

The allocation of the warehouses entails three aspects of great importance:

- The distances to be covered for the supply of products, from the unimpacted municipalities with the pre-positioned warehouse to the impacted municipalities, are based on a cost of \$1 per km traveled for model validation purposes;
- Compliance with the criteria established by the National System for the Integral Development of the Family, (SNDIF, for its acronym in Spanish) (SNDIF, 2011), and the National Civil Protection System, (SINAPROC, for its acronym in Spanish (SINAPROC, 2012), for a municipality to host a pre-positioned warehouse; for this, weight percentages are established for each criterion, in order to verify greater compliance with them, and;
- Logistical costs of storage, including costs related to ordering, maintaining, purchasing, and non-suppling of products.

For those reasons, the planners involved in the disaster must choose the position and quantity of supplies that minimize the costs. On the one hand, the cost of distances covered for the delivery of support in the shortest time. On the other hand, the total cost of the activities of the pre-positioned warehouses, without losing sight of the aspects of infrastructure, assembly, services, and access roads for the installation of the warehouse.

The objective function Eq. (8.1) is presented in order to describe in greater detail the parameters immersed in it.

$$\begin{aligned}\text{Min}\left\{f(Y) = \sum_{j=1}^{J}\sum_{i=1}^{I} w_i d_{ij}(Y_{ij})\right\} + \sum_{i=1}^{I}\sum_{j=1}^{J}\sum_{p=1}^{P}\left[\frac{Co^p_i\left(D^p_j\right)}{Q^p_{ij}}\right. \\ \left. + Cs_i\left(\frac{Q^p_{ij}}{2} + z_{cls} * S'^p_j\right) + C^p_i\left(D^p_j\right) + \frac{D^p_j}{Q^p_{ij}}\left(Cf^p_i + S^p_{\prime D_j} * E^p_{z_i}\right)\right]\end{aligned} \tag{8.1}$$

Where the location parameters are represented by w_i, where $i \in \{1,2,\ldots,\ I\}$ represents the *assigned weight* (according to the compliance level of the criteria established by the abovementioned secretariats) to the municipalities i that were not impacted by the hydrometeorological phenomenon, and by d_{ij}, where $i \in \{1,2,\ldots,\ I\}$ y $j \in \{1,2,\ldots,\ J\}$ represents the *distance* between impacted and unimpacted municipalities that could serve as locations for the pre-positioned warehouses.

Furthermore, the inventory levels are limited by parameters such as: FR^p_i, an indicator of the *level of service* of product p given by the pre-positioned warehouses located in municipalities *i;* $E^p_{z_i}$, the standardized function of loss of products p that are not supplied by the inventory of the pre-positioned warehouse located in municipality i and to which its supply corresponds; Z_{csl}, dimensionless number according to the *level of service to be provided*, i.e., the value of Z for the normal distribution with a defined Cycle Service Level (CSL); $S'^{p}_{D_j}$, is the adjusted standard deviation of products *p* belonging to each of the demands of the impacted municipality *j;* and L_i, is the delivery time of the products to the municipalities where there is a pre-positioned warehouse *i.*

The costs associated with the warehouses are calculated through formulas 8.2, 8.3, 8.4, and 8.5:

- *Order cost* from the pre-positioned warehouse located in municipality *i*, (Co_i^p), includes the administrative cost of order processing, pre-positioned warehouse transportation cost, and handling costs at the receipt of the product at the warehouse:

$$\frac{Co_i^p\left(D_j^p\right)}{Q_{ij}^p} \tag{8.2}$$

- *Holding cost* for product *p* inventory of the pre-positioned warehouse located in municipality $i, (Cs_i^p)$, comprises the costs of space, service, and risk:

$$Cs_i^p\left(\frac{Q_{ij}^p}{2} + z_{cls} * S'^{p}_{D_j}\right) \tag{8.3}$$

- The *purchase cost* of products *p* contained in the pre-positioned warehouse located in municipality $i, (C_i^p)$, in order to satisfy the affected demand:

$$C_i^p\left(D_j^p\right) \tag{8.4}$$

- *Shortage cost* for products *p* to be supplied by the pre-positioned warehouse located in municipality $i, (Cf_i^p)$, these are the costs associated with the non-availability of the demanded product, and it includes the request and the delivery from the permanent collection centers for its supply:

$$\frac{D_j^p}{Q_{ij}^p}\left(Cf_i^p + S_{D_j}^{\prime p} * E_{z_i}^p\right) \tag{8.5}$$

Variables: The proposed model involves the use of three variables, (a) Y_{ij}, binary decision variable, it represents that the impacted municipality j is assigned to municipality i, which was not impacted and served as the location for the pre-positioned warehouse, we indicate $Y_{ij} = 0$ if municipality j is not assigned to municipality i, and $Y_{ij} = 1$ otherwise, that is, in case of municipality j is assigned to municipality i; (b) Y_i, binary decision variable, represents the establishment of pre-positioned warehouses in any of the municipalities i, we indicate $Y_i = 1$ if the pre-positioned warehouse is established in i, and $Y_i = 0$ otherwise, i.e., if the pre-positioned warehouse is not established in the municipality i. (c) Q_{ij}^p is an integer type variable that represents the quantity of products p to be ordered from the pre-positioned warehouse located in municipality i, to supply the impacted municipalities j.

Table 8.2 presents the notation of the model, that is, the sets, indices, parameters, and decision variables that conform it to make practical the involvement of the reader.

Below is the mathematical model composed of 12 equations including the target function; these equations are explained in the paragraphs after the model.

Table 8.2 Notation of the mathematical model *Source* Authors

Notación de conjuntos	
I	Set of municipalities that were not impacted by a hydrometeorological natural phenomenon and that can serve as locations for the pre-positioned warehouse
J	Set of municipalities impacted by a hydrometeorological natural phenomenon
P	Set of products to be supplied depending on the type of demand D_j
Notation of subscripts and superscripts	
I	A number of municipalities in the set I
J	A number of municipalities in set J
P	Number of products in set P
Parameters	
A	A number of pre-positioned warehouses to be installed
w_i	The positional weight assigned to municipalities i, derived from the degree of compliance with the criteria established by the SNDIF and the SINAPROC for it to be the location of a pre-positioned warehouse
d_{ij}	Distance in kilometers between municipalities i that were not impacted and the impacted municipalities j
D_j^p	The demand of each sector of the population: products p to be sent to municipality j
FR_i^p	The service level of the pre-positioned warehouse located in municipality i gave to the affected people
$E_{z_i}^p$	Standardized loss function: products p that was not supplied by the inventory of the pre-positioned warehouse located in the municipality i

(continued)

Table 8.2 (continued)

Notación de conjuntos	
$Zcsl$	The dimensionless model according to the desired level of service to be provided, that is, the value of Z for the normal distribution with a defined cycle service level (CSL)
$S'^{p}_{D_j}$	The adjusted standard deviation of products p that pertains to each of the demands of impacted municipality j
$R^{p}_{D_j}$	Point of re-order: quantity of products p from which a supply order will be issued to cover the different types of demand of municipality Dj
L_i	Delivery time of products to the municipality that is serving as the location for pre-positioned warehouse i
Co^{p}_{i}	Order cost of product p in the pre-positioned warehouse located in the municipality i
Cs^{p}_{i}	Maintenance cost for the inventory of products p of the pre-positioned warehouse located in the municipality i
C^{p}_{i}	The purchase cost of the products maintained in the pre-positioned warehouse located in the municipality i
Cf^{p}_{i}	Cost for the lack of products p in the pre-positioned warehouse located in the municipality i
Decision variables	
Y_{ij}	Binary decision variable, it represents that the impacted municipality j is assigned to municipality i, which was not impacted and served as the location for the pre-positioned warehouse, we indicate $Y_{ij} = 0$ in case that municipality j is not assigned to municipality i, and $Y_{ij} = 1$ otherwise, that is, in case of municipality j is assigned to municipality i
Y_i	Binary decision variable, it represents whether the pre-positioned warehouse is established in the municipality i, we indicate $Y_i = 1$ if the pre-positioned warehouse is established in i, $Y_i = 0$ otherwise, that is, if the pre-positioned warehouse is not established in the municipality
Q^{p}_{ij}	Integer type variable, it represents the quantity of products p to be ordered from the pre-positioned warehouse located in the municipality i to supply the impacted municipalities j

8.4.1.3 Mathematical Model

$$\text{Min}\left\{f(Y)=\sum_{j=1}^{J}\sum_{i=1}^{I}w_i d_{ij}(Y_{ij})\right\}$$
$$+\sum_{i=1}^{I}\sum_{j=1}^{J}\sum_{p=1}^{P}\left[\frac{Co^{p}_{i}\left(D^{p}_{j}\right)}{Q^{p}_{ij}}+Cs_i\left(\frac{Q^{p}_{ij}}{2}+z_{cls}*S'^{p}_{j}\right)+C^{p}_{i}\left(D^{p}_{j}\right)+\frac{D^{p}_{j}}{Q^{p}_{ij}}\left(Cf^{p}_{i}+S'^{p}_{D_j}*E^{p}_{z_i}\right)\right] \tag{8.6}$$

Subject to:

$$\sum_{i=1}^{I}Y_{ij}=1 \quad i=1,2,\ldots,I \tag{8.7}$$

$$\sum_{i=1}^{I} Y_i = A \quad i = 1, 2, \ldots, I \tag{8.8}$$

$$Y_{ij} \le Y_i \quad i = 1, 2, \ldots, I \quad j = 1, 2, \ldots, J \tag{8.9}$$

$$S_{D_j}^{p} = S_{D_j}^{\prime p} \sqrt{L_i} \quad i = 1, 2, \ldots, I \quad j = 1, 2, \ldots, J \tag{8.10}$$

$$R_{D_j}^{p} = D_j^p L_i + z_{CSL} S_{D_j}^{\prime p} i = 1, 2, \ldots, I \quad j = 1, 2, \ldots, J \quad p = 1, 2, \ldots, P \tag{8.11}$$

$$Q_{ij}^{p} = \sqrt{2Co_i \frac{\sum_{j=1}^{J} Y_{ij}^p D_j^p}{Cs_i^p}} \quad i = 1, 2, \ldots, I \quad j = 1, 2, \ldots, J \quad p = 1, 2, \ldots, P \tag{8.12}$$

$$E_{z_i}^{p} = z\left[F_s^p(z) - 1\right] + f_s^p(z) \quad i = 1, 2, \ldots, I \quad p = 1, 2, \ldots P \tag{8.13}$$

$$FR_i^p = 1 - \frac{S_j^{\prime p} E_{z_i}^p}{Q_{ij}^p} \quad i = 1, 2, \ldots, I \quad p = 1, 2, \ldots, P \tag{8.14}$$

$$Y_i \in \{0, 1\} \tag{8.15}$$

$$Y_{ij} \in \{0, 1\} \tag{8.16}$$

$$Q_{ij}^p \in Z^+ \tag{8.17}$$

The following paragraphs describe each of the components of the model utilized.

Equation (8.1) represents the target problem function and is composed of two terms, through which it seeks to minimize the costs derived from the implementation of pre-positioned warehouses. The first term calculates the product of the *positional weight* factors of each municipality, multiplying the *distance* between the impacted and unimpacted Y_{ij} municipalities. It is worth remembering the previously established cost of \$1.00 per traveled km. The second term represents the sum of the *logistic costs* of the number of municipalities i and headquarters of pre-positioned warehouses. These places supply the municipalities j with the necessary products p to cover the needs of the people affected by the natural phenomenon. Therefore, the logistic costs originated by the implementation of a pre-positioned warehouse are related to the demand that each sector of the population has to meet D_j^p.

Equation (8.7) ensures that each impacted municipality j is assigned to single municipality i, which is the location of a pre-positioned warehouse.

Equation (8.8) ensures that the number of pre-positioned warehouses Y_i is equal to A, i.e., those determined at the planning stage.

Equation (8.9) ensures that each municipality affected by natural phenomenon j is assigned to single municipality i, which is the location of a pre-positioned warehouse.

Equation (8.10) allows the calculation of the adjusted standard deviation of each of the different demands belonging to the impacted municipalities $\left(S_{D_j}^{\prime p}\right)$, i t is worth remembering that the document mentions in previous paragraphs a normal behavior in the distribution of demand data.

Equation (8.11) reflects the calculation of the point of re-order of products (p) for each of the different types of demand in the impacted municipalities (D_j).

Equation (8.12) reflects the calculation of the order quantity per item (p) of the pre-positioned warehouses located in the municipalities (i) for the different stages of the human life cycle in the impacted municipalities (j).

Within Eq. (8.13), calculations of the loss function are made, i.e., the indicator of not enough products (p) to cover the total number of people affected by the hydrometeorological phenomenon.

Equation (8.14) calculates the Fill Rate (FR) indicator, which measures the level of service in the delivery of products (p) to affected people by the pre-positioned warehouses located in municipalities i, regardless of the total number of victims to be satisfied in municipality j.

Equations (8.15), (8.16), and (8.17) indicate the variable that represents whether the pre-positioned warehouse is established in the municipality i or not, where (Y_i) is of binary type. The variable (Y_{ij}) is a binary type and represents the assignment of affected municipalities to the unimpacted municipalities or pre-positioned warehouse. Moreover, the variable that represents the quantity of products p to be requested per order by the pre-positioned warehouse belonging to municipalities i in order to cover the demand of the impacted municipalities j, that is to say, $\left(Q_{ij}^{p}\right)$ is of the integer type.

8.4.2 *Location Data*

8.4.2.1 Positional Weight

The positional weight of the municipalities that may become the location of a pre-positioned warehouse is determined from different factors, which are presented in Table 8.3.

8.4.2.2 Distance Between Municipalities

The distances between municipalities are extracted using Google Maps® application. Figure 8.3 illustrates the position of the municipalities that integrate the *Olmeca* Region and *Los Tuxtlas* Region, which will allow the evaluation of the model.

Table 8.3 Weight factors *Source* National Institute for Federalism and Municipal Development, 2010; SCT, 2016; SEDESOL, 2015

Weight factor	Relative weight (%)	Description
Proximity to suppliers	20	Number of social development stores within the municipality
Road access	20	State or federal terrestrial communication road
Services	20	Constructions with drainage service, piped water, energy, telephone, internet
Classification according to the size of their location	10	Urban, medium urban, semi-urban, mixed, rural
Degree of marginalization	10	Too high, high, medium, low, too low
Terrain	20	A number of classrooms built per school, this data is taken as a reference, since due to it being a new pre-positioned warehouse, there is no construction for it

Fig. 8.3 Municipalities are belonging to the Olmeca Region and Los Tuxtla Region

8.4.3 Inventory Level Data

8.4.3.1 Demand

The calculation of the demand by the municipality is based on the emergency declarations issued by the National Disaster Fund (FONDEN, for its acronym in Spanish), located on the page of the National Civil Protection System (SINAPROC) in the section on Inputs authorized by emergency declaration (DGGR, 2017).

8.4.3.2 Products

A database is developed with the quantities and items needed to cover food, water, and equipment needs during the seven-day period, with the addition of vaccines and application elements, each of which is converted into kits for each affected person.

8.4.3.3 Costs

The costs considered for this document are described in Table 8.4. It should be noted that these data were taken from the pages of the following: (a) Office of the Federal Prosecutor for the Consumer (PROFECO, for its acronym in Spanish); (b) Tax Administration Service (SAT, for its acronym in Spanish); and (c) Ministry of Labor and Social Security (STPS, for its acronym in Spanish) (PROFECO, 2017; SAT, 2017; STPS, 2017).

Table 8.4 Information about the cost involved *Source* Authors

Cost	Aspects that comprise it
Order cost	Administrative and order processing cost + cost of transportation to the warehouse + cost of handling and reception in the warehouse. It includes the salaries of personnel, services, and fuels
Purchase cost	Food cost + equipment cost + cost per liter of water. It depends on the type of demand for which it is intended It includes formula and puréed food for babies, cereals, canned food, cookies, coffee, water, toiletries, clothing, shoes, and glasses
Shortage cost	Missing request cost + cost of transportation to the warehouse + cost of handling and reception in the warehouse. It includes salaries, services, and fuels
Maintenance cost	Space cost + service cost + risk cost. It includes rent, services, and insurance

8.4.4 Software Description

Lingo is an optimization modeling software for linear, nonlinear, quadratic, integer and another programming, and was created by Lindo Systems, Inc. (Lindo Systems, 2017).

8.4.5 Evaluation Scenarios

The model is evaluated through two scenarios, whose impact zones are the *Olmec* Region and *Los Tuxtlas* Region, which house 25 and four municipalities, respectively. They have a percentage of affectation by some hydrometeorological phenomenon of a contained unit of 40% and 100%. Mathematical programming evaluates the minimization of localization and inventory costs for two instances of two and three pre-positioned warehouses, respectively.

8.5 Results

Tables 8.5 and 8.6 show the results obtained through the programming of the model developed. In them, the optimal location for the installation of the pre-positioned warehouses is observed. In the selection of two locations, the optimal place is in the municipalities of (1) Jesus Carranza, and (2) Las Choapas; in the scenario of three locations, in order to further limit the distances, the results obtained are the municipalities of (1) Jesús Carranza; (2) Las Choapas, and (3) Uxpanapa.

Furthermore, it is also shown the number of products to be ordered, the point of re-order and the security inventory for the warehouse, which allows the survival of the victims, grouped in kits of (a) food; (b) equipment; (c) water; and (d) medicines for a one-week duration in consumables.

Additionally, a column is included for recording the non-supply, that is, the quantity of unsatisfied demand. It is noted that there are five types of demand for food and equipment items and a general demand for water and medicines. These are divided according to the human life cycle: conception, early childhood, childhood, adolescence, adulthood, and old age, and represented by Demand_1, Demand_2, Demand_3, Demand_4, and Demand_5, respectively.

As an example of this, we have the municipality of Jesús Carranza, which has an economic quantity of order $\left(Q_{ij}^{p}\right)$ of 4,224 food kits and 4,259 equipment kits, a point of re-order $\left(R_{j}^{p}\right)$ of 2,482 kits and maintains a security inventory of 1,391 kits, with a probability of not supplying two victims, with all of the above aimed at those affected in the early childhood phase.

Table 8.5 Results obtained by integrating two regions, two locations *Source* Authors

Type of demand	Municipality to host a pre-positioned warehouse	Economic order quantity $\left(Q_{ij}^{p}\right)$	Economic order quantity $\left(Q_{ij}^{p}\right)$	Point of re-order $\left(R_{j}^{p}\right)$	Security inventory	Not supplying
Demand_1 D_j^p (early childhood)	Jesús Carranza	4,224	4,259	2,482	1,391	2
	Las Choapas	5,603	4,713	1,198	92	1
Demand_2 D_j^p (childhood)	Jesús Carranza	8,264	8,264	5,314	2,907	5
	Las Choapas	9,174	9,174	3,761	1,233	2
Demand_3 D_j^p (adolescence)	Jesús Carranza	13,893	13,893	9,496	5,254	8
	Las Choapas	15,191	15,191	6,858	2,415	4
Demand_4 D_j^p (adulthood)	Jesús Carranza	12,355	12,355	8,225	4,478	7
	Las Choapas	13,641	13,641	5,996	2,054	3
Demand_5 D_j^p (old age)	Jesús Carranza	3,252	3,252	1,861	1,036	2
	Las Choapas	3,531	3,531	1,109	299	1
Demand_6 D_j^p (General)	Jesús Carranza	39,048	51,514	27,366	15,055	22
	Las Choapas	42,276	58,918	19,359	6,530	10

Weight*Distance.

$$\left\{f(Y) = \sum_{j=1}^{J}\sum_{i=1}^{I} w_i d_{ij}(Y_{ij})\right\} - 1,122,317,000.00$$

Cost of inventory

$$\sum_{i=1}^{I}\sum_{j=1}^{J}\sum_{p=1}^{P}\left[\frac{Co_i^p\left(D_j^p\right)}{Q_{ij}^p} + Cs_i\left(\frac{Q_{ij}^p}{2} + z_{cls} * S_j^{\prime p}\right) + C_i^p\left(D_j^p\right) + \frac{D_j^p}{Q_{ij}^p}\left(Cf_i^p + S_{D_j}^{\prime p} * E_{z_i}^p\right)\right].$$

$$= 648,312,980.00$$

A comparison of the term Weight * Distance assures that the optimal installation solution is two pre-positioned warehouses within the municipalities of Jesús Carranza and Las Choapas, derived from two attributes: (a) proximity to the impact zones, and (b) infrastructure. Regarding inventory costs, the optimal solution obtained is based on the installation of two pre-positioned warehouses, which will allow the most significant number of victims in the extracted regions to be supplied for the evaluation of the model. Table 8.7 shows the distances covered and the approximate delivery time to each of the impacted municipalities from the

Table 8.6 Results obtained integrating two regions, three locations *Source* Authors

Type of demand	Municipality to host a pre-positioned warehouse	The economic quantity of order $\left(Q_{ij}^{p}\right)$	The economic quantity of order $\left(Q_{ij}^{p}\right)$	Point of re-order $\left(R_{j}^{p}\right)$	Security inventory	No supplying
Demand_1 D_j^p (early childhood)	Jesús Carranza	4,924	4,259	2,482	1,391	2
	Las Choapas	4,424	3,760	1,500	624	1
	Uxpanapa	1,265	1,056	554	324	1
Demand_2 D_j^p (childhood)	Jesús Carranza	8,264	8,264	5,314	2,907	5
	Las Choapas	7,429	7,429	3,335	1,327	2
	Uxpanapa	2,023	2,023	1,311	791	2
Demand_3 D_j^p (adolescence)	Jesús Carranza	13,893	13,893	9,496	5,354	8
	Las Choapas	12,288	12,287	6,148	2,650	4
	Uxpanapa	3,448	3,448	2,316	1,372	2
Demand_4 D_j^p (adulthood)	Jesús Carranza	12,355	12,355	8,225	4,478	7
	Las Choapas	11,432	11,431	5,486	2,244	4
	Uxpanapa	2,607	2,607	1,513	813	2
Demand_5 D_j^p (old age)	Jesús Carranza	3,251	3,252	1,861	1,036	2
	Las Choapas	2,892	2,892	963	297	1
	Uxpanapa	699	699	358	214	1
Demand_6 D_j^p (General)	Jesús Carranza	39,048	51,515	27,366	15,054	22
	Las Choapas	34,892	47,181	17,410	7,120	11
	Uxpanapa	8,920	12,765	6,051	3,512	6

Weight*Distance.

$$\left\{f(Y)=\sum_{j=1}^{J}\sum_{i=1}^{I}w_i d_{ij}(Y_{ij})\right\}=1,156,429.00$$

Cost of inventory

$$\sum_{i=1}^{I}\sum_{j=1}^{J}\sum_{p=1}^{P}\left[\frac{Co_i^p\left(D_j^p\right)}{Q_{ij}^p}+Cs_i\left(\frac{Q_{ij}^p}{2}+z_{cls}*S_j^{\prime p}\right)+C_i^p\left(D_j^p\right)+\frac{D_j^p}{Q_{ij}^p}\left(Cf_i^p+S_{D_j}^{\prime p}*E_{z_i}^p\right)\right].$$

$$=681,148,463.60$$

pre-positioned warehouse locations. Delivery time of no more than 4 h can be observed.

The most notable characteristic of emergency problems is the urgency of time, so the reduction of this becomes the primary objective of the optimization. In other words, the delivery of necessary emergency and disaster relief and assistance supplies from the points closest to the disaster area should take place as soon as possible (PAHO, 2017). The data used can be seen in Appendix 8.A.

Table 8.7 Results of distance traveled/delivery time *Source* Authors

Affected municipality	Distance (Km)/ Delivery time Jesús Carranza	Distance (Km)/ Delivery time Las Choapas	Distance (Km)/ Delivery time Uxpanapa
Agua Dulce	172.34 km/2 H. 42 min	–	–
Coatzacoalcos	131.96 km/2 H. 9 min	–	–
Cosoleacaque	–	–	208.04 km/3 H. 54 min
Hidalgotitlan	–	104.24 km/2 H. 14 min	–
Ixhuatlán del Sureste	130.98 km/2 H. 2 min	–	–
Catemaco	–	197.21 km/3 H. 5 min	–
Jáltipan	–	–	189.89 km/3 H. 51 min
Minatitlán	111.31 km/1 H. 50 min	–	–
Hueyapan de Ocampo	–	158.64 km/2 H. 24 min	–
Moloacan	145.03 km/2 H. 36 min	–	–
Lázaro Cárdenas	137.55/2 H. 26 min	–	–
Pajapan	–	122.95 km/2 H. 18 min	–
San Andrés Tuxtla	–	207.97 km/3 H. 22 min	–
Santiago Tuxtla	–	225 km/3 H. 07 min	–
Distance/Maximum time	172.34 km/2 H. 42 min	225.41 km/3 H. 22 min	208.04 km/3 H. 54 min
Distance/Minimum time	111.31 km/1 H. 50 min	104.24 km/2 H. 14 min	189.89 km/3 H. 51 min

8.6 Conclusions and Future Work

The aim of this work is to establish the optimal location of several pre-positioned warehouses and their inventory levels during the preparation stage for the impact of a hydrometeorological phenomenon. Its purpose is to provide support to the responders in the event of a disaster, allowing them to attend in a timely and efficient manner to the people impacted, and thus maintain a certain level of quality of life of the affected people.

The model is evaluated in two regions of the State of Veracruz, Mexico. These regions have been severely damaged by natural hydrometeorological events. For this purpose, two scenarios are set up for the installation of two and three pre-positioned warehouses obtaining an optimal location within the municipalities of Jesús Carranza and Las Choapas for one region, and the municipalities of Jesús Carranza, Las Choapas and Uxpanapa, for the other one. In addition, the order quantities (Q), the point of re-order (R), the security supply (SS), and the probable number of people that cannot be supplied. The computational experiments show a variety of distance-feasibility relationships and expose how they affect location decisions and inventory levels. It is possible to observe the functionality of the model not only for the supply of a more significant number of victims but also for a minimum disbursement of resources. The optimal solution obtained by cost is that of the installation of two pre-positioned warehouses.

This document gives rise to the development of heuristic models with shorter run time and which can be used in the emergency stage before the disaster. Similarly, being the State of Veracruz, one with the most significant impact by natural hydrometeorological events, the model proposes a mapping by the municipality to allow the establishment of multi-warehouses, inventory levels for each of them, and delivery routing.

References

Bai, S., Liu, J. (2011). Research on resource response decision in emergency logistics. In *Paper Presented at International Conference on Web Information Systems and Mining IEEE, Sanya China*: pp. 378–381 June 2011. https://doi.org/10.1109/WISM.2010.150.

Barojas-Payán, E., Sánchez-Partida, D., Martínez-Flores, J. L., Gibaja-Romero, D. E. (2019). Mathematical model for locating a pre-positioned warehouse and for calculating inventory levels. *Journal of Disaster Research, 14*(4), 649–666. https://doi.org/10.20965/jdr.issn.1883-8030.

Campbell, A., & Jones, P. (2011). Prepositioning supplies in preparation for disasters. *European Journal of Operational Research, 209*(2), 156–165. https://doi.org/10.1016/j.ejor.2010.08.029.

CENAPRED National Center for Disaster Prevention. (2015). Disasters in Mexico: Social and economic impact. Available at: http://www.cenapred.unam.mx/es/Publicaciones/archivos/318-INFOGRAFADESASTRESENMXICO-IMPACTOSOCIALYECONMICO.PDF. Accessed April 2, 2019. [In Spanish].

City Council of Mexicali, B.C. (2011). Hydrometeorological hazards. Risk Atlas of the municipality of Mexicali. B.C. [In Spanish].

Civil Protection, Guanajuato. (2015). Hydrometeorological phenomenon, conceptual framework. State Government. Secretary of Public Security. [In Spanish].

Chamber of Deputies of the Congress of the Union. (2012). Last reform DOF 19-01-2018. *Official Journal of the Federation. General Law of Civil Protection*, Chapter I, article 2, section XLVIII: 5. [In Spanish].

Chamber of Deputies of the Congress of the Union. (2012). Last reform DOF 19-01-2018. *Official Journal of the Federation. General Law of Civil Protection*, Chapter I, article 2, section XVI: 3. [In Spanish].

Deqiang, F., & Xiaoming, T. (2011). *Research on location model of emergency logistics based on AHP/DEA*. Paper presented at International Conference on Information Management, Innovation Management, and Industrial Engineering, Shenzhen China, pp. 472–474, December 2011. https://doi.org/10.1109/ICIII.2011.394.

DGGR General Directorate for Risk Management, FONDEN Natural Disasters Fund. (2017). *Supplies authorized by emergency declaration*. Available at: http://www.proteccioncivil.gob.mx/es/ProteccionCivil/Insumos_autorizados_por_declaratoria_de_emergencia. Accessed May 30, 2017 [In Spanish].

FONDEN Natural Disasters Fund. (2012). *The Natural Disasters Fund in Mexico, a review. Mexico City*. Washington, DC. USA: Global Facility for Disaster Reduction and Recovery.

Government of the State of Veracruz, Secretariat of Civil Protection. (2011). *Municipal atlas of risk basic level Acajete*. Editor of the Government of the State of Veracruz of Ignacio de la Llave. [In Spanish].

Government of the State of Veracruz. Secretariat of Civil Protection. (2018). *Municipal Atlas of Risks*. Available at: http://www.veracruz.gob.mx/proteccioncivil/servicio/atlas-estatal-municipal/Accessed April 2, 2019. [In Spanish].

INEGI National Institute of Statistics and Geography. (2016). *Mexico in numbers*. Available at: https://www.inegi.org.mx/app/areasgeograficas/?ag=14# Accessed January 20th, 2017. [In Spanish].

Lindo Systems. Inc. (2017). *Lingo Optimization Modeling Software for linear, nonlinear, and integer programming*. Available at www.lindo.com/inde8.php/products/lingo-and-optimization-modeling. Accessed January 20, 2017.

Ministry of the Interior National Civil Protection System. National Center for Disaster Prevention. (2001). *National Atlas of Risks of the Mexican Republic. Diagnosis of hazards and identification of disaster risks in Mexico*. National Center for Disaster Prevention.

Mora, I., Morales, W., Rodríguez, S. (2015). *Atlas geological and hydrometeorological hazards of the State of Veracruz*. Secretary of Civil Protection, Government of the State of Veracruz of Ignacio de la Llave.

Nahmias, S. (2014). *Analysis of production and operations*, 6ª Edn. McGraw W-Hill/ InterAmerican Editors, S.A. de C.V., Mexico. pp. 205–2014.

National Institute for Federalism and Municipal Development. (2010). *National municipal information system: Municipalities in numbers*. Available at: http://www.snim.rami.gob.mx/. Accessed May 15, 2017.

OPS Pan American Health Organization, World Health Organization. (2001). Logistics and management of humanitarian supplies in the health sector. Washington, D.C. pp. 9–11.

Overstreet, E., Hall, D., Hanna, J., & Rainer, K. (2011). Research in humanitarian logistics. *Journal of Humanitarian Logistics and Supply Chain Management, 1*(2), 114–131. https://doi.org/10.1108/20426741111158421.

PAHO Pan American Health Organization, Department of Health Emergencies. (2017). *Supply management*. Available at: http://www.paho.org/disasters/inde8.php?option=com_content&view=article&id=748:supplymanagement&Itemid=661&lang=en. Accessed May 30, 2017.

PROFECO, Office of the Federal Prosecutor for the Consumer. (2017). Who's who in prices? Retrieved from https://www.profeco.gob.mx/precios/canasta/home.aspx?th=1. Accessed May 13, 2018). [In Spanish].

Rawls, C., & Turnquist, M. (2010). Prepositioning of emergency supplies for disaster response. *Transportation Research Part B, 44*(4), 521–534. https://doi.org/10.1016/j.trb.2009.08.003.

Sánchez, R., Cavazos, T. (2015). Chapter 1 Natural threats, society and disasters. *Living with nature: The problem of disasters associated with hydrometeorological and climatic phenomena in Mexico Editions ILCSA*. México: REDESClim, México, pp. 4–41.

SAT, Tax Administration Service (2017) Minimum wage. (2017). Available at: www.sat.gob.mx/informacion_fiscal/tabla_indicadores/Paginas/salarios_minimos.asp8. Accessed May 30, 2017 [In Spanish].

SCT Secretariat of Communications and Transport. (2016). *Draw your route*. Available at: http://app.sct.gob.mx/sibuac_internet/ControllerUI?action=cmdEscogeRuta. Accessed May 15, 2017.

Secretariat of Education. (2013). Geographic division: COPLADEVER veracruz planning committee for development. SEFIPLAN. *General directorate of educational evaluation and control*. Available at: http://www.sev.gob.mx/servicios/anuario/2011/regiones.php. Accessed February 20, 2017 [In Spanish].

SEDESOL Secretariat of Social Development. (2015). *Rural supply program*. Available at: http://catalogo.datos.gob.mx/dataset/programa-de-abasto-rural-tiendas-2015. Accessed May 15, 2017.

SINAPROC, National System of Civil Protection Mexico. (2012). *Collection center*. Available at: http://sismos.gob.mx/en/sismos/Centro_de_acopio. Accessed September 20, 2016. [In Spanish].

SNDIF, National System for the Integral Development of the Family. (2011). Unit of attention to the vulnerable population, General Directorate of Food and Community Development. *Operational manual, attention to the population at risk or emergency condition APCE*, 2011.

STPS, Ministry of Labor and Social Security. (2017). *Minimum wage*. Available at: www.gob.mx/cms/uploads/attachment/file/175865/Tabla_de_salarios_minimos_vigentes_a_partir_de_01_enero_2017.pdf. Accessed May 30, 2017. [In Spanish].

Timperio, G., Panchal, G. B., De Souza, R., Goh, R., Samvedi, A. (2016). Decision making framework for emergency response preparedness: A supply chain resilience approach. *Paper presented at Management of Innovation and Technology (ICMIT) 2016. International Conference on IEEE, Bangkok, Thailand*, pp. 78–82. October 2016. https://doi.org/10.1109/ICMIT.2016.7605011.

Tomasini, R., Van-Wassenhove, L. (2009). *Humanitarian logistics*. London: Palgrave Macmillan. https://doi.org/10.1057/9780230233485.

Vela, J. (2014). *Health logistics in catastrophes*. Logística sanitaria en catástrofes. SANT0108. In: IC Editorial.

WHO World Health Organization Electronic library of scientific documentation on nutritional measures (eLENA). (2017). *Life cycle*. Available at: http://www.who.int/elena/life_course/es/. Accessed January 30, 2017.

Chapter 9
A Hybrid Capacitated Multi-facility Location Model for Pre-positioning Warehouses and Inventories in Regions at Risk in Mexico

Erika Barojas-Payán, Diana Sánchez-Partida, Santiago-Omar Caballero-Morales, and José-Luis Martínez-Flores

Abstract During the period 2012–2016, 99.0% of the municipalities of the State of Veracruz in Mexico have been subject to emergency declarations by the National Fund for Disaster Relief derived from the impact of hydro-meteorological phenomena. The present work develops a hybrid capacitated multi-facility location model based on (1) the p-median algorithm, (2) the Nearest Neighbor (NN) algorithm, (3) the Greedy Randomized Adaptive Search Procedure (GRASP) algorithm, and (4) the continuous review (q, R) inventory model with uncertain demand. To support the organization of humanitarian relief efforts and maintain the quality of life of specific affected people in this region. The hybrid model identifies multiple feasible locations for the establishment of pre-positioned warehouses and their inventories. In contrast to other works, the proposed model extends on the characteristics of the required inventories to be stored. These are classified into food, clothing, personal hygiene, water, and personal health, and each one is classified depending on the life stages of the person: (a) early childhood; (b) childhood; (c) adolescence; (d) adult, and (e) third age. This classification is performed to provide products according to the specific needs of the affected people according to their age. As a result, 20 feasible locations throughout the State of Veracruz were identified, and for each pre-positioned warehouse, economic order quantities, reorder points, and safety stocks were defined.

Electronic supplementary material The online version of this chapter (https://doi.org/10.1007/978-3-030-67295-9_9) contains supplementary material, which is available to authorized users.

E. Barojas-Payán (✉) · D. Sánchez-Partida · S.-O. Caballero-Morales · J.-L. Martínez-Flores
Department of Logistics and Supply Chain Management, Universidad Popular Autónoma del Estado de Puebla (UPAEP University), 17 Sur 901 Barrio de Santiago, CP 72410 Puebla, Mexico
e-mail: erika.barojas@upaep.edu.mx

D. Sánchez-Partida et al. (eds.), *Disaster Risk Reduction in Mexico*,
https://doi.org/10.1007/978-3-030-67295-9_9

Keywords Multi-facility location problem · Pre-positioned warehouses · GRASP · Continuous review · Hydro-meteorological phenomena · Humanitarian logistics

9.1 Introduction

During the period between the years 1996 and 2015, 1.3 million people died worldwide due to natural disasters. Mainly, disasters related to climate and water were prevalent according to the Emergency Events Database (EM-DAT). Although in the last decade the occurrence of natural phenomena has decreased, the lethality of storms has increased by a factor of three (from 64 deaths per event between 1996 and 2005 to 180 deaths per event between 2006 and 2015) (CRED & UNISDR, 2016).

Mexico is among the world's most vulnerable countries to the effects of climate change due to its geographical location and social structures. Its territory is made up of two mountain systems, the *Sierra Madre Occidental* and the *Sierra Madre Oriental.* These are part of the Neo-volcanic axis, which is a chain of volcanoes that belong to the so-called fire belt of the Pacific. Likewise, due to its geographic and topographic characteristics, a vast region of the Mexican territory has abundant water resources. Its vast expanses of coastal areas surrounded by the Pacific Ocean, the Gulf of Mexico, the Sea of Cortes and the Caribbean Sea, make Mexico a Bio-oceanic country (INECC, 2016; SEMAR, 2014).

It makes Mexico susceptible to be affected by diverse destructive agents such as disturbing phenomena of geological and hydro-meteorological origin. Its assets, such as production plants, public services, and the environment, can be severely affected by these phenomena, increasing the magnitude of the damage as the density of the population increases and concentrates in regions at risk. In recent decades, natural phenomena have left considerable damage to both human lives and economic wealth. It leads to recognizing the essential need to establish strategies and programs focused on preventing and reducing the adverse effects of these disturbing phenomena (INECC, 2016; SEMAR, 2014).

Notably, the State of Veracruz is a narrow strip of slightly curved land, which extends from northwest to southeast to the center of the Gulf of Mexico. Its territory has a coastline of 745 km and represents 3.7% of the Mexican territory. It consists of 212 municipalities and 8,112,505 inhabitants, according to the population census of 2015.

Although much of the territory of Veracruz is made up of slightly curved lands, part of it is crossed by one of the neo-volcanic mountain systems. In this region, there are towering mountains and volcanoes where two of them are active: the San Martín Tuxtla and the Pico de Orizaba which is the highest of Mexico and one of the last glaciers in the national territory (PC, 2017; INEGI, 2013).

Derived from its geographical location the State of Veracruz is frequently exposed to multiple natural disturbing phenomena (PC, 2011). During the last years, it has been affected by disturbing agents such as (a) severe rain and fluvial flood (2014); (b) severe rain, rain flood and river flood (2015); (c) severe rain and fluvial flood (2016), and (d) severe rain, fluvial flood and the hurricane Katia (2017), for which Veracruz received emergency funds of +300 million of pesos by the Natural Disasters Fund (FONDEN, by its acronym in Spanish) (SEGOB, 2017). These events have led to significant loss of human lives and infrastructure.

The Mexican government is performing actions through the MX Plan to make relief efforts more efficient by reducing response times, avoiding duplication of efforts, and focusing more accurately on the most affected regions. Its objective is to protect the life and heritage of all people within the Mexican territory before, during and after a contingency. This plan is aligned to: (a) Plan DN—III by the Ministry of National Defense (SEDENA, by its acronym in Spanish); (b) Marina Plan by the Secretariat of the Navy (SEMAR, by its acronym in Spanish); (c) National Response Plan for the Civilian Population; (d) response plans by dependencies and entities of the Public Administration, and (e) volunteers and associations of the civil society (PF, 2016; Presidency of the Republic, 2014; SEMAR, 2014).

As support to these collaborative efforts, the present work is focused on providing a humanitarian relief plan which consists of pre-positioning specific supplies at strategic places. This plan can reduce the effect of disturbing phenomena by supplying and storing survival kits in advance, before the disaster event occurs. This has the advantage of reducing response times and avoiding duplicate supply efforts.

To accomplish this plan, a capacitated hybrid multi-facility location model is developed. This model integrates: (a) p-median algorithm; (b) the Nearest Neighbor (NN) algorithm; (c) the Greedy Randomized Adaptive Search Procedure (GRASP) algorithm, and (d) an inventory control method based on the continuous review (q, *R*) model with uncertain demand. The integration of (a)–(c) determines the set of feasible municipalities for the establishment of pre-positioned warehouses and the municipalities affected by the disturbing phenomenon that must be supplied. Then, the integration of (d) determines the optimal inventory levels to supply food products, clothing and personal hygiene, water, and health for the affected people depending on their age stage: early childhood, childhood, adolescence, adulthood, or third age (old age).

The present work is structured as follows: in Sect. 9.2 a literary review on the logistic aspect of facility location for humanitarian relief and inventory management facilities is presented; then, the development and evaluation of the model are presented and discussed in Sects. 9.3–9.5; finally, in Sect. 9.6 the conclusions and future work are presented.

9.2 Literature Review

Many countries are prone to be affected by disturbing phenomena. This is why aid organizations, both local and international, face a unique mix of challenges that characterize the design and management of assistance plans: unpredictability of the demand and short response/delivery times. The estimated time window usually is 72 h for the dispatch of a wide range of supplies and human, technological and economic resources (Balcik & Beamon, 2008).

In this context, Humanitarian Logistics (HL) is focused on the achievement of this task. However, there are factors that affect the required decision process. On one hand, there is a lack of knowledge about the dynamism of the demand. On the other hand, the service level must balance between equity and efficiency, particularly for human well-being. In addition, there is uncertainty related to available resources (Van Wassenhove & Pedraza, 2012). A successful humanitarian operation mitigates the urgent needs of the population, with a sustainable reduction of their vulnerability in the shortest time and with the least amount of resources (Van Wassenhove, 2006). Likewise, HL must respect the basic three humanitarian principles: neutrality, impartiality, and humanity (Tomasini & Van Wassenhove, 2009). In addition, International Humanitarian Organizations (IHOs) must respect survival mechanisms, social structures, economies and local markets (Wassenhove & Pedraza, 2012).

The *risk of disaster* is determined by three fundamental factors: danger, degree of exposure, and vulnerability of the exposed goods. Danger lies within the disturbing phenomena themselves. However, the probability that danger will materialize in some damage to the community depends of the degree of exposure and its vulnerability (PC, 2017). For reducing the impact caused by the materialization of disturbing phenomena, HL performs a set of activities classified into four phases:

1. *Mitigation*: plans and mechanisms such as training to reduce the vulnerability of the population, where organizations as government and associations play an essential role.
2. *Preparation*: operations or strategies that must be planned before a disaster occurs, incorporating past experiences in order to improve response actions.
3. *Response*: all operations planned in the previous phases are carried out in order to reduce the number of deaths.
4. *Reconstruction*: rehabilitation process for the infrastructure affected by the disaster (Cozzolino, 2012).

In these activities, different entities contribute to their achievement. As presented by Santiago and Oliveira (2014) there are critical factors that help logistic processes in cases of environmental disasters. They identified that suppliers, donors, beneficiaries, and distribution centers, play important roles which have presence in the supply chain after the disaster.

Diaz and Gaytan (2014) identified the relationship between flood risk assessment and the design of HL processes for humanitarian relief in Tabasco (Mexico). They predicted the time and severity of floods to perform an efficient planning of the logistic network for delivery of supplies before the disaster event.

Extended research, focused on the location of supplies and the determination of inventory levels to cover the highest number of affected people, has been reported in the HL literature. Boonmee et al. (2017) reported on deterministic, stochastic, dynamic and robust facility location problems related to HL. For each problem the type of facility location, data modeling, disaster, decisions, goals, constraints and solution methods, were evaluated for applications in the real world.

Likewise, the research of Safeer et al. (2014) used a classification-based review methodology to identify cost functions and constraints for emergency logistics operations. Additionally, Dufour et al. (2017) presented the cost-benefit analysis of adding a regional distribution center in Kampala, Uganda to the already existing network of The United Nations Humanitarian Response Depot (UNHRD) to efficiently respond to humanitarian crises in East Africa. A methodology based on fieldwork, simulation, network optimization, and statistical analysis was launched. With 5000 scenarios generated, the addition of the center represented a cost reduction of 21%.

Meanwhile, Rodríguez et al. (2018) developed a multi-objective system for multi-organizational decision making in the disaster preparedness phase. A cartographic model to avoid the selection of facilities susceptible to floods, and a bi-objective optimization model, led to determine the location of emergency facilities, stock pre-positioning, resource allocation, and distribution. With an application in the City of Acapulco (Mexico) the system led to potential benefits in terms of cost and service level when compared to current contingency plans. In the same way, Morales and Akhavan (2012) presented a linear optimization model for the location of emergency warehouses. The model minimized costs and it was restricted to cover a certain percentage of affected people. By using Monte Carlo simulation, the performance of the model for HL was evaluated, obtaining an improvement of more than 27% in average travel hours.

The work of Maharjan and Hanaoka (2017) determined the number and locations of the warehouses to be placed for a support chain in Nepal to respond to sudden onset disasters. The proposed model introduced indexes for development, disaster safety and accessibility of transportation to reflect the socioeconomic, climatic and topographic characteristics of Nepal.

In the work of Gösling and Geldermann (2014) a framework of reference was presented to support the decision of practitioners in the field of HL to compare available Operations Research (OR) models. Three ways to compare OR models were introduced: (a) based on the decision of the support; (b) based on the decision of criticism and considered metrics, and (c) based on their adjacent methodologies and assumptions. As examples, two facility location models were developed as reference test instances for other works.

A last-mile distribution model for HL was proposed by Maghfiroh and Hanaoka (2017). It incorporated flexible vehicle routing processes to minimize unmet demand. As results it determined the routing sequence, the quantities of relief supplies, the assignment of the demand point to each vehicle, and the decision to serve or reject an additional demand within a work period. On the other hand, Santoso et al. (2017) reported a Rescue Team Assignment and Scheduling Problem (RUASP) model which considered the relationship between capacity requirements of incident locations and the capabilities of rescue teams in the handling time, and fuzzy logic for the travel times. The model was solved through the Greedy Randomized Adaptive Search Procedure (GRASP) algorithm approach to obtain results in reasonable time.

Among the different forms of preparedness for disaster relief management, it is considered that the pre-purchase of stocks in a pre-positioned warehouse is the best way to maximize the effectiveness of humanitarian aid supply chains. However, there have been very few studies that consider the business-centric application of the multi-criteria location problem for the pre-positioning of warehouses for humanitarian aid organizations. The study by Sae et al. (2013) empirically identified the critical factors to be considered for the selection of locations for a humanitarian aid warehouse. The results indicated that cooperation is the most important factor when selecting the location of the facility in humanitarian aid, followed by national stability, cost, logistics, and location.

The main objective of the research of Landa et al. (2016) was to relocate the distribution sites for the economic support provided by PROSPERA in La Perla (municipality of Veracruz). This was proposed by using the p-median algorithm. Three different scenarios were created in order to compare to the current distribution plan. Through a comparison of the number of beneficiaries by travelled kilometers, the overall distances were minimized by 25.0%. Meanwhile, the research developed by Bai (2016) studied a new fuzzy prepositioning emergency supply model considering a credibility value-at-risk (CVaR) metric.

Ahmadi et al. (2015) designed a variable neighborhood search algorithm to solve a last mile model which determined the locations of local distribution centers and routing for product distribution after an earthquake. The results showed that unmet demands can be reduced significantly at the expense of a higher number of local vehicles and deposits. Additionally, the research of Özdamar and Demir (2012) described a hierarchical cluster and a route procedure (HOGCR) to coordinate the routing of vehicles in large-scale evacuation and distribution activities after the disaster. HOGRC is a multilevel clustering algorithm that groups demand nodes into smaller clusters at each planning level, allowing the optimal solution of cluster routing problems. Routing problems were presented as trained network flow models, which CPLEX solved to optimality and independently in a parallel computing platform. Its performance was evaluated through the use of large-scale scenarios with satisfactory results.

Barzinpour et al. (2014) developed a multiobjective model for distribution considering two objectives: minimize the total costs and maximize the satisfaction rate in the sense of being fair while distributing the items. The model simultaneously determined the location of relief distribution centers and the allocation of affected areas to relief distribution centers. Furthermore, the authors developed an efficient solution approach based on a genetic algorithm to solve the proposed mathematical model. The computational results showed that the proposed genetic algorithm provided relatively good solutions in a reasonable time when compared to exact methods.

The research developed by Caballero et al. (2018) provided an efficient metaheuristic to determine the most appropriate location for support centers in Veracruz (Mexico). The metaheuristic was based on the K-Means Clustering (KMC) algorithm which was extended to integrate capacity restrictions on the support centers, a genetic algorithm to estimate the most suitable number of support centers, a variable number of assigned locations to centers to add flexibility to the allocation task, and a random-based decision model to improve the final assignments. As a result, it was determined that 260 support centers were required to provide relief for 3837 communities at risk.

The research of Döyen et al. (2011) reported a two-stage stochastic programming model for HL. Its objective was to minimize the total cost of facility location, inventory holding, transportation, and shortage. The decisions were made considering the pre- and post-disaster rescue centers, the number of relief items to be stocked at the pre-disaster rescue centers, the amount of relief item flows at each echelon, and the amount of relief item shortage. The deterministic equivalent of the model was formulated as a mixed-integer linear programming model and solved by a heuristic method based on Lagrangean relaxation. Results on randomly generated test instances showed that the proposed solution method exhibited good performance up to 25 scenarios.

Rawls et al. (2010) developed an emergency response planning tool to determine the location and quantities of various types of emergency supplies to be pre-positioned. It was performed considering uncertainty about if or where a natural disaster will occur. The authors presented a two-stage stochastic mixed-integer program (SMIP) to provide an emergency response pre-positioning strategy for hurricanes or other disaster threats. Due to the computational complexity of the problem, a heuristic algorithm referred to as the Lagrangian L-shaped method (LLSM) was developed to solve large-scale instances of the problem. The tool was tested with a case study focused on hurricane threats in the Gulf Coast area of the US. In the same field, Rawls and Turnquist (2011) extended the model described in (Rawls and Turnquist, 2010, 44:521–534) with additional service quality constraints. The added constraints ensured that the probability of meeting all demand is at least α and that the demand is met with supplies whose average shipment distance is no higher than a specific limit. A case study using hurricane threats was used to illustrate the model and how the additional constraints modify the pre-positioning strategy.

Finally, the study by Deng et al. (2017) presented a two-stage stochastic capacitated location-allocation (LA) problem for emergency logistics. To solve the model, the authors proposed an improved particle swarm optimization algorithm with the Gaussian cloud operator, the Restart strategy, and the adaptive parameter strategy.

As presented, there have been many approaches to HL. However, no specific inventory is considered for each type of affected people. Thus, the separation concerning the type of demand is an area of opportunity which has not been

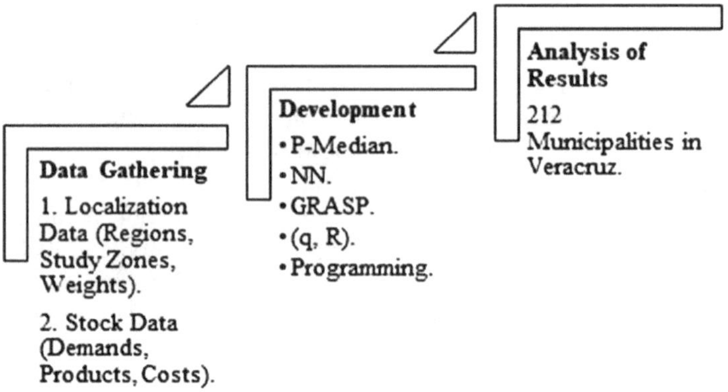

Fig. 9.1 Solution methodology

explored. Also, the differentiation between suitable and non-suitable locations for the distribution centers is not widely considered. Thus, the present work, which is based on the *disaster preparedness phase*, is focused on the development of a hybrid model that integrates two heuristic methods and two exact methods. It leads to the localization of feasible locations obtained for the construction of pre-positioned warehouses. The estimation of inventory levels depends on the type of demand to be supplied (early childhood, childhood, adolescence, adulthood, and third age).

For the development and evaluation of the hybrid model, the methodology showed in Fig. 9.1 was followed.

9.3 Data Gathering

9.3.1 Location Data

9.3.1.1 Regions of Veracruz

The State of Veracruz is made up of 212 municipalities, which are grouped into ten regions, as presented in Fig. 9.2. The north of Veracruz, which borders with the States of Tamaulipas and San Luis Potosi, is within the region of the *Huasteca Alta* (I). It is followed by the *Huasteca Baja* (II) region, which is limited by the State of Hidalgo. In that order, and bordering the State of Puebla, are the *Totonac* (III) region, the *Nautla* (IV) region, the *Capital* (V) region and *Las Altas Montañas* (VII) region. Between the *Capital*, *Altas Montañas*, and *Papaloapan* regions, the *Sotavento* (VI) region is located. The *Papaloapan* (VIII) region is adjacent to the State of Oaxaca, limiting the *Tuxtlas* (IX) region and ending with the *Olmeca* (X) region, which has its limits with the States of Oaxaca and Tabasco.

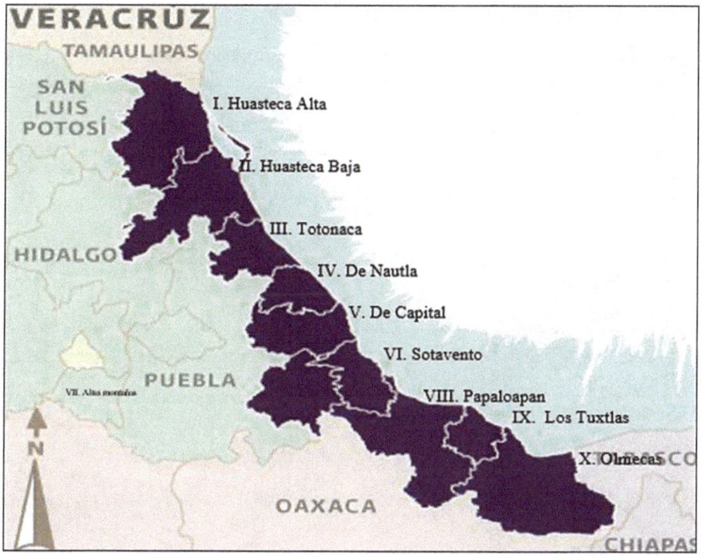

Fig. 9.2 Regions of the State of Veracruz (INAFED, 2010a, b)

Table 9.1 Emergency declarations 2012–2016 (SINAPROC, 2017)

Number of municipalities	Region	Number of declarations	Percentage
11	Nautla	86	11.08
15	Totonaca	39	4.86
18	Huasteca Baja	73	9.09
15	Huasteca Alta	40	4.86
33	Capital	154	19.05
57	Altas Montañas	174	21.42
4	Los Tuxtlas	23	2.99
25	Olmeca	123	15.19
22	Papaloapan	47	5.60
12	Sotavento	47	5.85

During the period between the years 2012 and 2016, 211 of the 212 municipalities belonging to the entity have received emergency declarations according to the FONDEN database. Table 9.1 shows the number of municipalities per region, the number of declarations that have been issued in one or several municipalities per region, and the percentage of declarations by region concerning the total.

As presented, the region with the highest number of declarations is *Las Altas Montañas,* with 21.42%, followed by *Capital* with 19.05% and *Olmeca* with 15.19%. These regions also have the highest number of municipalities.

Table 9.2 Declarations from 2016 per zone (SINAPROC, 2017)

Zones	Regions	Number of municipalities	Number of affected municipalities	Number of declarations
North	Huasteca Alta	15	4	4
	Huasteca Baja	18	10	15
	Nautla	11	9	17
	Totonaca	15	3	1
Center	Capital	33	17	26
	Altas Montañas	57	49	53
South	Los Tuxtlas	4	4	6
	Olmeca	25	10	17
	Papaloapan	22	7	7
	Sotavento	12	10	10
Total		212	123	

9.3.1.2 Study Zones

Based on the *regionalization* of Veracruz and the lack of municipality space for the establishment of the pre-positioned warehouses, a classification was made between affected and non-affected municipalities based on the presence of hydro-meteorological events by region from the declarations of emergency in the year 2016. As presented in Table 9.2, this led to classify the regions into zones.

The *North Zone* consists of the *Huasteca Alta*, *Huasteca Baja*, *Nautla* and *Totonaca* regions with a total of 59 municipalities (where 26 have been affected, and 33 have not been affected). In the *Center Zone,* the *Capital* and *Las Altas Montañas* regions are considered with a total of 90 municipalities (where 66 have been affected, and 24 have not been affected). Finally, in the S*outh Zone*, *Los Tuxtlas*, *Olmeca*, *Papaloapan* and *Sotavento* regions are considered with a total of 63 municipalities (where 31 have been issued at least one emergency declaration). From this data, a total of 123 affected municipalities, 156 declarations, and 89 non-affected municipalities were identified (these can become headquarters for pre-positioned warehouses).

9.3.1.3 Weight Factors for Non-affected Municipalities

To obtain feasible locations, places within short distances were not only considered. As recommended by the National Civil Protection System (SINAPROC, by its acronym in Spanish), we also considered attributes such as direct accesses to road infrastructure and closeness with suppliers and services (i.e., piped water, energy, telephone, and internet) to establish the support centers. For this, the weighting of these properties was proposed to describe the suitability of each non-affected municipality. Table 9.3 shows the percentages assigned to each of these properties.

Table 9.3 Weight factors (INAFED, 2010a, b; SCT, 2016; SEDESOL, 2015)

Weight factor	Relative weight (%)	Description
Road access	25	Road communication (state or federal type)
Suppliers	25	Number of social development stores within the municipality
Services	25	Constructions with drainage service, piped water, energy, telephone, internet
Infrastructure	25	Number of available rooms or schools. This data is taken as a reference because new pre-positioned storage does not require pre-existing buildings

Finally, the geographic coordinates of the affected and non-affected municipalities were obtained from the database of the National Institute of Statistics and Geography of 2012 (INEGI, 2012). From this, the hybrid model obtains the Euclidean distance between the described municipalities.

9.3.2 *Stock Data*

9.3.2.1 Demand and Supplies

The average of victims was estimated for all municipalities within each region by using the FONDEN 2016 database available on the website of SINAPROC.

Likewise, and with the purpose of being able to cover the needs of all human beings during their different stages of life (i.e., early childhood, childhood, adolescence, adulthood and old age), a database was developed to allow the efficient supply of food, equipment, water and medicines to the affected population. Food and equipment were grouped into kits, while water and medicines were estimated individually.

9.3.2.2 Cost

The purchase costs of products considered for the purposes of the research were extracted from (a) the Federal Consumer Attorney's Office (PROFECO, by its acronym in Spanish); (b) the Secretariat of Tax Administration (SAT, by its acronym in Spanish); and (c) the Secretariat of Labor and Social Security (STPS, by its acronym in Spanish) (PROFECO, 2018; SAT, 2018; STPS, 2018). These costs were classified into:

1. *Ordering cost*: integrated by an administrative and order processing cost; a transportation cost to the warehouse; and a handling and receiving cost in the warehouse.
2. *Purchasing cost*: the addition of monetary expenditures for the acquisition of supplies such as food, equipment, and liters of water.
3. *Stockout cost*: it involves the costs of the missing request, transportation cost to the warehouse and the handling and reception costs in the warehouse.
4. *Holding cost*: use of space cost, service, and loss of the goods.

9.4 Development

For solving the problem, a hybrid model that integrates the *p-median* algorithm and the metaheuristics of the *Nearest Neighbor* (*NN*) and *Greedy Randomized Adaptive Search Procedure* (GRASP) for the identification of suitable locations for the establishment of pre-positioned warehouses, was developed. The hybrid model also uses the continuous review (q, R) inventory model with uncertain demand for establishing the economic lot size for each product, the reorder point, safety stock, and shortage. In the following sections, each of these tools and their entry into the model is described.

9.4.1 P-Median Problem

"The p-median problem is a decision problem that consists of a set F of m facilities, a set C of n clients, and a matrix D of dimension $n \times m$ for the distances d_{ij} between each client $i \in C$ and the facility $j \in F$. The objective is to minimize the sum of the distances while connecting each one of the clients with the sets $S \subseteq F$ of p facilities.

The mathematical formulation of the problem is the following:

$$\mathbf{min} \sum_{i=1}^{n} \sum_{j=1}^{m} d_{ij} x_{ij} \tag{9.1}$$

$$s.t. \sum_{j=1}^{m} x_{ij} = 1, \quad \forall_i, \tag{9.2}$$

$$x_{ij} \leq y_j, \quad \forall_{ij}, \tag{9.3}$$

$$\sum_{j=1}^{m} y_j = p, \tag{9.4}$$

$$x_{ij}, y_j \in \{0, 1\} \tag{9.5}$$

The p-median problem determines the optimal location of the p facilities and allocation of clients to each facility in such a way that the total cost is minimized. In the mathematical formulation, (9.1) represents the objective function of the model, indicating that it seeks to minimize the summation of the distances d_{ij} for each client or demand location; (9.2) makes sure that each client is assigned to a facility; (9.3) makes sure that each client is attended by precisely one facility; (9.4) indicates that p facilities must be assigned, and (9.5) indicates the variable domain (binary)" (Landa et al., 2016).

It is important to mention that the p-median problem is of NP-hard computational complexity. Thus, large problems cannot be solved to optimality. Due to this situation, metaheuristics are commonly used to provide suitable solutions (near-optimal) to problems in a reasonable time.

9.4.2 Algorithm

The NN metaheuristic is one of the basic constructive methods for the p-median problem. Table 9.4 presents the steps of the *NN*.

For a given reference location (vertex representing a facility), clustering starts by assigning it the closest locations (vertexes representing client locations). This process is iterated until all locations are uniquely assigned to each reference location. Although this procedure is fast, it can lead to low quality results as some client locations can be assigned to far away facility locations. Due to this situation,

Table 9.4 Nearest neighbor algorithm (Ríos & González, 2000)

Initialization
Select a vertex j *at random.*
Do $t = j$ *and* $W = V \setminus \{j\}$.
While
$(W \neq 0)$
Take $j \in W$ / $Ctj = min\ \{Cti \mid i \in W\}$
Connect t *to* j
Do $W = W \setminus \{i\}$ *and* $t = j$

Table 9.5 Steps of the GRASP algorithm (Tabli, 2009)

Input: Number of iterations.
Repeat
s = Random-Greedy(seed); /* apply a randomized greedy heuristic */
s′ = Local − Search(s); /* apply a local search algorithm to the solution*/
Until Stopping criteria /* e.g. a given number of iterations */
Output: Best solution found.

for this work the NN was extended to add some characteristics of the GRASP metaheuristic, which is another method for solving NP-hard problems.

9.4.3 GRASP Algorithm

The *GRASP* metaheuristic seeks to improve initial solutions. It is an iterative greedy heuristic to solve combinatorial optimization problems. Each iteration of the GRASP algorithm contains two steps: (a) construction, where it creates a feasible solution using a greedy random algorithm, and (b) local search, where a local search heuristic is applied from the constructed solution. Table 9.5 presents the steps of the GRASP algorithm (Feo & Resende, 1995; Tabli, 2009).

While the GRASP is also an approximate solving method, due to the randomness of its local search, it can add flexibility to fix search mechanisms as those of the NN algorithm.

9.4.4 Capacitated NN + GRASP for P-Median Problem

The clustering (groups of client locations assigned to a specific facility) are restricted based on (a) municipalities which are not affected as venues for the establishment of pre-positioned warehouses; (b) a warehouse can supply more than one affected municipality; (c) an affected municipality must be supplied by a single warehouse; and (d) the maximum amount of demand that the store can supply is 55,247 affected people (this is the most significant number of people who has been affected in a municipality).

With these restrictions, the proposed hybrid model solves the p-median problem by performing the following steps:

1. Initial solution by NN → S_0 (constructed solution). Minimization of the objective function (see Sect. 4.1) is performed considering an integrated distance metric, which is based on the Euclidean distance between each facility and

its assigned clients, and w_i, which represents the weight assigned to each location as described in Table 9.3. i and j index to identify affected and non-affected municipalities. Thus, d_{ij} (where $i \in \{1, 2, \ldots, I\}$ and $j \in \{1, 2, \ldots, J\}$) represents the weighted Euclidean distance between affected and non-affected municipalities, which can be the headquarters for the pre-positioned warehouses.

2. The GRASP algorithm modifies S_0 by performing a random exchange of clients between clusters (local search). This process is iterated T times.
3. At each iteration, compliance of restrictions is verified as exchange can lead to non-valid solutions (e.g., clusters with total demand larger than the capacity of 55,247).

9.4.5 *Inventory Model (q, R)*

Para realizar el cálculo de los niveles de inventario, se programa la sección del modelo plasmado por Barojas-Payán et al. (2019), que corresponde a dicho proceso, en el cual se presenta "an extension of the continuous review model (q, R) with uncertain demand indicates that, in most cases, the demand is not known with utmost certainty De la Fuente et al. (2008). The same happens with other variables, such as the supply time. If an *Economic Order Quantity Model* (*EOQ*) is implemented in these conditions, using as demand the mean of the expected values, the natural thing is that at one point there will be a break in the available inventory, since they were produced in that period by a demand superior to the mean. A level of inventory above what would be strictly necessary to prevent the stockout. It is called Safety Stock (SS), and it serves to mitigate fluctuations.

From the above, obtaining the inventory levels for each type of demand within each pre-positioned warehouse is carried out through an extension to the *model (q, R) of continuous review with uncertain demand, based on the service level.* The mathematical formulation is the following:

$$\mathbf{Min} = \sum_{i=1}^{I}\sum_{j=1}^{J}\sum_{p=1}^{P}\left[\frac{Co_i^p D_j^p}{Q_{ij}^p} + Cs_i\left(\frac{Q_{ij}^p}{2} + z_{cls} * S_{D_j}^{\prime p}\right) + C_j^p D_j^p + \frac{D_j^p}{Q_{ij}^p}\left(Cf_i^p + S_{D_j}^{\prime p} * E_{z_i}^p\right)\right] \tag{9.6}$$

The objective function (9.6) represents the sum of the *logistic costs* of the number of municipalities i headquarters of pre-positioned warehouses, which supply the municipalities j, with the necessary products p to cover the needs of the people affected by the natural hydro-meteorological phenomenon. Therefore, the logistic costs originated by the implementation of a pre-positioned warehouse are related to the demand of each sector of the population D_j^p.

Table 9.6 Notation of mathematical model *Source* Authors

Notation of sets	
I	Set of municipalities that were not affected by a hydro-meteorological natural phenomenon and that can serve as locations for the pre-positioned warehouse
J	Set of municipalities affected by a hydro-meteorological natural phenomenon
P	Set of products to be supplied depending on the type of demand D_j
Notation of subscripts and superscripts	
i	A number of municipalities in the set I
j	A number of municipalities in set J
p	Number of products in set P
Parameters	
$\boldsymbol{D_j^p}$	The demand of each sector of the population: products p to be sent to municipality j
$\boldsymbol{FR_i^p}$	The service level of the pre-positioned warehouse located in municipality i gave to the affected people
$\boldsymbol{E_{z_i}^p}$	Standardized loss function: products p that was not supplied by the inventory of the pre-positioned warehouse located in the municipality i
Z_{csl}	The dimensionless metric according to the service level to be provided, that is, the value of Z for the normal distribution with a defined cycle service level (CSL)
$S_{D_j}^{\prime p}$	Adjusted standard deviation of products p that pertains to each of the demands of the affected municipality j
$\boldsymbol{R_{D_j}^p}$	Point of re-order: quantity of products p from which a supply order will be issued to cover the different types of demand of municipality D_j
$\boldsymbol{L_i}$	Delivery time of products to the municipality that is serving as the location for pre-positioned warehouse i
$\boldsymbol{Co_i^p}$	Order cost of product p in the pre-positioned warehouse located in the municipality i
$\boldsymbol{Cs_i^p}$	Maintenance cost for the inventory of products p of the pre-positioned warehouse located in the municipality i
$\boldsymbol{C_i^p}$	The purchase cost of the products maintained in the pre-positioned warehouse located in the municipality i
$\boldsymbol{Cf_i^p}$	Cost for the lack of products p in the pre-positioned warehouse located in the municipality i
Decision variables	
$\mathbf{Q_{ij}^p}$	Integer type variable, it represents the quantity of products p to be ordered from the pre-positioned warehouse located in the municipality i to supply the affected municipalities j

Table 9.6 presents the extended notation of the model, that is, the sets, indexes, parameters, and decision variables that comprise it, to make practical the involvement of the reader.

Then, the costs associated to the warehouses are computed as follows:

- *Ordering cost* from the pre-positioned warehouse located in municipality i (Co_i^p), includes the administrative cost of order processing, pre-positioned warehouse transportation cost, and handling costs at the reception of the product at the warehouse:

$$\frac{Co_i^p D_j^p}{Q_{ij}^p} \tag{9.7}$$

- *Holding cost* for product p inventory of the pre-positioned warehouse located in municipality i (Cs_i^p), it comprises the costs of space, service, and risk:

$$Cs_i^p\left(\frac{Q_{ij}^p}{2} + z_{cls} * S_{D_j}^{\prime p}\right). \tag{9.8}$$

- *Purchasing cost* of products p contained in the pre-positioned warehouse located in municipality i (C_i^p), in order to satisfy the affected demand:

$$C_i^p D_j^p \tag{9.9}$$

- *Stockout cost* for products p to be supplied by the pre-positioned warehouse located in municipality i (Cf_i^p), these are the costs associated with the non-availability of the demanded product, and it includes the request and the delivery from the permanent collection centers for its supply:

$$\frac{D_j^p}{Q_{ij}^p}\left(Cf_i^p + S_{D_j}^{\prime p} * E_{z_i}^p\right) \tag{9.10}$$

With this information, the objective function (9.6) is subject to the following restrictions:

$$S_{D_j}^{\prime p} = S_{D_j}^p \sqrt{L_i}, \quad i = 1, 2, \ldots, I,\ j = 1, 2, \ldots, J \tag{9.11}$$

$$R_{D_j}^p = D_j^p * L_i + z_{CSL} * S_{D_j}^{\prime p}, \quad i = 1, 2, \ldots, I,\ j = 1, 2, \ldots, J,\ p = 1, 2, \ldots, P \tag{9.12}$$

$$Q_{ij}^p = \sqrt{2Co_i \frac{\sum_{j=1}^{J} Y_{ij}^p D_j^p}{Cs_i^p}}, \quad i = 1, 2, \ldots, I,\ j = 1, 2, \ldots, J,\ p = 1, 2, \ldots, P \tag{9.13}$$

$$E_{z_i}^p = z\left[F_s^p(z) - 1\right] + f_s^p(z), \quad i = 1, 2, \ldots, I,\ p = 1, 2, \ldots P \tag{9.14}$$

$$FR_i^p = 1 - \frac{S_j^{\prime p} E_{z_i}^p}{Q_{ij}^p}, \quad i = 1, 2, \ldots, I,\ p = 1, 2, \ldots, P \tag{9.15}$$

$$Q_{ij}^p \in Z^+ \tag{9.16}$$

Equation (9.11) enables the estimation of the adjusted standard deviation of each of the different demands belonging to the impacted municipalities ($S_{D_j}^{\prime p}$). Revealing that demand has normal behavior in the distribution of data. Equation (9.12) reflects the calculation of the re-order point of products (p) for each of the different types of demand in the affected municipalities (D_j). Equation (9.13) determines the order quantity per item (p) of the pre-positioned warehouses located in the municipalities (i) for the different stages of the human life cycle in the affected municipalities (j).

Equation (9.14) calculates the loss function, which estimates the likelihood of not having enough products (p) to supply the total number of victims. Equation (9.15) calculates the Fill Rate (FR), which measures the service level for the delivery of products (p) to affected people by the pre-positioned warehouses located in municipalities i, regardless of the total number of victims to be satisfied in municipality j.

Equation (9.16) defines the variable that represents the quantity of products p to be requested per order by the pre-positioned warehouse belonging to municipalities i in order to cover the demand of the affected municipalities j, that is to say, ($\mathrm{Q}_{\mathrm{ij}}^{\mathrm{p}}$) is of the integer type".

The purpose of adding mathematical programming is to obtain the optimal quantity to be requested from each of the items (food, equipment, water, medicines). As well as the stock point that marks the request for a new order, based on time of delivery, to reduce not only the costs associated with the store but the number of people not supplied derived from the lack of products.

9.4.6 *Software*

GNU Octave is a high-level language, primarily intended for numerical computations. It provides a convenient command-line interface for solving linear and nonlinear problems numerically, and for performing other numerical experiments using a language that is mostly compatible with Matlab. It may also be used as a batch-oriented language (Eaton, 2018).

Due to these features, Octave was used for the implementation of the hybrid algorithm with the processes described in Sects. 9.4.4 and 9.4.5.

9.5 Application and Results

A first run was performed with the p-median problem and the NN algorithm as a solving method. It led to the results presented in Table 9.7, where two scenarios are considered: with and without the positional weight (see Sect. 9.3.1.3). Out of a total of 59 municipalities contained in the North Zone (26 have been affected and 33 not affected), the model estimated the establishment of 7 pre-positioned warehouses if the municipal attributes mentioned in previous sections are considered, and no facility in case they are omitted. In the same way, within the Center Zone, which

Table 9.7 Location results considering NN (PPW = Pre-positioned warehouses with/without positional weight)

Zone	Region	PPW	Municipalities per region	
			With positional weight	Without positional weight
North	Huasteca Alta	2	Naranjos de Amatlán—Pueblo Viejo	
	Huasteca Baja	1	Cerro Azul	
	Totonaca	2	Coatzintla—Gutiérrez Zamora	
	Nautla	2	San Rafael—Vega de Alatorre	
Center	Capital	4/10	Banderilla—Emiliano Zapata—Jilotepec—Naolinco	Emiliano Zapata—Jalcomulco—Tlalnelhuayocan—Jilotepec—Actopan—Coacoatzintla—Naolinco—Tonayan—Landero y Coss—Chiconquiaco
	Altas Montañas	3/8	Atoyac—Fortín—Omealca	Comapa—Chocaman—Fortín—Atoyac—Paso del Macho—Camarón de Tejeda—Rafael Delgado—Omealca
South	Sotavento	1/2	Puente Nacional	Puente Nacional—Manlio Fabio Altamirano
	Papaloapan	3	Carlos A. Carrillo—Cosamaloapan Lerdo de Tejada	
	Los Tuxtlas	0	–	
	Olmeca	2	Acayucan—Las Choapas	

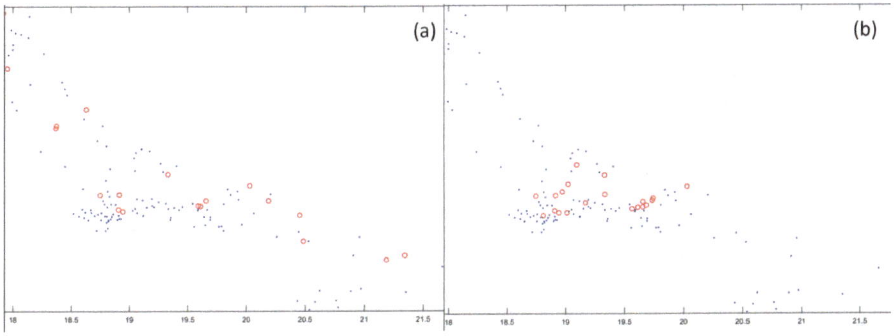

Fig. 9.3 Location of pre-positioned warehouses considering NN (with/without positional weight)

Fig. 9.4 Location of pre-positioned warehouses considering NN (map with positional weight)

consists of 90 municipalities (66 have been affected and 24 not been affected), the feasible result with positional weight is the establishment of 7 pre-positioned warehouses and 18 if that weight is omitted. The South Zone houses 63 municipalities (31 have been affected, and 32 not affected), and the method proposed to establish six warehouses taking into consideration their compliance with governmental attributes and two if not. In both cases, 20 feasible locations were determined.

Figure 9.3 presents the graphical comparison of the results obtained: 3a) shows the location of the 20 warehouses considering the positional weight of each non-affected municipality, and 3b) shows the 20 warehouses without considering the positional weight. Both figures show the affected municipalities.

Figure 9.4 shows the 20 municipalities located for the installation of the pre-positioned warehouses, distributed throughout the State of Veracruz in the Google Maps® application.

Table 9.8 presents the details of the municipalities assigned to each of 20 pre-positioned warehouses. This consists of information regarding the number of municipalities to be supplied, the name of each of the municipalities, the distance between the municipality where the warehouse is located and the affected municipalities, and the total demand to supply.

Likewise, Fig. 9.5 shows the location of 6 of the 20 municipalities located for the establishment of pre-positioned warehouses and their respective municipalities to supply. These options are determined through the minimum distance of travel while considering the number of victims of each one of them to determine if its addition to a specific cluster is feasible. The municipalities presented are Pueblo Viejo, Cerro Azul, Banderilla, Omealca, Acayucan, and Las Choapas.

The addition of the GRASP algorithm to the NN was performed to reduce the distances within the clusters of the affected municipalities to supply. As part of the

Table 9.8 Extended results considering NN (NM = number of municipalities)

Host municipality for the pre-positioned warehouse	NM	Municipalities to supply	Distance (Km)	Demand
Naranjos	3	Ilamatlán, Zontecomatlán de López y Fuentes, Ozulama de Mascareñas	383.60	16,595
Pueblo Viejo	2	Tampico, Panuco	106.00	38,344
Cerro Azul	7	Benito Juárez, Huayacocotla, Tlachichilco, Tuxpán, Chicontepec, Tantoyuca, Álamo de Temapache	758.10	54,061
Coatzintla	5	Zacualpan, Texcatepec, Espinal, Papantla, Poza Rica	456.80	49,835
Gutiérrez Zamora	5	San Andrés Tenejapa, Aquila, Coetzala, Tlilapan, Tenochtitlán	1493.00	6716
San Rafael	14	Teocelo, Texhuacán, Los Reyes, Rafael Lucio, Colipa, Tenampa, Huiloapan de Cuauhtémoc, Tomatlán, Tepatlaxco, Nautla, Jalancingo, Las Vigas de Ramírez, Tlapacoyan, Martínez de la Torre	2234.30	54,949
Vega de Alatorre	13	Sochiapa, Alpatláhuac, Yecuatla, Tlacolulan, Misantla, Ayahualulco, Zentla, Xalapa Enríquez, Calcahualco, Juchique de Ferrer, Totutla, Maltrata, Tlaltetela	2040.80	55,073
Omealca	8	Tlaquilpan, Soledad Atzompa, Ixhuatlán del Café, Cuitláhuac, Tehuipango, Tezonapa, Zongolica, Tierra Blanca	457.10	55,009
Banderilla	5	Villa Aldama—Perote—Coatepec—Cosautlán de Carvajal—Tlacotepec de Mejía	274.60	54,901
Emiliano Zapata, Dos Ríos	7	Yanga, Coscomatepec, Rio Blanco, Huatusco, Camerino Z. Mendoza, Mariano Escobedo, Nogales	758.40	54,957
Jilotepec	2	Altotonga, Las Minas	116.50	54,857
Naolinco	9	Naranjal, Atzacán, Acultzingo, Ixhuatlancillo, Tatatila, Acajete, Atzalan, Úrsulo Galván, Xico	1268.00	55,219
Atoyac	7	Ixhuacán de los Reyes, La Antigua, La Perla, Soledad de Doblado, Paso de Ovejas, Medellín de Bravo, Boca del Rio	659.70	55,144
Fortín	5	Tequila, Amatlán de los Reyes, Ixtaczoquitlán, Orizaba, Córdoba	85.10	55,155
Puente Nacional	1	Veracruz	54.90	55,218

(continued)

Table 9.8 (continued)

Host municipality for the pre-positioned warehouse	NM	Municipalities to supply	Distance (Km)	Demand
Carlos A. Carrillo	9	Xoxocotla, Ixhuatlán del Sureste, Hidalgotitlan, Pajapan, Nanchital de Lázaro Cárdenas del Río, Agua Dulce, Catemaco, Juan Rodríguez Clara, Isla	1429.00	55,204
Cosamaloapan de Carpio	11	Magdalena, Atlahuilco, Cuichapa, Jamapa, Mixtla de Altamirano, Tlacotalpan, Carrillo Puerto, Ignacio de la Llave, Cotaxtla, Tlalixcoyan, Tres Valles	1307.50	54,528
Lerdo de Tejada	6	Astacinga, Alvarado, Hueyapan de Ocampo, Santiago Tuxtla, Cosoleacaque, San Andrés Tuxtla	681.70	54,513
Acayucan	1	Minatitlán	49.70	55,247
Las Choapas	3	Jáltipan, Moloacan, Coatzacoalcos	195.90	54,257

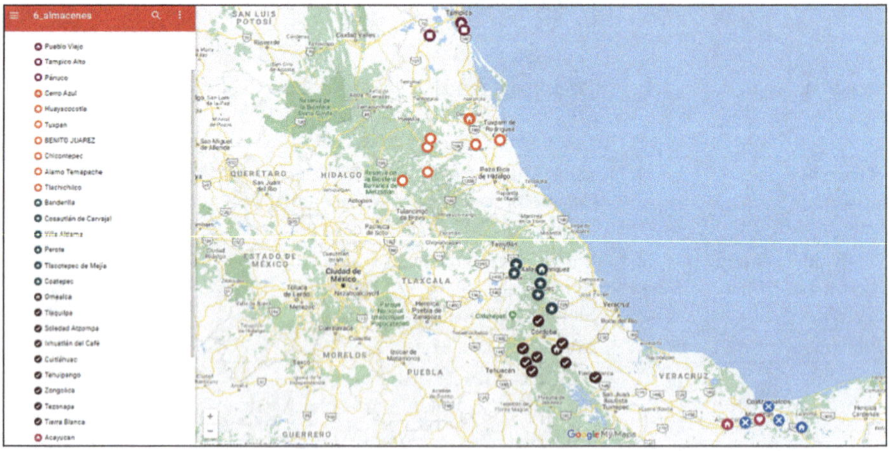

Fig. 9.5 Location of 6 pre-positioned warehouses and municipalities to supply considering NN

evaluation, a search process of T = 10,000 iterations and with a processing time of 2.28 s, changes were obtained within 11 clusters, and a decrease in distance of approximately 366.5 km was obtained.

Table 9.9 shows the details of the revised clusters, the sum of the distances between the municipality where the pre-positioned warehouse is located and the affected municipalities to be supplied, and the demand to be supplied from the integration of the algorithm, only for the municipalities that were altered. It should

Table 9.9 Extended results considering NN + GRASP (NM = number of municipalities)

Host municipality for the pre-positioned warehouse	NM	Municipalities to supply	Distance (Km)	Demand
Cerro Azul	7	*Zacualpan*, Huayacocotla, Tuxpan, Chicontepec, Tantoyuca, Álamo de Temapache, Ilamatlán	802.10	51,768
Coatzintla	5	*Benito Juárez*, Texcatepec, Espinal, Papantla, Poza Rica.	316.80	52,128
Gutiérrez Zamora	5	San Andrés Tenejapa, *Huatusco*, *Tequila*, *Totutla*, *La Perla*	1461.00	26,425
San Rafael	14	Teocelo, Texhuacán, Los Reyes, Rafael Lucio, Colipa, Tenampa, Huiloapan de Cuauhtémoc, *Astacinga*, Tepatlaxco, Nautla, Jalancingo, Las Vigas de Ramírez, Tlapacoyan, Martínez de la Torre	2290.30	54,642
Vega de Alatorre	13	Sochiapa, Alpatláhuac, Yecuatla, Tlacolulan, Misantla, Ayahualulco, Zentla, Xalapa Enríquez, Calcahualco, Juchique de Ferrer, *Tenochtitlán*, *Cuitláhuac*, *Ixhuatlancillo*	1936.40	54,969
Omealca	8	Tlaquilpan, Soledad Atzompa, Ixhuatlán del Café, Maltrata, Tehuipango, Tezonapa, Zongolica, Tierra Blanca	511.00	53,511
Emiliano Zapata, Dos Ríos	7	Yanga, Coscomatepec, Rio Blanco, *Aquila*, Camerino Z. Mendoza, Mariano Escobedo, Nogales	712.30	47,492
Naolinco	9	Naranjal, Atzacán, Acultzingo, Tlaltetela, Tatatila, Acajete, Atzalan, Úrsulo Galván, Xico	1122.40	53,989
Atoyac	7	Ixhuacán de los Reyes, La Antigua, *Tlilapan*, Soledad de Doblado, Paso de Ovejas, Medellín de Bravo, Boca del Rio	653.90	50,001
Fortín	5	*Coetzala*, Atzacán, Aquila, Ixtaczoquitlán—Córdoba	76.70	50,886
Lerdo de Tejada	6	*Tomatlán*, Alvarado, Hueyapan de Ocampo, Santiago Tuxtla, Cosoleacaque, San Andrés Tuxtla	643.70	54,820

be noted that, within the column of municipalities to be supplied, those that were added to the headquarters to be supplied are underlined.

In Fig. 9.6 the location of the six warehouses taken as an example to evaluate the NN method was the same as the last. The change of municipalities caused by the

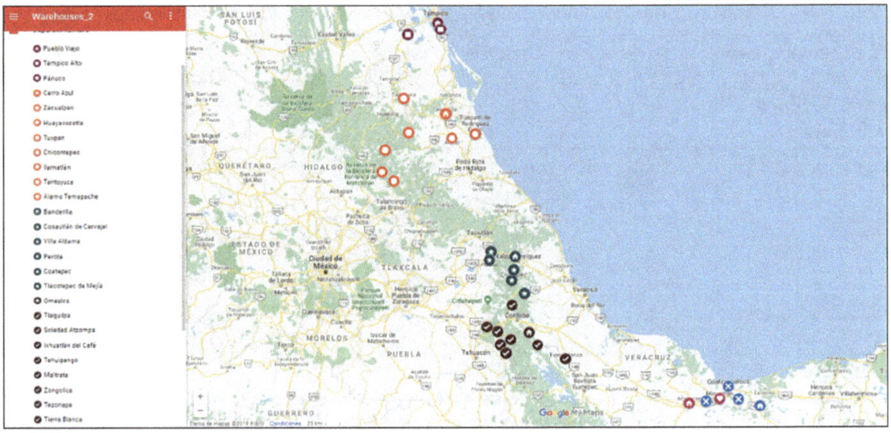

Fig. 9.6 Location of six pre-positioned warehouses and municipalities to supply considering NN + GRASP

GRASP method can be observed in the following municipalities: in Cerro Azul, there was a change from the municipality of Benito Juárez to Zacualpan while in Omealca there was a change from the municipality of Cuitlahuac to Maltrata,

Regarding the inventory, Table 9.10 presents the results for the six warehouses presented in Figs. 9.5 and 9.6. Information regarding the economic order quantity (Q_{ij}^{p}) of each type of demand and product to be supplied; the reorder point of each product on demand ($R_{D_j}^{p}$); the safety stock of the warehouses; and affected people not supplied ($E_{z_i}^{p}$) is presented. The data used in this research can be seen in Appendix 9.A.

9.6 Conclusions and Future Work

Pre-positioning of emergency supplies is a means to increase preparedness for natural disasters (Rawls et al. 2010). Critical decisions in pre-positioning are the locations and capacities of emergency distribution centers, as well as allocations of inventories for multiple relief commodities to those distribution locations. The location and allocation decisions are complicated by uncertainty about if or where a natural disaster will occur (Rawls et al. 2011). The hybrid model presented in this research combines the NN and GRASP methods for solving the location problem, which is modeled by the p-median problem. Also, it considers the non-linear inventory model of continuous review (q, R) to determine the appropriate inventory levels for each pre-positioned warehouse.

The hybrid model was evaluated through its application in the 212 municipalities that constitute the State of Veracruz, of which, during 2016, 123 of them have been

Table 9.10 Food and water inventory levels

Type of demand	Host municipality for the pre-positioned warehouse	The economic order quantity ($\mathbf{Q}_{ij}^{p}$)	Reorder point ($\mathbf{R}_{j}^{p}$)	Safety stock	Non-supplied people
Demand $\mathbf{D}_{j}^{p}$ (early childhood)	Pueblo Viejo	1825	146	612	1
	Cerro Azul	2062	1361	301	1
	Banderilla	2988	1502	646	1
	Omealca	2446	250	313	1
	Acayucan	2764	150	936	2
	Las Choapas	2246	153	547	2
Demand $\mathbf{D}_{j}^{p}$ (childhood)	Pueblo Viejo	4015	1606	1361	2
	Cerro Azul	5010	2516	716	2
	Banderilla	6208	1791	1359	2
	Omealca	5235	1811	664	1
	Acayucan	5850	146	1987	3
	Las Choapas	4987	1920	1215	3
Demand $\mathbf{D}_{j}^{p}$ (adolescence)	Pueblo Viejo	6663	2369	2236	4
	Cerro Azul	7050	3243	1052	2
	Banderilla	8845	2404	1893	3
	Omealca	7750	2524	1020	2
	Acayucan	10,520	197	3550	6
	Las Choapas	9326	3101	2238	6
Demand_d $\mathbf{D}_{j}^{p}$ (adulthood)	Pueblo Viejo	5144	2193	612	1
	Cerro Azul	6174	2799	301	1
	Banderilla	5400	1833	646	1
	Omealca	5570	2111	313	1
	Acayucan	6949	183	936	2
	Las Choapas	6570	2364	547	2
Demand $\mathbf{D}_{j}^{p}$ (old age)	Pueblo Viejo	1637	778	524	1
	Cerro Azul	1983	1008	301	1
	Banderilla	2196	667	278	1
	Omealca	1483	753	218	1
	Acayucan	2076	87	710	2
	Las Choapas	1481	725	381	2
Demand $\mathbf{D}_{j}^{p}$ (General)	Pueblo Viejo	20,816	6716	6907	11
	Cerro Azul	23,004	9730	3394	5
	Banderilla	25,320	6164	5405	8
	Omealca	22,990	6710	3036	5
	Acayucan	30,282	333	10,242	15
	Las Choapas	25,977	7878	6295	10

affected by a disturbing phenomenon of hydro-meteorological type. In the first stage, the problem was solved with the NN algorithms. As result, 20 feasible locations were obtained for the establishment of pre-positioned warehouses and the municipalities to be supplied by each one of them were determined. In a second stage, the GRASP algorithm was added to the NN algorithm. This led to decrease the distance between affected municipalities by 366.5 km. Then, in a third stage, the mathematical model of the (q, R) inventory method was extended to provide the inventory levels for the different people affected by the phenomena. It provided the order quantities to be requested, the reorder point, the safety stock, and the non-supplied people. Demand was assumed to consist of product kits (divided into four classes: food, equipment, water and medications) according to the life stages to which the affected people belong (early childhood, childhood, adolescence, adult or third age). This provided the support to serve the most significant number of affected people within the shortest time with products according to their stage of life.

Quick response to urgent relief right after natural disasters through efficient HL is vital to the alleviation of disaster impact in the affected areas, which remains challenging in the field of logistics and related study areas (Jian et al., 2010). This is why the work of Abushaikha and Schumann (2016) was aimed to assess the suitability of mobile phone technology for humanitarian aid operations and the creation of new HL concepts. The introduction of the concept of mobile telephony will be used in future work in order to reduce the non-supply and the response time derived from traffic problems, security or lack of infrastructure.

It should be mentioned that both location and routing should be considered vital in the phases before and after the disaster, both in aspects of distribution of supplies and in the transfer of victims to shelters and care centers. Given the above, future work is also focused on proposing multiple routing for the delivery of supplies and the inclusion of temporary collection centers and warehouses for the post-disaster phase.

References

Abushaikha, I., & Schumann, D. (2016). Mobile phones: Established technologies for innovative humanitarian logistics concepts. *Procedia Engineering, 159,* 191–198. https://doi.org/10.1016/j.proeng.2016.08.157.

Ahmadi, M., Seifi, A., & Tootoni, B. (2015). A humanitarian logistics model for disaster relief operation considering network failure and standard relief time: A case study on San Francisco district. *Transportation Research Part E: Logistics and Transportation Review, 75,* 145–163. https://doi.org/10.1016/j.tre.2015.01.008.

Bai, X. (2016). Optimal decisions for prepositioning emergency supplies problem with type-2 fuzzi variables. *Discrete Dynamics in Nature and Society*, 1–17. doi: http://d9.doi.org/10.1155/2016/9275192.

Balcik, B., & Beamon, M. (2008). Facility location in humanitarian relief. *International Journal of Logistics Research and Applications, A Leading Journal of Supply Chain Management, 11,* 101–121. https://doi.org/10.1080/13675560701561789.

Barojas-Payán, E., Sánchez-Partida, D., Martínez-Flores, J. L., & Gibaja-Romero, D. E. (2019). Mathematical model for locating a pre-positioned warehouse and for calculating inventory levels. *Journal of Disaster Research, 14*(4), 649–666. doi: https://doi.org/10.20965/jdr.issn.1883-8030.

Barzinpour, F., Saffarian, M., Makoui, A., & Teimoury, E. (2014). Metaheuristic algorithm for solving biobjective possibility planning model of location allocation in disaster relief logistics. *Journal of Applied Mathematics 2014*, 1–18. doi: http://d9.doi.org/10.1155/2014/239868.

Boonmee, C., Arimura, M., & Asada, T. (2017). Facility location optimization model for emergency humanitarian logistics. *International Journal of Disaster Risk Reduction, 24*, 485–498. https://doi.org/10.1016/j.ijdrr.2017.01.017.

Caballero, S., Barojas, E., Sánchez, D., & Martínez, J. (2018). Extended GRAPS-capacited K-means clustering algorithm to establish humanitarian support centers in large regions at risk in Mexico. *Journal of Optimization, 118*, 1–14. https://doi.org/10.1155/2018/3605298.

CRED & UNISDR (2016). *ArmResearch poverty & death: disaster mortality 1996–2015*. Brussels, Belgium: Centre for Research on the Epidemiology of Disasters (CRED) & United Nations Office for Disaster Risk Reduction (UNISDR).

Cozzolino, A. (2012). Humanitarian logistics and supply chain management. In *Humanitarian logistics, Springerbriefs in Business* (pp. 5–16). Berlin, Heidelberg: Springer. doi: https://doi.org/10.1007/978-3-642-30186-5_2.

De la Fuente, D., Parreño, J., Fernández, I., Pino, R., Gómez, A., & Puente, J. (2008). *Ingeniería de organización en la empresa: Dirección de Operaciones*. Asturias: Universidad de Oviedo.

Deng, Y., Zhu, W., Tang, J., & Qin, J. (2017). Solving a two-stage stochastic capacitated location-allocation problem with an improved PSO in emergency logistics. *Mathematical Problems in Engineering, 2017*, 1–15. https://doi.org/10.1155/2017/6710929.

Diaz, C., & Gaytan, J. (2014). Flood risk assessment in humanitarian logistics process design. *Journal of Applied Research and Technology, 12*, 976–984. https://doi.org/10.1016/S1665-6423(14)70604-2.

Döyen, A., Aras, N., & Barbarosoğlu, G. (2011). A two-echelon stochastic facility location model for humanitarian relief logistics. *Optimization Letters, 6*, 1123–1145. https://doi.org/10.1007/s11590-011-0421-0.

Dufour, Ê., Laporte, G., & Rancourt, M. (2017). Logistics service network design for humanitarian response in East Africa. *Omega, 74*, 1–14. https://doi.org/10.1016/j.omega.2017.01.002.

Eaton, J. -W. (2018). *GNU octave*. Available from https://www.gnu.org/software/octave/about.html. Accessed on April 14, 2018.

Feo, T., & Resende, M. (1995). Greedy randomized adaptive search procedures. *Journal of Global Optimization*, 109–133. doi: https://doi.org/10.1007/BF01096763.

Gösling, H., & Geldermann, J. (2014). A framework to compare OR models for humanitarian logistics. *Procedia Engineering, 78*, 22–28. https://doi.org/10.1016/j.proeng.2014.07.034.

INAFED, National Institute for Federalism and Municipal Development, SEGOB, Secretary of the Interior. (2010a). *Encyclopedia of municipalities and delegations of Mexico*. Available from http://siglo.inafed.gob.mx/enciclopedia/EMM30veracruz/regionalizacion.html. Accessed on May 3, 2017 (in Spanish).

INAFED, National Institute for Federalism and Municipal Development SEGOB, Secretary of the Interior. (2010b). *National municipal information system: Municipalities in numbers*. Available from http://www.snim.rami.gob.mx/. Accessed on May 3, 2017 (in Spanish).

INECC, National Institute of Ecology and Climate Change. (2016). *Vulnerability to climate change*. Available from https://www.gob.mx/inecc/acciones-y-programas/vulnerabilidad-al-cambio-climatico-actual. Accessed on April 10, 2018 (in Spanish).

INEGI. (2012). ITER30XLS10. National Institute of Statistic and Geography (INEGI). Census and Counts, 2012. Government of the United Mexican States.

INEGI. (2013). Knowing Veracruz by Ignacio de la Llave. National Institute of Statistics and Geography (INEGI), 2013. Government of the United Mexican States.

Jian, H., Bing, Y., Jie, W., Zhi, W., & Zhi, H. (2010). GIS-based safe area discovery for emergency logistics. In *2010 2nd IEEE International Conference on Information Management and Engineering*. Chengdu, China. doi: https://doi.org/10.1109/ICIME.2010.5477642.

Landa, I., Sánchez-Partida, D., Barojas, E., & Martínez-Flores, J. L. (2016). Optimization model for the relocation on of distribution sites of economic support in areas of high marginalization. *DYNA Management Research Article, 4*(1), 1–15. doi: https://d9.doi.org/10.6036/MN8104.

Maghfiroh, M., & Hanaoka, S. (2017). Las Mile distribution in humanitarian logistics under stochastic and dynamic consideration. In *2017 IEEE International Conference on Industrial Engineering and Engineering Management (IEEM)* (pp. 1411–1415). Singapore, Singapore. doi: https://doi.org/10.1109/IEEM.2017.8290125.

Maharjan, R., & Hanaoka, S. (2017). Warehouse location determination for humanitarian relief distribution in Nepal. *Transportation Research Procedia, 25,* 1151–1163. https://doi.org/10.1016/j.trpro.2017.05.128.

Morales, R., & Akhavan, R. (2012). A location model for storage of emergency supplies to respond to technological accidents in Bogota. In C. Laroque, J. Himmelspach, R. Pasupathy, O. Rose, & A. M. Uhrmacher (Eds.) *Proceedings of the 2012 Winter Simulation Conference* (pp. 2413–2424).

Özdamar, L., & Demir, O. (2012). A hierarchical clustering and routing procedure for large scale disaster relief logistics planning. *Transportation Research Part E: Logistics an Transportation Review, 48,* 591–602. https://doi.org/10.1016/j.tre.2011.11.003.

PC, Secretary of Civil Protection. (2011). *A Municipal atlas of basic level risks, Ixtaczoquitlán.* Secretariat of Civil Protection (CP), 2011. Government of the State of Veracruz, Mexico (in Spanish).

PC, Secretariat of Civil Protection. (2017). *State and municipal Atlas.* Available from http://www.veracruz.gob.mx/proteccioncivil/servicio/atlas-estatal-municipal. Accessed on April 11, 2018 (in Spanish).

PF, Federal Police. (2016). *Rescue and support actions for the population.* Available from https://www.gob.mx/policiafederal/acciones-y-programas/acciones-de-rescate-y-apoyo-a-la-poblacion . Accessed on April 15, 2018 (in Spanish).

Presidency of the Republic. (2014). *Do you already know what the MX Plan is?* Available from https://www.gob.mx/presidencia/articulos/ya-sabes-que-es-el-plan-m9. Accessed on April 15, 2018.

PROFECO, Federal Consumer Procurator's Office. (2018). *Who's who in the prices.* Available from https://www.profeco.gob.mx/precios/canasta/home.aspx?th=1. Accessed on May 13, 2018 (in Spanish).

Rawls, C. -G., & Turnquist, M. -A. (2010). Pre-positioning of emergency supplies for disaster response. *Transportation Research Part B, 44,* 521–534. https://doi.org/10.1016/j.trb.2009.08.003.

Rawls, C. -G., & Turnquist, M. -A. (2011). Pre-positioning planning for emergency response with service quality constraints. *OR Spectrum, 33,* 481–498. https://doi.org/10.1007/s00291-011-0248-1.

Ríos, M. -R., & González, J. -L. (2000). Investigation of operations in action: Heuristics for the solution of the TSP. *Engineering* (pp 15–20).

Rodríguez, O., Pavel, P., & Brewster, C. (2018). Disaster preparedness in humanitarian logistics: A collaborative approach for resource management in floods. *European Journal of Operational Research, 264,* 978–993. https://doi.org/10.1016/j.ejor.2017.01.021.

Sae, R., Hyun, J., & Chul, H. (2013). Warehouse location decision factors in humanitarian relief logistics. *The Asian Journal of Shipping and Logistics, 29,* 103–120. https://doi.org/10.1016/j.ajsl.2013.05.006.

Safeer, M., Anbuudayasankar, S. -P., Balkumar, K., & Ganesh, K. (2014). Analyzing transportation and distribution in emergency humanitarian logistics. In *12th Global Congress on Manufacturing and Management, GCMM 2014. Procedia Engineering 97,* 2248–2258. doi: https://doi.org/10.1016/j.proeng.2014.12.469.

Santiago, M., & De Oliveira, R. (2014). Humanitarian logistics: Empirical evidences from a natural disaster. *Procedia Engineering, 78,* 102–111. https://doi.org/10.1016/j.proeng.2014.07.045.

Santoso, A., Prayogo, D., & Parung, J. (2017). Model development of rescue assignment and scheduling problem using grasp metaheuristic. In *2017 IEEE International Conference on Industrial Engineering and Engineering Management (IEEM)* (pp. 1407–1410). Singapore, Singapore. https://doi.org/10.1109/IEEM.2017.8290124.

SAT, Secretary of tax administration. (2018). *Minimum wage 2018.* Available from http://www.sat.gob.mx/informacion_fiscal/tablas_indicadores/paginas/salarios_minimos.asp9. Accessed on May 13, 2018 (in Spanish).

SCT, Secretariat of Communications and Transportation. (2016). *Draw your route.* Available from http://app.sct.gob.mx/sibuac_internet/ControllerUI?action=cmdEscogeRuta. Accessed on May 14, 2017 (in Spanish).

SEDESOL. (2015). Directory of stores. Secretary of Social Development (SEDESOL), 2015. Government of the United Mexican States.

SEGOB. (2017). Declarations of natural disaster. Secretary of the Interior (SEGOB), 2017. Government of the United Mexican States.

SEMAR. (2014). Marine plan to help the population in cases and zones of emergency or disaster. Secretary of the Navy of Mexico (SEMAR), 2014. Government of the United Mexican States.

SINAPROC, National System of Civil Protection. (2017). *Supplies authorized by emergency declaration.* Available from http://www.proteccioncivil.gob.mx/es/ProteccionCivil/Insumos_autorizados_por_declaratoria_de_emergencia. Accessed on April 20, 2018 (in Spanish).

STPS, Secretary of Labor and Social Security. (2018). *Minimum wages in effect as of January 2018.* Available from https://www.gob.mx/cms/uploads/attachment/file/285013/TablaSalariosMinimos-01ene2018.pdf. Accessed on May 13, 2018 (in Spanish).

Talbi, E. G. (2009). *Metaheuristics, from design to implementation* (pp. 164). University of Lille—CNRS—INRIA. Wiley A John Wiley & Sons, Inc., Publication.

Tomasini, R., & Van Wassenhove, L. (2009). From preparedness to partnerships: Case study research on humanitarian logistics. *International Transactions in Operational Research, 16*(5), 549–559. https://doi.org/10.1111/j.1475-3995.2009.00697.x.

Van Wassenhove, L. (2006). Blackett Memorial lecture. Humanitarian aid logistics: Supply chain management in high gear. *Journal of the Operational Research Society, 57*(5), 475–489.

Van Wassenhove, L., & Pedraza, A. (2012). Using OR to adapt supply chain management best practices to humanitarian logistics. *International Transactions in Operational Research, 19,* 307–322. https://doi.org/10.1111/j.1475-3995.2010.00792.x.

Chapter 10
Risk Analysis of Unmanned Aerial Systems to Supply Survival Kits in Search-and-Rescue (SAR) Operations

Diana Sánchez-Partida, Georgina G. Rosas-Guevara, José Luis Martínez-Flores, and Azgad Casiano-Ramos

Abstract In a "during disaster" phase, the response activities are executed during the emergency period or immediately after the event occurs. In most disasters, this period passes very quickly, and the human being is found vulnerable. Many of the registered cases the people can lose life in the rescue process, whether they cannot access the first aid within the first hours. Until now, many researchers and practitioners have determined that a survival kit must be light to transport and safeguard life. In this research, a solution for the delivery of survival kits to vulnerable people during a disaster is proposed by using Unmanned Aerial Systems (UAS) under coordinated response with strategic support centers. An operational and technical risk analysis of Unmanned Aerial Vehicles (UAV) technology is made in order to determine its feasibility for the mission from a technical and legal point of view. A study case is developed in the mountain region of Veracruz, Mexico. It leads to particular attention on UAV design issues such as range, autonomy, sustainable, and communication levels by considering broader implications to create necessary humanitarian operations.

Keywords Unmanned aerial systems (UAS) · Unmanned aerial vehicles (UAV) · Risk analysis · AHP · Humanitarian aid · Search-and-Rescue (SAR)

Electronic supplementary material The online version of this chapter (https://doi.org/10.1007/978-3-030-67295-9_10) contains supplementary material, which is available to authorized users.

D. Sánchez-Partida · J. L. Martínez-Flores
Department of Logistics and Supply Chain Management, UPAEP Universidad, 17 Sur 901, Barrio de Santiago, CP 72410 Puebla, Puebla, Mexico

G. G. Rosas-Guevara
Department of Innovation, HappenIoT, CP 72760 Cholula, Puebla, Mexico
e-mail: grosas@happeniot.com

A. Casiano-Ramos (✉)
Faculty of Electronic Sciences, Benemérita Universidad Autónoma de Puebla, Puebla, Mexico
e-mail: azgadcara@camautsa.com

D. Sánchez-Partida et al. (eds.), *Disaster Risk Reduction in Mexico*,
https://doi.org/10.1007/978-3-030-67295-9_10

10.1 Introduction

Mexico is located in a region of intense seismic and volcanic activity. As a consequence, earthquakes or eruptions are originated. In addition to this, the country is located in the intertropical zone, threatened by hurricanes originated in the Pacific and Atlantic oceans following Mexican Geological Service (SGM 2019), and the Ministry of Environment and Natural Resources—(SEMARNAT 2019). In this context, the country is subject to a great variety of natural phenomena that can cause disasters.

Disasters can be defined as sudden, calamitous events that disrupt the activities of a society or a community and cause human, material, economic, or environmental losses that exceed the recovery capacity of the affected community or society using only its resources (Natarajarathinam et al., 2009). That is why the Mexican authorities recognized that in order to deal with them in a better way, it is necessary to have an operational protection scheme to support the population. With this objective, the Mexican National Civil Protection System (SINAPROC) created many civil protection plans. One of these plans encourages the population to make an emergency backpack with the primary care as acting in the prevention or "before disaster" phase, in order to reduce risks (SINAPROC, 2019).

This primary care or survival kit contains essential elements that victims need right after a disaster occurs, such as food, vaccines, first aid items, medicines, among others. However, it is well known that sometimes the magnitude or the rapidity with which the disaster occurs can exceed the ability to react in the phase of an emergency. However, people can forget the emergency backpack when they want to escape from some calamity, for instead get trapped under rubble; or on the roof of their homes in the case of a flood as occurred in August 29th, 2005 in the United States due to Hurricane Katrina recorded in National Oceanic and Atmospheric Administration (Ahued-Ortega, 2018; NOAA, 2019). For that reason, a strategic plan to supply the survival kits to trapped people is necessary.

In the last years, Unmanned Aerial Systems (UAS) have been considered as a feasible alternative to support humanitarian and disaster relief operations (Van Tilburg, 2017; Turki et al. 2019). UAS can be of great importance when some disaster has disrupted the supply chain through the destruction or obstruction of roads. The aerial supply has already been proposed in other investigations like (Garnica et al., 2016), which found the optimum number and location for first aid distribution centers in Chiapas, Mexico, decreasing response time for emergency services. They propose the use of a V/STOL (Vertical/Short Take-Off and Landing) aircraft to establish an efficient air supply to the affected areas from the first-aid distribution centers. The operational advantages of this aircraft are enormous that are used in the field of military, humanitarian, and rescue operations, or even in general aviation (Páscoa et al., 2013).

Similarly, in De Oliveira et al. (2017), the authors use fixed-wing Unmanned Aerial Vehicles (UAV) and Geographic Information Systems (GIS) in a procedure for structuring direct response distribution networks of humanitarian aid after the occurrence of a natural disaster. They show the effectiveness of their scheme in speeding up the humanitarian operations after a flood occurred in Duque de Caixas, Brazil, in 2013. The recent use of UAVs in other disaster management scenarios such as earthquakes, forest fires, the spread of hazardous materials, and nuclear accidents has been reported in Chowdhury et al. (2017) and Restas (2015); and specific indicators for the operational, economic, and risk value of the particular UAV implementation models have been proposed in Haidari et al. (2016) and Vidyadharan et al. (2017).

This work is organized as follows. In Sect. 10.2, the literature review is presented. In Sect. 10.3, The motivation and the selected region for the study are described. In Sect. 10.4, the flight operational and technical criteria of different UAVs configurations that affect the Search-and-Rescue (SAR) mission are analyzed. In Sect. 10.4, are described the external factors like environmental of the mission. In Sect. 10.5, the results of this analysis are presented, and some legal issues for the implementation of this model in Mexico are discussed. Moreover, for last, in Sect. 10.6, general conclusions about this work are given.

10.2 Literature Review

When a small plane crashes in a remote area or a fishing boat is lost at sea, or a hurricane devastates a region, or a person simply gets lost while he or she is hiking, Search-and-Rescue (SAR) teams must scan vast areas in search for evidence of victims. For this purpose, UAVs equipped with remote sensing devices can be programmed to fly predefined pat patterns, possibly at low altitudes, and produce various types of images (thermal, optical, and so forth) captured from their privileged point-of-view. Ideally, this imagery is transmitted real-time back to a ground control station via a data link (Everaerts, 2008; Molina et al., 2014).

The UAVs are not only used to detect and rescue people but are used to deliver medicines improved vaccine availability and produced logistics cost savings considering range and payload within currently available UAV specifications (Haidari et al., 2016). The payload is an essential criterion because a larger payloads result in slower operating speeds. However, in some cases, it can be challenging to control, and with less maneuverability, mainly due to their high susceptibility to ambient weather conditions (Villa et al., 2016).

Whatever, the use of UAVs for the so-called “wilderness SAR” is rapidly evolving. For example, Alotaibi et al. (2019) consider the use of a team of multiple unmanned aerial vehicles (UAVs) to accomplish a search and rescue (SAR) mission in the minimum time possible while saving the maximum number of people. The novelty of Layer SAR involves simulating real disasters to distribute SAR tasks among UAVs. The performance of LSAR is compared, in terms of percentage of rescued survivors and rescue and execution times, with the max-sum, auction-based, and locust-inspired approaches for multi UAV task allocation (LIAM) and opportunistic task allocation (OTA) schemes.

Półka et al. (2017) work conducted to gather end-user requirements in purpose to create a tailor-made solution supporting traditional activities of SAR teams. The target group of the system has been extended, as other public services might benefit from the system implementation as well. The system used combines Digital Cellular Technologies and Global Navigation Satellite DCT—EGNSS technology to track victims.

Performing a feasibility analysis of the use of the system in SAR missions is preponderant because each UAV has its operational and technical criteria, which are affected by the conditions of use.

So, Burke et al. (2019) made a pilot study with a thermal-equipped drone for SAR applications in Morecambe Bay. In a variety of realistic SAR scenarios, it was found that the UAV could detect humans who would need rescue, both by the naked eye and by a simple automated method. Besides, they explore the current advantages and limitations of thermal UAV systems and outline the future path to a useful system for deployment in real-life SAR.

Also, Witczuk et al. (2018) analyzed the feasibility of the UAVs and thermal infrared (TIR) imaging. They achieve the right level of area coverage. However, the main challenges of the method were difficulties in species identification due to a relatively low resolution of TIR cameras, regulations limiting drone operations to the visual line of sight, and high dependence on weather.

Rudol and Doherty (2008), the technique has been implemented and tested on-board the UAVTech autonomous unmanned helicopter platform as a part of a complete autonomous mission. Moreover, the detected human positions are geolocated, and a map of points of interest is built. Such a saliency map can, for example, be used to plan medical supply delivery during a disaster relief effort. The results of the flight-tests were the video sequences collected onboard an unmanned aerial vehicle. The technique presented uses two video sources (thermal and color) and allows for high rate human detection at more considerable distances than in the case of using the video sources separately with standard techniques. The high processing rate is essential in case of video collected onboard a UAV in order not to miss potential objects as a UAV flies over it.

Moreover, Silvagni et al. (2017) present a multipurpose UAV (unmanned aerial vehicle) for mountain rescue operations. The multi-rotors based flying platform and its embedded avionics are designed to meet environmental requirements for mountainous terrain such as low temperatures, high altitude, and strong winds, assuring the capability of carrying different payloads (separately or together) such as avalanche beacon (ARTVA) with automatic signal recognition and path following algorithms for the rapid location of the snow-covered body; camera (visible and thermal) for search and rescue of missing persons on snow and in woods during the day or night; payload deployment to drop emergency kits or specific explosive cartridge for controlled avalanche detachment. The resulting small (less than 5 kg) UAV is capable of fully autonomous flight (including take-off and landing) of a pre-programmed, or easily configurable, custom mission.

In this research, an operational and technical risk analysis of commercial UAS technology is made. The evaluation considers the operational and technical criteria, a risk analysis developed by NASA, in order to evaluate to determine its feasibility for the Search-and-Rescue (SAR) mission for supplying the survival kits to the victims during a disaster.

10.3 Problem Description

This work is based on previous research by Caballero-Morales et al. (2018). It was aimed to provide an efficient metaheuristic to determine the most appropriate location for support centers in the State of Veracruz, which is one of the most affected regions in Mexico. The objective was to minimize the distance between the support centers and the affected regions. The metaheuristic was based on the -Means Clustering (KMC) algorithm which was extended to integrate (a) the associated capacity restrictions of the support centers, (b) a micro Genetic Algorithm GA to estimate a search interval for the most suitable number of support centers, (c) a variable number of assigned elements to centers in order to add flexibility to the assignation task, and (d) a random-based decision model to improve the final assignments further. These extensions on the KMC algorithm led to the GRASP-Capacitated -Means Clustering (GRASP-CKMC) algorithm, which was able to provide very suitable solutions for the establishment of 260 support centers for 3837 communities at risk in Veracruz, Mexico. Validation of the GRASP-CKMC algorithm was performed with well-known test instances and metaheuristics. The validation supported its suitability as an alternative to standard metaheuristics such as Capacitated -Means (CKM), Genetic Algorithms (GA), and Variable Neighborhood Search (VNS).

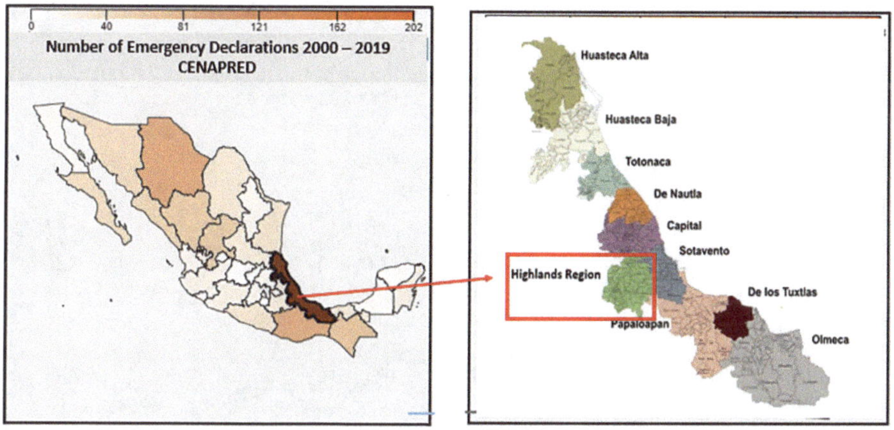

Fig. 10.1 Regions of Veracruz, Mexico selected for study for its high risk of disaster

Based on the research above, the affected communities of Veracruz selected for the study were that belong to the Highlands regions located in the mountainous region, as can be seen in Fig. 10.1.

The Highlands region was chosen because, according to the National Atlas of Risks (ANR, 2019), it is one of the regions with the highest degree of Danger, Exposure, and Vulnerability to electrical storms, floods, slope susceptibility, and flammable substances. Then, based on the database of emergency declarations of the Natural Disasters Fund (FONDEN, 2019) corresponding to the period 2012–2016, three communities were selected. One, for the highest density of the affected population (Orizaba) and the other two for the frequency of impact (Ixtaczoquitlán and Unión y Progreso) of some disturbing phenomenon. In Table 10.1, the selected communities and the type of phenomena that are sensitive to, as well as its location, the number of emergency declarations, and the number of victims for declaration are shown.

The assumptions considered are shown below:

- The three communities selected can be inaccessible at some time because of the damage of the roads by landslides or floods.
- The demand was determined based on three estimations. First, was obtained the total number of rescued victims per hydrometeorological event. This event not only affects one state but a set of nearby states, according to the fourth, fifth, and sixth report of the Government of the Republic 2012–2018, (2019). Second, the number of rescued victims was calculated per state considering the number of emergency and disaster declarations of that event recorded by the National Center for Disaster Prevention (CENAPRED) 2000–2019 (CENAPRED, 2019); giving an average of 1072, a minimum of 360 and a maximum of 21, 594 rescued victims for the state of Veracruz. There is a significant difference

Table 10.1 Information about the communities selected for the study

High risk of	Communities	Latitude	Longitude	#Emergency declarations	#Victims per declaration
Electrical storms, floods, slope susceptibility, and flammable substances	Ixtaczoquitlán	18.86277	−97.06222	6	23,574
Electrical storms, floods, and slope susceptibility	Orizaba	18.85861	−97.09666	5	60,425
Electrical storms and slope susceptibility	Unión y Progreso	18.87277	−97.11916	6	14,730

Table 10.2 Information about the strategic points that support the victims in a rescue operation

Center	Communities assigned	Latitude	Longitude	#Victims per event per community in recue operations
Center No. 247	Ixtaczoquitlán	18.88868	−97.05019	13
Center No. 162	Orizaba	18.78618	−97.03319	4
Center No. 173	Orizaba	18.78783	−97.10085	4
Center No. 174	Unión y Progreso	18.89287	−97.13071	4

between the number of rescued victims due to there are events with low and high magnitude. Third, the average of rescued victims was distributed among the ten regions of Veracruz according to the number of municipalities impacted with adverse effects following the emergency declarations from 2012 to 2016 reports made by the Natural Disasters Fund (FONDEN, 2019). Thus, to the Highlands region correspond an average of 25 victims that need a survival kit, which was distributed through the three selected communities based on the risk found in the ANR, see Table 10.2.

- The communities can be supplied by the strategic points identified in Caballero-Morales et al. (2018); these can be seen in Table 10.2. Two centers must supply the Orizaba community due to the large number of people to attend. These points can be the temporary shelters that have previously been designated to provide support to the victims. These shelters are roofed installations that have security protocols and the authorization of operation by the Civil Protection of the State. If they do not exist or if they are insufficient, they must search for facilities that meet the requirements and thus can be authorized.
- The survival kit will be the Personal First Aid Kit recommended by the American Red Cross (ARC, 2019a, b). The kit considered 72 h of emergency items, and it is prepared for minor injuries and illness. The kit contains essential first-aid supplies to treat most common injuries, including cuts, scrapes, swelling, sprains, strains, and more housed in a clear carrying tube. The kit also contains the American Red Cross Emergency First Aid Kit Card that provides emergency action information and necessary caregiving steps for a range of situations. Taking into account the work developed by (Valenzuela et al., 2016), an emergency blanket, an emergency food bar, and water purifier straw are also considered. Four more elements are suggested in the kit to guarantee the rescue of the victim. A Rescue Flash Signal Mirror, visible for more than 10 miles, reflective aiming aid gives the power to signal to planes, helicopters, and distant SAR teams; Oral Rehydration Salts that control the effects of stomach illness because the most common causes of morbidity and mortality in this situation are a diarrheal disease (Jafari et al., 2011).

  Moreover, is necessary to add two bags; both where water can be stored, one where dirty water is stored to filter, and the other where clean water is stored and the oral rehydration salts can be prepared as well as a Card that provides information about the use each item suggested. Therefore, the water purifier straw is proposed that contains a bag. All the items are considered from ARC and Lifestraw company (Lifestraw, 2019). In appendix 10.A, it is possible to see the characteristics of each item. The estimated weight of the survival kit is 2.58 lbs. Moreover, the total dimensions are Width 8”, Height 8” and Depth 12.49”.

10.4 Benchmarking of Operational and Technical Criteria of Commercial UAVs

The initial time after a disaster is crucial, and the rapid and effective delivery of these items is essential to minimize casualties of those who were affected (Bozorgi-Amiri et al., 2012). Hence, the pre-positioning of these rescue resources is a strategy to reduce the delivery time and increase the preparedness requiring additional investments before the emergency occurs (Natarajarathinam et al., 2009). Therefore, it is essential to have an efficient logistics system that can provide promptly.

As pointed out by the author De Oliveira et al. (2017), UAV can facilitate the coordination logistics by yielding timely geographical information of the affected region to the decision-makers. For the case of SAR operations, a set of predisposed UAVs must be capable of transporting the survival kits from the available support centers to the affected locations. In this context, the operational and technical criteria of the UAS are critical since the current UAS technology still has some flight limitations. For example, the maximum flight range of a UAV could avoid the coverage of the affected locations or limit the operation to a one-way trip; or the lost communication with a telemetry ground station would put the mission or persons at risk. Moreover, the inability of the aircraft to sense and avoid obstacles that could provoke more problems than it wishes to solve is critical too.

In Table 10.3, some modern UAVs, commercially known as drones, are compared concerning its flight range. Note that the maximum round trip achieved by the listed aircraft is approximately 9–48 km, with an estimated delivery time of 8.5–30 min, respectively, by considering an approximated survival kit weight of 5 kg. These characteristics will let to conclude if the current operational and technical criteria of commercial UAVs will be feasible for making the SAR mission or if the UAVs require to be improved.

Table 10.3 The estimated flight range of different UAV technologies for transporting a survival kit

Drone brand	Model	Aircraft type	Energy source	Max speed (km/min) (No wind)	Estimated flight time (min) @ 5 kg load	Estimated flight distance (km) @ 5 kg load
DJI	Matrice 600	Hexacopter	LiPo 6S battery	1.08	17	18.36
Drone Tools	Condor	VTOL Hexacopter	LiPo/Lition battery	0.96	26	24.96
DJI	S1000+	Octacopter	Lipo 6S battery	1.32	28.5	37.62
Harris Aerial	Carrier H6 Hybrid	Hexacopter	Gas-electric hybrid	0.9	90	81
Intelligent Energy	DJI Matrice 100 Adaptation	Quadricopter	Fuel cell	1.32	NA (500 g max payload)	NA
WingCopter	178 Heavy Lift	VTOL Hybrid	Batteries	0.83–2.5	52	68.1
eBlimp	e2400B	Air Blimp	Lithium batteries	1.625	60	97.5

10.5 Methodology and Results

In this section show, this methodology is based on Agency Risk Management Procedural Requirements and NASA/SP-2011-3422, NASA Risk Management Handbook (NASA, 2011). Which consists of to elaborate a risk matrix that is a graphical representation of the likelihood and consequence scores of a risk.

It is sometimes called a “5 × 5 Matrix” because it contains five rows and five columns. The rows of a risk matrix show likelihood scores, while the columns show the consequence scores. The likelihood is a measure of the possibility that a consequence is realized. This probability accounts for the frequency of the consequence and the timeframe in which the consequence can be realized. For some purposes, it can be assessed qualitatively. For other purposes, it is quantified in terms of frequency of probability. A consequence is the quantitatively or qualitatively expressed the outcome of a risk that may lead to degraded performance concerning one or more performance measures.

A Priority Score can represent each cell in a Risk Matri10. A risk is a potential for performance shortfalls that may be realized in the future concerning achieving explicitly established and stated performance requirements. The performance shortfalls are related to the mission execution in safety, technical, and cost.

Table 10.4 Risk identification methods

Formal	Informal
System safety assessments—fault tree analysis, hazard analysis, failure modes, and effects analysis	Brainstorming
Quantitative risk assessments	Test and verification
System and software engineering	Pause and learn sessions
Program planning and control—cost and schedule risk analysis	Experience—previous analysis, lessons learned, historical data
Models and simulations	

Table 10.5 Risk likelihood criteria

Likelihood		
Score	Likelihood of occurrence (p)	
5	Near certainty	$p > 80\%$
4	Highly likely	$60\% < p \leq 80\%$
3	Likely	$40\% < p \leq 60\%$
2	Low likelihood	$20\% < p \leq 40\%$
1	Not likely	$p \leq 20\%$

The first step is *identifying the risk* that realizes as is shown in the following Table 10.4.

The second step is *analyzing criterion risk.* It is determined with Table 10.5, to assign a likelihood score from 1 to 5.

The consequence score is determined by assessing the consequence of the risk and assigning a consequence score from 1 to 5 based on the criteria. The risks must be analyzed and scored on each separate consequence category as performance, safety, asset, schedule, and cost.

Finally, it is built the priority risk. Once the likelihood score, consequence score, impact timeframe, and impact horizon have been determined, risk can be prioritized by determining its Priority Score. The Priority Score is the score assigned via the Risk Matrix cell into which the risk falls. The Risk Matrix cell into which the risk falls is based on the risk's likelihood and consequence scores.

10.5.1 Evaluation of UAS Operational and Technical Criteria in a SAR Mission

UAS flight operational and technical criteria are synthesized as four operational and technical criteria for the mission. These criteria are identified as critical for the transport and delivery of the survival kits. The name and meaning of each attribute are explained in the following lines.

Table 10.6 AHP matrix for the determination of the relevance of the operational and technical criteria in UAS survival kit delivery

	Telemetry	Payload		Range	Autonomy
Telemetry	1	9		6	6
Payload	1/9	1		1/3	1/3
Range	1/6	3		1	2
Autonomy	1/6	3		½	1

Table 10.7 Weights results from the operational criteria

Criteria	Weighting
Telemetry	0.67524198
Payload	0.15916402
Range	0.11254596
Autonomy	0.05304804

Telemetry criterion refers to the ability of a remote operator to take complete control of the aircraft from a ground station and to visualize the operational scenario using visual sensors, e.g., cameras, installed on board of the UAV all the time and locations of the mission.

The payload criterion refers to the ability of the aircraft to transport and deliver the desired package volume and weight without damaging it or requiring structural modifications to the aircraft.

The range criterion refers to the maximum distance allowed by the aircraft to travel before the energy source is needed.

The autonomy criterion refers to the sense and avoids the capability of the aircraft without the need for the intervention of a human operator during the flight.

With the operational attributes described above, a standard procedure for decision making, known as Analytical Hierarchy Process (AHP) was applied in order to determine the relevance of each attribute for the mission by assigning different levels of importance to each one, according to the experience of the decision-makers we proceed to make a judgment relationship. Table 10.6 shows the AHP matrix for the four critical criteria in UAS survival kit delivery. The Saaty scale and the procedure were used on base to Chap. 2-X of this book based on (Saaty, 1994).

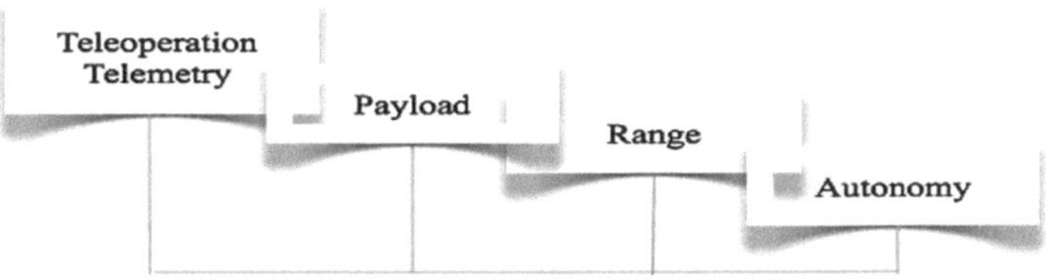

Fig. 10.2 Degree of the importance of UAS operational and technical criteria for a SAR mission

According to the proposed operational and technical criteria for the mission, Saaty Consistency Ratio was CR = 0.0484668 < 10%, which indicates that the expert criteria in the AHP matrix were consistent. The corresponding CR and weights are shown in Table 10.7.

According to Table 10.7, the results of the hierarchy of the UAS criteria are shown in Fig. 10.2. Here, the telemetry criterion has been determined by the experts as the most important, since total control of the aircraft is needed to guarantee safety during all the mission. Secondly, the payload criterion is recognized for its elemental aim to transport goods to the affected persons. The range is evaluated as the third criterion due to the possibility of performing shorter scale missions. Lastly, autonomy is considered as the last criterion for the success of the mission since basic sense and avoid capabilities are actively compensated by the teleoperation capability of a human operator by assuming an ideal telecommunication link.

10.5.2 Risk Analysis of the UAS-Assisted SAR Mission

In order to evaluate the risk of the UAS-assisted SAR mission, externalities to the operational and technical criteria of the UAS are incorporated, and its probability of occurrence, which is determined as shown in Table 10.8. The operational availability factor of UAVs in the predefined centers must be considered due to the sudden occurrence of a disaster or the obstruction of the predefined distribution centers; not all the UAVs may be available to start the mission. The weather is a critical factor determining the mission's success. It directly affects the telemetry criterion of a fully operational UAV by opposing to its secure teleoperation. Direct link communication in the proposed scenario becomes almost impossible due to the orography and the normal adverse situation of communication systems in the region. Thus, the geography factor mainly contributes to the risk for the UAV when flying in the mountains region where numerous and diverse obstacles are present. Furthermore, finally, the available communication networks factor such as the last

Table 10.8 External (non-operational) factors that affect the UAS-assisted SAR mission

Code	Factors	Description	% Probability (0–5)	% Consequence (0–5)
H1	Operational availability	Available UAVs in the strategic facilities	4	2
H2	Weather	Conditions to rain and wind	4	4
H3	Geography	Physical obstacles	3	2
H4	Available communication networks	Terrestrial and Satellite IoT networks	2	5

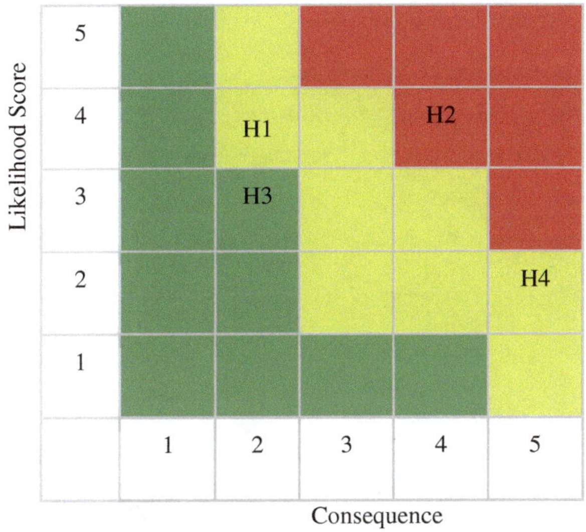

Fig. 10.3 Risk matrix for the UAS-assisted mission by considering external (nonoperational) factors

generation Internet of Things (IoT) networks would be essential for communicating with the UAV to preserve its telemetry capabilities.

In conjunction, the level of impact of the risk factors as determined in the risk matrix of Fig. 10.3, showing that weather is the most critical external factor for the success of the UAS-assisted SAR mission. Available communication networks are a common risk factor due to the new growth of satellite IoT infrastructure along with the possibility of adopting radio communication boosting equipment. Operational availability is considered at the same risk level due to a lower probability of occurrence but with a high impact on the mission's success. Finally, geography is a lower risk factor, being compensated by the autonomy and telemetry operational criteria, where total flight control is possible.

UAS systems have been demonstrated to be an appropriate platform for in situ inspection systems to sample closer to the source. According to with evaluation of UAS for SAR mission, the risk factors as determined by weather (H2), which can be mitigated with forecast analysis in real-time after disaster events and using tools developed for us that support operations of the UAS with in-depth intelligence. Following shown an example from Table 10.2 for rescue zones: centers 247, 162/173, and 174 using UAV Forecast (https://www.uavforecast.com/) that help determine fly conditions.

There are four centers of rescue for a mission with a maximum range of 100 km and altitude during the flight of 2000 m. UAS should be capable of providing command and control links and real-time during the whole flight. The next Fig. 10.4 is shown some characteristics in the prediction of the weather on a specific date. However, weather predictions can be obtained around disaster date. Some parameters o index that is analyzed for functional fly are maxim wind, wind altitude, minimum temperature, maxim temperature, gust, visibility, cloud cover, visible satellites.

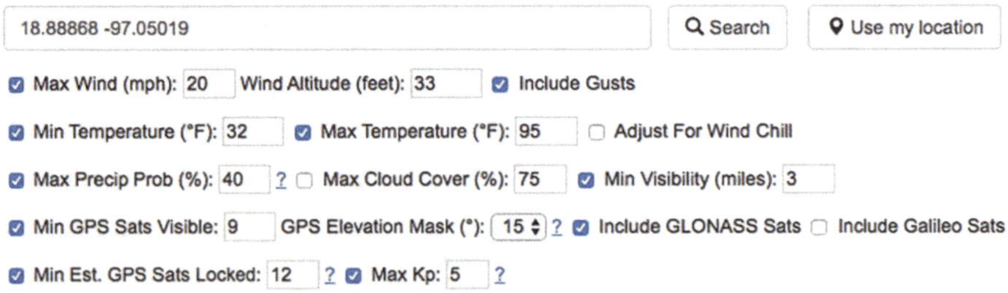

Fig. 10.4 Parameters of the forecast weather of Center No. 247—Ixtaczoquitlán (18.88868, −97.05019)

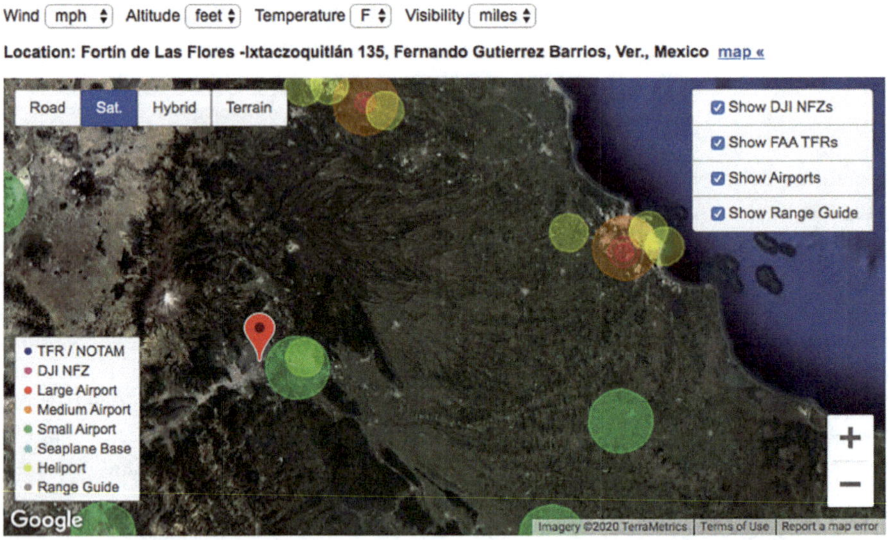

Fig. 10.5 Results of the forecast weather of Center No. 247—Ixtaczoquitlán (18.88868, −97.05019)

Precip Pro is the probability of there being any kind of precipitation, including rain, snow, hail, and sleet. The *Kp* index measures geomagnetic disruption caused by solar activity, on a scale from 0 (calm) to 9 (major storm). Anything at or below 3 or 4 is usually safe for flying. The elevation of a satellite is how high above the horizon it appears to be at a given point in time at their location. GPS satellites that are at low elevation are close to the horizon and are more likely to be obscured by nearby buildings or hills, making it more difficult for their GPS receiver to get a lock. The GPS Elevation Mask lets count how many satellites are visible from their location, by excluding all satellites whose elevations are below the threshold value. The default value (5°) is suitable for most locations.

Thursday 2020-03-05: sunrise 06:43, sunset 18:35									
Time	Gusts	Temp	Precip Prob	Cloud Cover	Visibility	Visible Sats	Kp	Est. Sats Locked	Good To Fly?
00:00	4 mph	67°F	-	-	10 miles	13	2	12.7	yes
01:00	4 mph	67°F	-	-	10 miles	13	2	12.2	yes
02:00	4 mph	66°F	1%	-	10 miles	15	2	13.0	yes
03:00	4 mph	65°F	1%	-	10 miles	11	2	9.9	no
04:00	4 mph	64°F	1%	7%	10 miles	16	2	13.7	yes
05:00	4 mph	63°F	-	17%	10 miles	20	2	18.5	yes
06:00	4 mph	63°F	-	28%	10 miles	19	2	19.0	yes
07:00	4 mph	63°F	1%	42%	10 miles	19	2	19.0	yes
08:00	4 mph	66°F	2%	59%	10 miles	17	2	17.0	yes
09:00	5 mph	68°F	3%	71%	10 miles	13	2	13.0	yes
10:00	6 mph	71°F	5%	77%	10 miles	13	2	13.0	yes
11:00	8 mph	72°F	7%	79%	9 miles	8	2	8.0	no
12:00	9 mph	73°F	8%	81%	7 miles	11	2	11.0	no
13:00	9 mph	71°F	7%	86%	7 miles	15	2	14.4	yes
14:00	8 mph	69°F	5%	91%	9 miles	11	2	11.0	no

Current Conditions as of Wednesday 2020-03-04 14:22 CST									
Time	Gusts	Temp	Precip Prob	Cloud Cover	Visibility	Visible Sats	Kp	Est. Sats Locked	Good To Fly?
14:22	7 mph	91°F	-	-	10 miles	13	3	12.7	yes
Wednesday 2020-03-04: sunrise 06:44, sunset 18:34									
Time	Gusts	Temp	Precip Prob	Cloud Cover	Visibility	Visible Sats	Kp	Est. Sats Locked	Good To Fly?
15:00	8 mph	90°F	-	-	10 miles	11	2	10.8	no
16:00	7 mph	87°F	-	-	10 miles	11	2	11.0	no
17:00	7 mph	84°F	-	-	10 miles	11	2	11.0	no
18:00	6 mph	81°F	-	-	10 miles	12	2	12.0	yes
19:00	5 mph	78°F	-	-	10 miles	12	2	11.9	no
20:00	4 mph	74°F	-	-	10 miles	13	2	12.1	yes
21:00	4 mph	71°F	1%	-	10 miles	12	3	10.8	no
22:00	4 mph	69°F	-	-	10 miles	15	3	13.0	yes
23:00	4 mph	68°F	-	-	10 miles	13	3	11.9	no

Fig. 10.6 Current conditions of the weather and the forecast of Center No. 247—Ixtaczoquitlán (18.88868, −97.05019)

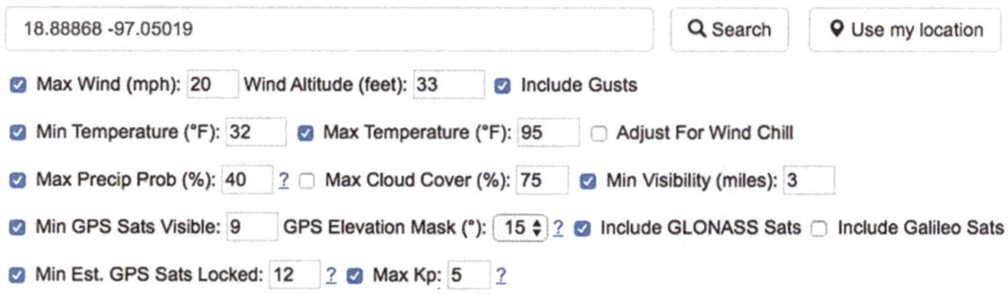

Fig. 10.7 Parameters of the forecast weather of Center No. 162—Orizaba (18.78618, −97.03319)

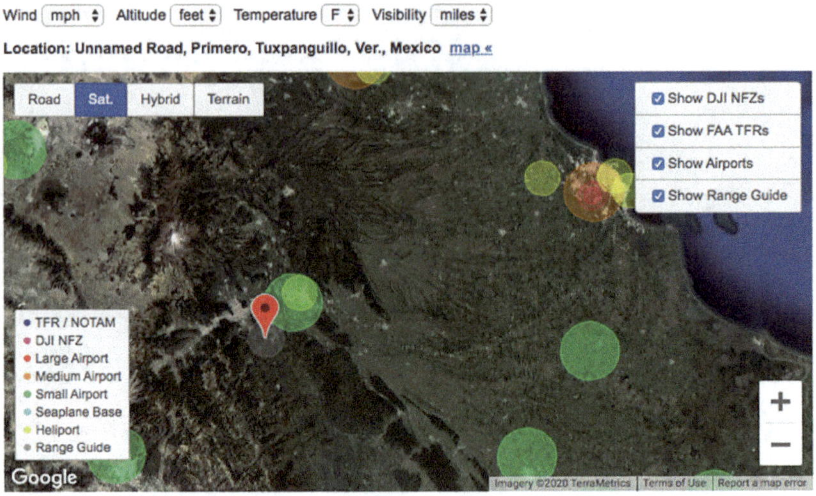

Fig. 10.8 Results of the forecast weather of Center No. 162—Orizaba (18.78618, −97.03319)

Current Conditions as of Wednesday 2020-03-04 14:41 CST									
Time	Gusts	Temp	Precip Prob	Cloud Cover	Visibility	Visible Sats	Kp	Est. Sats Locked	Good To Fly?
14:41	7 mph←	89°F	-	-	10 miles	12	3	11.7	no

Wednesday 2020-03-04: sunrise 06:44, sunset 18:34									
Time	Gusts	Temp	Precip Prob	Cloud Cover	Visibility	Visible Sats	Kp	Est. Sats Locked	Good To Fly?
15:00	7 mph←	89°F	-	-	10 miles	11	2	10.8	no
16:00	7 mph←	86°F	-	-	10 miles	11	2	11.0	no
17:00	6 mph←	83°F	-	-	10 miles	11	2	11.0	no
18:00	6 mph←	81°F	-	-	10 miles	12	2	12.0	yes
19:00	5 mph←	78°F	-	-	10 miles	12	2	11.9	no
20:00	4 mph→	74°F	-	-	10 miles	13	2	12.1	yes
21:00	4 mph→	71°F	1%	-	10 miles	12	3	10.8	no
22:00	4 mph↘	69°F	-	-	10 miles	15	3	13.0	yes
23:00	4 mph↗	69°F	-	-	10 miles	13	3	11.8	no

Thursday 2020-03-05: sunrise 06:43, sunset 18:35									
Time	Gusts	Temp	Precip Prob	Cloud Cover	Visibility	Visible Sats	Kp	Est. Sats Locked	Good To Fly?
00:00	4 mph→	68°F	-	-	10 miles	13	2	12.7	yes
01:00	4 mph→	68°F	-	-	10 miles	13	2	12.2	yes
02:00	4 mph→	67°F	1%	-	10 miles	15	2	13.0	yes
03:00	4 mph→	66°F	1%	-	10 miles	11	2	9.9	no
04:00	4 mph→	65°F	1%	2%	10 miles	16	2	13.7	yes
05:00	4 mph→	64°F	-	5%	10 miles	20	2	18.5	yes
06:00	4 mph↗	63°F	-	12%	10 miles	19	2	19.0	yes
07:00	4 mph↑	63°F	1%	32%	10 miles	19	2	19.0	yes
08:00	4 mph↖	65°F	2%	57%	10 miles	17	2	17.0	yes
09:00	5 mph←	67°F	3%	75%	10 miles	13	2	13.0	yes
10:00	6 mph←	69°F	4%	80%	10 miles	13	2	13.0	yes
11:00	8 mph←	71°F	6%	79%	9 miles	8	2	8.0	no
12:00	9 mph←	72°F	6%	79%	7 miles	11	2	11.0	no
13:00	9 mph←	71°F	5%	83%	8 miles	15	2	14.4	yes
14:00	8 mph←	68°F	4%	88%	10 miles	11	2	11.0	no

Fig. 10.9 Current conditions of the weather and the forecast of Center No. 162—Orizaba (18.78618, −97.03319)

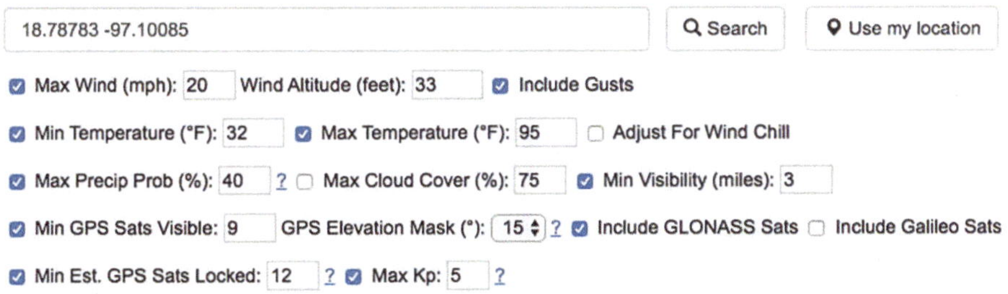

Fig. 10.10 Parameters of the forecast weather of Center No. 173—Orizaba (18.78783, −97.10085)

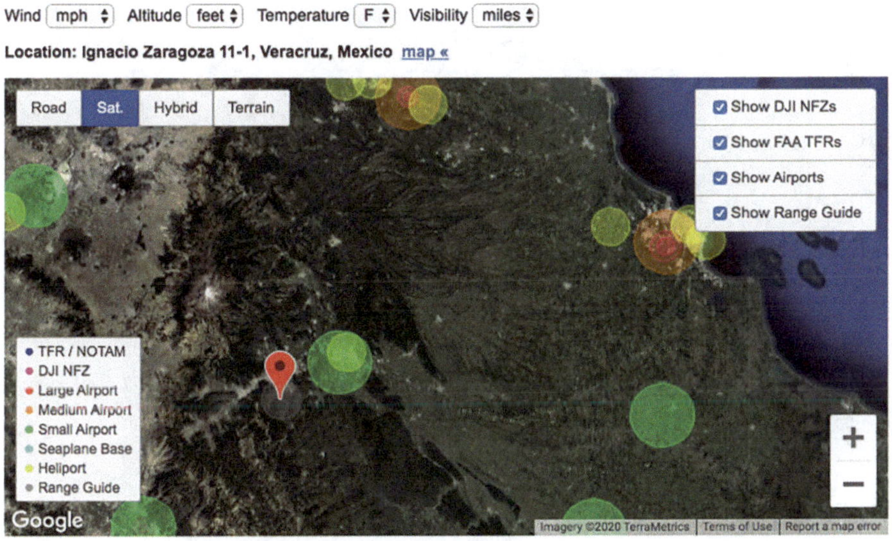

Fig. 10.11 Results of the forecast weather of Center No. 173—Orizaba (18.78783, −97.10085)

Figure 10.5 is shown the results; even within the simulations, it is possible to see the different range guide of the airports or heliports or other UAVs. Also, the current weather conditions and the forecast are presented, see Fig. 10.6. It is the same for the rest of the centers' simulation Figs. 10.7, 10.8, 10.9, 10.10, 10.11, 10.12, 10.13, 10.14 and 10.15.

Finally, bias can be introduced in several ways, including sampling a region or strategic center, a refined model UAS forecast using model intelligent, an mechanical design fit for SAR applications, an feed for mission planner, and a communication system integral that to ensure control and mission of the UAV. Therefore, this work is viable for effective mission/path planning, and validation targeted at ensuring a UAS-assisted SAR implementation.

Time	Gusts	Temp	Precip Prob	Cloud Cover	Visibility	Visible Sats	Kp	Est. Sats Locked	Good To Fly?
Current Conditions as of Wednesday 2020-03-04 14:48 CST									
14:48	7 mph←	88°F	-	-	10 miles	11	3	10.9	no
Wednesday 2020-03-04: sunrise 06:44, sunset 18:35									
15:00	7 mph←	87°F	-	-	10 miles	11	2	10.8	no
16:00	7 mph←	84°F	-	-	10 miles	11	2	11.0	no
17:00	7 mph←	82°F	-	-	10 miles	11	2	11.0	no
18:00	6 mph←	79°F	-	-	10 miles	12	2	12.0	yes
19:00	5 mph←	75°F	-	-	10 miles	12	2	11.9	no
20:00	4 mph→	71°F	-	-	10 miles	13	2	12.1	yes
21:00	4 mph→	68°F	1%	-	10 miles	13	3	11.5	no
22:00	4 mph↘	66°F	-	-	10 miles	15	3	13.0	yes
23:00	4 mph↗	65°F	-	-	10 miles	13	3	11.9	no
Thursday 2020-03-05: sunrise 06:43, sunset 18:35									
00:00	4 mph→	65°F	-	-	10 miles	13	2	12.7	yes
01:00	4 mph→	64°F	-	-	10 miles	13	2	12.2	yes
02:00	4 mph→	63°F	1%	-	10 miles	15	2	13.0	yes
03:00	4 mph→	62°F	1%	-	10 miles	11	2	9.9	no
04:00	4 mph→	62°F	1%	2%	10 miles	17	2	14.5	yes
05:00	4 mph→	61°F	-	6%	10 miles	20	2	18.5	yes
06:00	4 mph↘	60°F	-	14%	10 miles	19	2	19.0	yes
07:00	5 mph↙	61°F	1%	32%	10 miles	19	2	19.0	yes
08:00	5 mph←	63°F	2%	56%	10 miles	17	2	17.0	yes
09:00	6 mph←	65°F	2%	73%	10 miles	13	2	13.0	yes
10:00	7 mph←	68°F	4%	79%	10 miles	13	2	13.0	yes
11:00	8 mph←	70°F	5%	79%	10 miles	8	2	8.0	no
12:00	9 mph←	71°F	5%	80%	9 miles	11	2	11.0	no
13:00	9 mph←	69°F	5%	84%	9 miles	15	2	14.4	yes
14:00	9 mph←	67°F	4%	90%	10 miles	11	2	11.0	no

Fig. 10.12 Current conditions of the weather and the forecast of Center No. 173—Orizaba (18.78783, −97.10085)

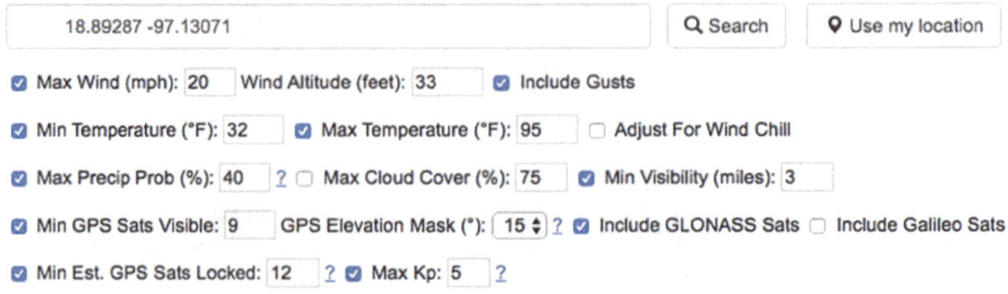

Fig. 10.13 Parameters of the forecast weather of Center No. 174—Unión y Progreso (18.89287, −97.13071)

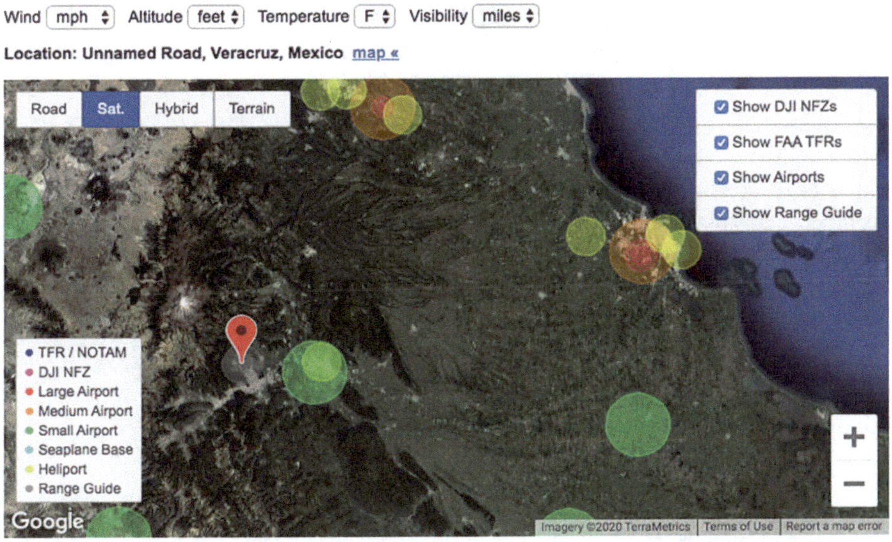

Fig. 10.14 Results of the forecast weather of Center No. 174—Unión y Progreso (18.89287, −97.13071)

Time	Gusts	Temp	Precip Prob	Cloud Cover	Visibility	Visible Sats	Kp	Est. Sats Locked	Good To Fly?
Current Conditions as of Wednesday 2020-03-04 15:32 CST									
15:32	8 mph↖	85°F	1%	-	10 miles	10	2	10.0	no
Wednesday 2020-03-04: sunrise 06:44, sunset 18:35									
16:00	8 mph↖	83°F	1%	-	10 miles	11	2	11.0	no
17:00	7 mph↖	81°F	1%	-	10 miles	11	2	11.0	no
18:00	6 mph↖	77°F	-	-	10 miles	12	2	12.0	yes
19:00	6 mph←	74°F	-	-	10 miles	12	2	11.9	no
20:00	5 mph↓	70°F	1%	-	10 miles	13	2	12.1	yes
21:00	4 mph↘	67°F	1%	-	10 miles	13	3	11.5	no
22:00	4 mph↘	65°F	-	-	10 miles	15	3	13.0	yes
23:00	4 mph→	64°F	-	-	10 miles	13	3	11.9	no
Thursday 2020-03-05: sunrise 06:44, sunset 18:35									
00:00	4 mph→	63°F	-	-	10 miles	13	2	12.7	yes
01:00	4 mph→	62°F	-	-	10 miles	13	2	12.2	yes
02:00	4 mph↘	62°F	1%	-	10 miles	15	2	13.0	yes
03:00	4 mph↘	62°F	1%	-	10 miles	11	2	9.9	no
04:00	4 mph↘	61°F	1%	7%	10 miles	17	2	14.6	yes
05:00	4 mph↘	60°F	1%	18%	10 miles	20	2	18.5	yes
06:00	4 mph↘	60°F	1%	28%	10 miles	19	2	19.0	yes
07:00	4 mph↓	61°F	1%	41%	10 miles	19	2	19.0	yes
08:00	5 mph←	64°F	1%	54%	10 miles	17	2	17.0	yes
09:00	6 mph←	67°F	2%	66%	10 miles	13	2	13.0	yes
10:00	7 mph↖	69°F	3%	75%	10 miles	13	2	13.0	yes
11:00	8 mph↖	70°F	5%	83%	10 miles	8	2	8.0	no
12:00	9 mph↖	71°F	7%	88%	9 miles	11	2	11.0	no
13:00	9 mph↖	70°F	6%	91%	9 miles	15	2	14.4	yes
14:00	9 mph↖	67°F	6%	93%	9 miles	11	2	11.0	no
15:00	8 mph↖	64°F	6%	94%	9 miles	10	2	10.0	no

Fig. 10.15 Current conditions of the weather and the forecast of the Center No. 174—Unión y Progreso (18.89287, −97.13071)

10.5.3 Legal Issues for the UAS-Assisted SAR Implementation

Nowadays, The General Directorate of Civil Aeronautics (DGAC) is an administrative unit dependent on the Ministry of Communications and Transportation, which is responsible for ensuring that air transport participates in the process of sustained and sustainable growth and of the approval of legal use of the UAV in Mexico. There is a big space for the growing of opportunities in a country where UAV applications are limited by regulations focused on UAV flight security for civilian and commercial type users, such as the restriction of an operator direct line-of-sight or a maximum teleoperation distance of 457 meters.

10.6 Conclusions

This research was carried out in the pre-disaster phase because it is analyzed a risk analysis to know if these communities affected by some phenomena can be serviced by a UAV from the strategic points previously selected. Critical operational and technical criteria were obtained from the flight performance specifications of the UAS, and external risk factors were incorporated in order to assess the risk for the mission success. In conclusion, the current commercial UAVs must be improved in their operational and technical characteristics for overcoming the SAR mission. Therefore, UAS-assisted is viable for effective SAR mission implementation. Finally, legal issues for the implementation of the proposed model in Mexico were discussed. The use of UAV for humanitarian applications in Mexico is not established yet. However, this research may be a start point to suggest some essential aspects of its application.

Acknowledgements The first author would like to thanks Erika Barojas-Payán Ph.D. and Santiago-Omar Caballero-Morales Ph.D. for all the support and growth of this research line.

References

Ahued-Ortega, A. (2018). Terremoto en México: la respuesta en salud del gobierno de la Ciudad de México. *Salud Publica Mex, 60*(supl 1), S83–S89. https://doi.org/10.21149/9327. (in Spanish).

Alotaibi, E. T., Alqefari, S. S., Koubaa, A. (2019). LSAR: Multi-UAV collaboration for search and rescue missions. *IEEE Access, 7*(2019), 55817–55832. https://doi.org/10.1109/access.2019.2912306.

American Red Cross - ARC (2019a). *Make a first aid kit*. Retrieved February 19, 2019, from https://www.redcross.org/get-help/how-to-prepare-for-emergencies/anatomy-of-a-first-aid-kit.html.

American Red Cross - ARC (2019b). *Deluxe personal first aid kit*. Retrieved February 19, 2019, from https://www.redcross.org/store/deluxe-personal-first-aid-kit/329164.html?utm_source=RCO&utm_medium=Referral&utm_term=Tweezers&utm_campaign=Anatomy_of_a_First_Aid_Kit.

Burke, C., McWhirter, P. R., Veitch-Michaelis, J., McAree, O., Pointon, H. A. G., Wich, S., & Longmore, S. (2019). Requirements and limitations of thermal drones for effective search and rescue in marine and coastal areas. *Drones, 3*, 78. https://doi.org/10.3390/drones3040078.

Bozorgi-Amiri, A., Jabalameli, M. S., Mohammad, A. M., Alinaghian, J., & Heydari, M. (2012). A modified particle swarm optimization for disaster relief logistics under uncertain environment. *International Journal of Advanced Manufacturing Technology, 60,* 357–371. https://doi.org/10.1007/s00170-011-3596-8.

Caballero-Morales, S. O., Barojas-Payan, E., Sanchez-Partida, D., & Martinez-Flores, J. L. (2018). Extended GRASP-capacitated K-means clustering algorithm to establish humanitarian support centers in large regions at risk in Mexico. *Journal of Optimization 2018*(3605298), 14. https://doi.org/10.1155/2018/3605298.

Chowdhury, S., Emelogu, A., Marufuzzaman, M., Nurre, S. G., & Bian, L. (2017). Drones for disaster response and relief operations: A continuous approximation model. *International Journal of Production Economics, 188,* 167–184.

De Oliveira, S. L., De Mello, B. R. A., & Gouvea, C. V. B. (2017). The use of UAV and geographic information systems for facility location in a post-disaster scenario. *Transportation Research Procedia, 27,* 1137–1145.

Everaerts, J. (2008). The use of unmanned aerial vehicles (UAVs) for remote sensing and mapping. *The International Archives of the Photogrammetry, Remote Sensing and Spatial Information Sciences, 37*(2008), 1187–1192.

Garnica, E. A., Flores, J. L., Benítez, D. P., & Morales, S. O. C. (2016). Localisation of first-aid centres in South-Eastern Mexico. *International Journal of Supply Chain and Operations Resilience, 2*(2), 166. https://doi.org/10.1504/IJSCOR.2016.078185.

Government of the Republic 2012–2018. *Fourth, fifth and sixth report of the Government of the Republic 2012–2018*. Retrieved February 20, 2019 from http://www.presidencia.gob.mx/informe/. (in Spanish).

Haidari, L. A., Brown, S. T., Ferguson, M., Bancroft, E., Spiker, M., Wilcox, A., et al. (2016). The economic and operational value of using drones to transport vaccines. *Vaccine, 34,* 4062–4067.

Jafari, N., Shahsanai, A., Memarzadeh, M., & Loghmani, A. (2011). Prevention of communicable diseases after disaster: A review. *Journal of Research in Medical Sciences: the Official Journal of Isfahan University of Medical Sciences, 16*(7), 956–962.

LifeStraw. *Lifestraw flex with gravity bag*. Retrieved February 26, 2019, from https://www.lifestraw.com/products/lifestraw-flex-multi-use-water-filter-with-gravity-bag.

Mexican Geological Service. (SGM). *Evolución de la tectónica en México*. Retrieved February 14, 2019, from https://www.sgm.gob.mx/Web/MuseoVirtual/Riesgos-geologicos/Evolucion-tectonica-Mexico.html. (in Spanish).

Ministry of Environment and Natural Resources. (SEMARNAT). Retrieved February 14, 2019, from https://www.gob.mx/semarnat/articulos/mexico-territorio-vulnerable-ante-huracanes. (in Spanish).

Molina, P., Parés, M. E., Colomina, I., & Vitoria, T. (2014). Drones to the rescue! unmanned aerial search missions based on thermal imaging and reliable navigation. White paper.

NASA. (2011). Risk management handbook. Washington, D.C. 20546. NASA/SP-2011-3422 Version 1.0 November 2011.

Natarajarathinam, M., Capar, I., & Narayanan, A. (2009). Managing supply chains in times of crisis: A review of literature and insights. *International Journal of Physical Distribution and Logistics Management, 39*(7), 535–573.

National Center for Disaster Prevention. (CENAPRED). *Declaratorias sobre emergencia, desastre y contingencia climatológica a nivel municipal entre 2000 y 2019*. Retrieved February 25, 2019, from https://datos.gob.mx/busca/dataset/declaratorias-sobre-emergencia-desastre-y-contingencia-climatologica. (in Spanish).

National Civil Protection System. (SINAPROC). *Emergency backpack and first aid kit*. Retrieved February 14, 2019, from https://www.gob.mx/profeco/documentos/mochila-de-emergencia-y-botiquin-de-primeros-auxilios. (in Spanish).
Natural Disasters Fund. (FONDEN). Retrieved February 19, 2019, from https://www.gob.mx/segob/documentos/fideicomiso-fondo-de-desastres-naturales-fonden. (in Spanish).
National Oceanic and Atmospheric Administration. (NOAA). *Extremely powerful hurricane katrina leaves a historic mark on the northern Gulf coast: a killer hurricane our country will never forget*. Retrieved February 19, 2019, from https://www.weather.gov/mob/katrina.
National Risk Atlas. (ANR). Retrieved February 19, 2019, from http://www.atlasnacionalderiesgos.gob.mx.
Páscoa, J. C., Dumas, A., & Trancossi, M. (2013). A review of thrust-vectoring in support of a V/STOL non-moving mechanical propulsion system. *Central European Journal of Engineering, 3*(3), 374–388. https://doi.org/10.2478/s13531-013-0114-9.
Półka, M., Ptak, S., & Kuziora, L. (2017). The use of UAV's for search and rescue operations. *Procedia Engineering 192*(2017), 748–752, ISSN 1877-7058. https://doi.org/10.1016/j.proeng.2017.06.129.
Restas, A. (2015). Drone applications for supporting disaster management. *World Journal of Engineering Technology, 3,* 316–321.
Rudol, P., & Doherty, P. (2008). Human Body detection and geolocalization for uav search and rescue missions using color and thermal imagery. *IEEE Aerospace Conference, Big Sky, MT, 2008,* 1–8. https://doi.org/10.1109/AERO.2008.4526559.
Saaty, T. L. (1994). How to make a decision: the analytic hierarchy process. University of Pittsburg, USA.
Silvagni, M., Tonoli, A., Zenerino, E., & Chiaberge, M. (2017). Multipurpose UAV for search and rescue operations in mountain avalanche events. *Geomatics, Natural Hazards and Risk, 8*(1), 18–33. https://doi.org/10.1080/19475705.2016.1238852.
Turki, E., Saleh, S., & Koubaa, A. (2019). LSAR: Multi-UAV collaboration for search and rescue missions. *IEEE Access 7*(2019), 55817–55832. https://doi.org/10.1109/access.2019.2912306.
Valenzuela, O., Báez, M., Chancey, E., Garzón, E., Quevedo, M., Sánchez-Partida, D., & Martínez-Flores, J. L. (2016). Rapid response center for disasters, inventory management in Southeastern Mexico. In *the Global Conference on Business & Finance Proceedings 11*(1), 73. Institute for Business & Finance Research.
Van Tilburg, C. (2017). First report of using portable unmanned aircraft systems (drones) for search and rescue. *Wilderness & Environmental Medicine, 28*(2), 116–118. https://doi.org/10.1016/j.wem.2016.12.010.
Vidyadharan, A., Philpott III, R., Kwasa, B. J., & Bloebaum, C. (2017). Analysis of autonomous unmanned aerial systems based on operational scenarios using value modeling. *Drones 1*(5), 1–17. https://doi.org/10.3390/drones1010005.
Villa, T. F., Gonzalez, F., Miljievic, B., Ristovski, Z. D., & Morawska, L. (2016). An overview of small unmanned aerial vehicles for air quality measurements: Present applications and future prospectives. *Sensors (Basel, Switzerland), 16*(7), 1072. https://doi.org/10.3390/s16071072.
Witczuk, J., Pagacz, S., Zmarz, A., & Cypel, M. (2018). Exploring the feasibility of unmanned aerial vehicles and thermal imaging for ungulate surveys in forests—preliminary results. *International Journal of Remote Sensing, 39*(15–16), 5504–5521. https://doi.org/10.1080/01431161.2017.1390621.

Chapter 11
Donation Management in Disaster Relief Operations: A Survey

Irais Mora-Ochomogo, Marco Serrato, Jaime Mora-Vargas, Raha Akhavan-Tabatabaei, and Isabel Serrato

Abstract Donations play a crucial part in disaster relief operations since many of Humanitarian Organizations operate exclusively with donated resources. Donations can be made in-cash or in-kind, and depending on the situation, both present some advantages and disadvantages to their management in times of crisis. In this chapter, we explore some of the characteristics of both types of donations, the impact each one has on the general operations, and some challenges that they could present on the Humanitarian Organizations. Finally, a survey of the current state of the art is performed. This research presents a total of 24 papers that discuss different areas concerning donation management and prediction. The papers included are classified mainly according to the type of donation, and research focus, the latter being either Donations' Analysis and Modeling or Donations Management and Material Convergence. The literature review clearly reflects the lack of research that has been made towards donations. Considering the relevance they have on real-life operations, it is motivated to continue this line of research.

Keywords Financial donations · In-kind donations · Disaster relief

11.1 Introduction

Natural disasters such as earthquakes, tornadoes, hurricanes, floods, among others, represent a latent threat for every country in the world. Due to climate change and other factors, statistics show that they continue on the rise not only in number but

---

I. Mora-Ochomogo (✉) · J. Mora-Vargas · I. Serrato
Department of Industrial Engineering, Tecnologico de Monterrey, Av Lago de Guadalupe KM 3.5, Margarita Maza de Juárez, 52926 Cd López Mateos, Mexico
e-mail: irais.mora@tec.mx

M. Serrato
The University of Chicago, 5801 S Ellis Ave, Chicago, IL 60637, USA

R. Akhavan-Tabatabaei
Sabanci School of Management, Orta Mahalle, 34956 Tuzla, İstanbul, Turkey

D. Sánchez-Partida et al. (eds.), *Disaster Risk Reduction in Mexico*,
https://doi.org/10.1007/978-3-030-67295-9_11

also in damage costs and the number of victims or people affected by the disasters. This situation presents a challenge for the communities and the humanitarian organizations to be better prepared and react faster to natural disasters.

Disasters have been given several definitions throughout the literature, including "a disruption that physically affects a system as a whole and threatens its priorities and goals" (Van Wassenhove, 2006). Moreover, "a non-routine event that exceeds the capacity of the affected area to respond to it in such a way as to save lives; to preserve property; and to maintain the social, ecological, economic, and political stability of the affected region" (Pearce, 2000).

On the other hand, humanitarian organizations such as the International Federation of the Red Cross and Red Crescent define a disaster as "A sudden, calamitous event that seriously disrupts the functioning of a community or society and causes human, material, and economic or environmental losses that exceed the community's or society's ability to cope using its resources. Though often caused by nature, disasters can have human origins" (IFRC, 2018). Relief Web International defines a disaster as "A serious disruption of the functioning of a community or a society causing widespread human, material, economic or environmental losses which exceed the ability of the affected community or society to cope using its resources" (Relief Web, 2008).

Four phases have been defined through the literature to address disasters: Mitigation and Preparedness take place before a disaster strikes and aim to reduce or eliminate the negative impact it could have on the communities. Once the disaster happens, the response and recovery phases take place.

The Response phase happens immediately after the disaster strikes. This phase includes actions taken to save lives and prevent further property damage in an emergency. The response is putting the preparedness plans into action. Some of the activities that take place in this phase include activating emergency plans, evacuation of areas at risk, search and rescue, medical attention, and emergency supplies start to arrive.

This stage can be subdivided into four phases, each requiring a different level of resources to be delivered to the affected population, as shown in Fig. 11.1 (Balcik et al., 2008).

In the activities of supplies delivery, most of the resources come from different sources of donations, either financial or in-kind. These sources can be the government, businesses, companies, organizations, and the general public. In-kind donations are received through collection centers that are enabled in different areas outside the affected community.

Different types of humanitarian organizations sustain their operations, mainly with contributions made by the public, and the organizations' goodwill and disaster relief operations are a perfect reflection of that. Besides the fact that some of these organizations could get government grants, donations have always been the backbone of disaster relief operations, since they directly determine the available resources to work with and fulfill their purpose; nevertheless, the research community often ignores the fact that donations are uncertain and very limited.

Understanding the behavior of the donations, especially after a disaster occurs, benefits the Non-Governmental Organizations (NGOs) to be better prepared and

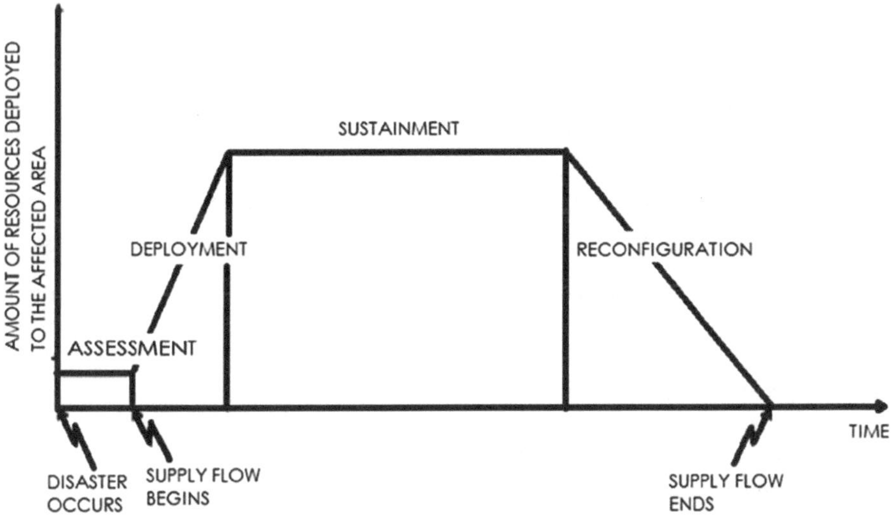

Fig. 11.1 Amount of resources deployed trough the response stage. *Source* Balcik and Beamon (2008)

make better decisions on the use and management of the available and expected resources.

The objective of this chapter is to present a general setting of the importance of donations' management and the challenges that each type represents the operations of humanitarian organizations. It is complemented with a survey of state of the art regarding this topic.

The remainder of this chapter is organized as follows, financial donations, along with their characteristics and challenges, are described in Sect. 11.2. In-kind or material donations are addressed in Sect. 11.3; in this section, collection centers are also described since they represent an essential factor to be considered with material donations management. Section 11.4 presents a literature review with the research that has been made to understand and model donations in disaster relief operations. Finally, Sect. 11.5 presents the conclusions of this research.

11.2 Financial Donations

According to the Giving USA report, Americans gave $427.71 billion to charity in 2018. This number includes different charities and non-profit organizations that are clustered in the following categories: religion, education, human services, foundations, health organizations, public society benefit organizations, arts, culture and humanities, international affairs, and environmental and animal organizations. Donations used for disaster relief are included in the clusters of human services, public-society, and foundations. The human services and public-society sectors combined represented 16.8% of the total donations in 2018, with almost 72 billion dollars (Giving USA, 2019).

This report on its 2018 edition also provides information on the source of all these donations. Individuals represent by far the largest source of donations in the USA, with an estimated $292.09 billion, followed by foundations with $75.86 billion. Even though the percentage of donations made by individuals is decreased compared to 2017, it is still on record values and reminds of the importance of the general population in these situations.

There have been studies on the determinant features that influence the donations. Okten and Weisbrod (2000) compare for-profit versus non-profit organizations. They present an analysis of whether the aspects that benefit the for-profit market, such as price and advertising, apply in the same manner to fundraising and non-profit organizations. The result of this research confirms that there is a different economic relationship towards the traditional market factors in the private sector and charitable giving.

Other research made by McKenzie (2011) includes analysis of the relationship between government grants and public donations to non-profits. It is established that they have a significant and inverse relationship following the occurrence of a natural disaster. Public donations increase by a factor more significant than government grants because government grants move inversely with public donations. In the case of industries, almost all saw a decrease in donations to non-profits following the occurrence of a natural disaster.

On the other hand, McKenzies' study showed that hurricanes generate a more considerable increase in donations to non-profits than other types of natural disasters, which may be related to media coverage, as the study revealed as well that the mean media coverage for hurricanes is higher than for other types of disasters.

Finally, in recent years, technology plays a more critical role in every aspect of our lives, and donations are not an exception. A research made by the Blackbaud Institute found that online giving has been a more common choice for donors in recent years and now represents around 8.5% of the total fundraising revenue (MacLaughlin, 2019). It is a result of the increasing popularity and development of broadband, dial-up, online banking, cell phones, smartphones, and social media.

In the same line, crowdfunding has become more popular. Crowdfunding is defined as the use of small amounts of capital from a large number of individuals to finance new business (Smith, 2019). In the case of disaster relief funding, crowdfunding works the same way; there are different platforms that have been used for these purposes, such as YouCaring, GoFundMe, and CrowdRise. These platforms have provided individuals with a personalized approach to fundraising using social media, as well as a way to quickly raise cash for disaster recovery with relatively little effort (Giving USA, 2018).

11.2.1 Challenges of Financial Donations

One of the main challenges when receiving financial aid is that the decision-makers in the organizations must make sure to use those resources efficiently, and they must report to their donors in what the money was spent. It puts humanitarian organizations on the spot and places additional pressure to optimize the use of resources for the benefit of the affected people but also to fulfill some of the donors' agenda, i.e., in the cases, the donors asked their donations to be used for a specific purpose. Using the money for administrative and another operational cost is often seen by donors as misused, leading the organizations to cut expenses in these areas even though they are necessary.

Money management can also derive a series of criticism by the media and the general public. One example is the case of the American Red Cross and the ProPublica (2018) series of articles criticizing different disaster operations and their expense reports.

11.2.2 How to Vet a Charity

On the other hand, after a disaster strikes, there is chaos and uncertainty; and donors want to bring their good intentions to action. However, there are many times that these situations are used to scam people with fake stories and even false organizations that are allegedly using the money to help the affected communities.

The online donations platforms and social media make it easier for these scams to reach more people all around the globe. To make sure donors are making accurate decisions with their money, the Wall Street Journal (Saunders, 2015) suggests that donors first research the charities in websites such as GuideStar, BBB Wise Giving Alliance, Charity Navigator, among others.

11.3 In-Kind Donations and Collection Centers

In many countries, the population's primary response to disasters is to give material or in-kind donations, and they end up constituting a sizeable portion of the available resources for the NGOs (Sewordor et al., 2019).

These donations are especially useful when the area affected by a disaster has completely collapsed, and there is the necessity to provide all the basic needs from outside. Meanwhile, for the humanitarian organizations and NGOs, this represents different challenges in terms of economy and logistics.

11.3.1 Challenges of Material Donations

Managing in-kind donations presents a series of challenges from different angles, like the economical and the logistical.

In the economic area, there is the collecting, packing, and transporting costs that the humanitarian organizations must absorb to carry out their activities. An essential factor to consider is that the donations could be made anywhere around the world, and thus the farther the donations are made, the more expensive it becomes to transport them. Also, when we consider the handling of a large number of items, renting an appropriate space and the corresponding transport must be considered among the expenses. Finally, since the organizations receive products, these administrative and logistical costs are not covered.

When talking about the logistical challenges, material donations imply that collection centers must be addressed. The Pan-American Health Organization (2001) classifies warehouses in four types: General Warehouse Offices, where products are stored for an extended period waiting to be sent to the disaster zone or a secondary warehouse. Slow Rotation Warehouses, where products of non-urgent distribution, in reserve or not frequent consumption are stored, such as spare parts, specialized equipment, or tools. Fast Rotation Warehouses, which are those that manage a daily or frequent shipment of products. They are the most common type of warehouses in the operations field and usually store the products of prompt distribution for the affected population. At last, Collection Centers are defined as sites-enabled for the reception of goods (patios, offices, and so forth). They should seek prompt shipment of the collected products to another warehouse. They can be used for separating and sorting donations, only sending materials that are considered useful and already categorized to the warehouses.

Meanwhile, the Mexican National Civil Protection System (2010) defines Collection Centers as reception locations enabled by public, private, and social authorities and organizations to receive aid supplies for people that have been affected by a disaster. Most of the time, these facilities are located outside the area that has been affected. The Mexican Red Cross is the NGO with more presence in the country; they work with around 80% of in-kind donations that are received through these facilities (Oxenhaut, 2015). In the USA, the percentage of in-kind donations is approximately 42%, and it tends to decrease with the distance of the donors (Destro & Holguín-Veras, 2011).

Several humanitarian organizations like the International Federation of the Red Cross, the Scouts Association, Doctors without Borders, and many others work with different predefined aid kits that were designed to satisfy a family's needs for a certain amount of time. It means that specific in-kind donations are required so they can complete the kits to send them to the affected area.

The Mexican Red Cross works with three principal types of aid kits. The most common is the food kit; it has products to eat for a family of five members for one week. There is also the personal hygiene kit, which contains products that are estimated to last one month for a family of five members as well. Moreover, the household cleaning kit, that is given only once to each family in the affected area.

As an example of the operations that take place in a collection center, we present the Mexican Red Cross' seven operational areas:

- Reception and unload: This area receives aid supplies. These donations can come from corporations, institutions, individuals, or local Red Cross facilities.
- Selection: This area oversees the classification of supplies that fulfill requirements such as expiration date and size, among others.
- Classification: In this area, products for the different kinds of kits are separated for packing.
- Packing: The different types of kits are assembled with their predetermined products.
- Label and documentation: Once the kits are assembled and packed, it is essential to label and fill the corresponding documentation before they can exit the facility.
- Load and exit: The shipment is loaded in the corresponding mean of transportation and sent to the affected area.
- Waste management: This area is in charge of giving proper disposal to a different kind of waste resulting from operations of the Collection Center.

Mexican National Civil Protection System (2010) identifies three critical considerations that become challenges in Collection Center's operations:

- Lack of knowledge of the type of products that will be received
- All the donations will be mixed
- The reception timing varies.

These considerations directly contribute to the high level of uncertainty in the supply of donations. Therefore, it makes it difficult for organizations to plan appropriately or to make more accurate decisions since not all the information is available.

Another logistical inconvenience when we talk about international in-kind donations is that they have to go through customs processes upon arrival in the affected country. Stablished humanitarian organizations, when the corresponding emergency status is issued, can introduce their donations to the country more easily. On the other hand, there have been cases when donations are retained in customs for long periods or even indefinitely.

Finally, material convergence is a crucial concern with in-kind donations since it includes the supplies and equipment sent by all of the entities that respond to a disaster, including governments, relief agencies, companies, churches, local community groups, and individuals (Holguín-Veras et al., 2012). Also, this becomes a significant problem when the donations are not useful, many of them do not meet the culture's customs or preferences, are not suitable for consumption (like in the case of food or medicines) or are merely non-priority items at the time and are consuming logistic resources of the organizations. It is estimated that between 50 and 70% of the in-kind donations received are non-priority (Jaller & Holguín-Veras, 2011).

One of the main reasons that this happens is because donors do not know the real needs, and large companies see an opportunity to position their brand or reduce inventories (Suárez-Moreno et al., 2016), some people use these disasters to get rid of what they do not need and do not think of what is really needed in the affected area. Therefore, educating the media, leaders, and the general population can help reduce this problem (Edwards, 2009).

For this, the Pan-American Health Organization, with the sponsorship of UNICEF, OCHA, World Food Program, Oxfam, IFRC, CRID, and Spain's government, issued the campaign "Saber Donar" (2008) were the key considerations that donors and organizations must consider when donating items were explained.

Meanwhile, the Center for International Disasters (2017) states that in-kind donations are beneficial when they meet the following criteria:

- Items are requested explicitly by a charitable or local organization working on-site.
- Items are available in enough quantity to serve affected populations.
- Items are easily integrated into existing relief and development programs.
- Items are monitored for effectiveness.
- Transportation costs are paid for by the donor.
- The recipient organization has a distribution plan.

11.3.2 Why Do People Make in-Kind Donations?

Even considering the several challenges previously mentioned, a significant number of people still prefer this type of donations. One of the reasons is that the donors want to make concrete donations and know what they are giving (Fessler, 2013). With this, there is no uncertainty in what the cash will be spent.

Another of the main reasons is that donors have a certain level of distrust in the organizations and do not believe monetary donations will get to the final donor (Pérez Romero, 2004). The fear of corruption and mismanagement of funds causes donors to want to deliver in-kind donations to the humanitarian organizations or even directly to the beneficiaries.

11.4 Literature Review

This section provides a literature review on papers that address the donations' uncertain behavior, i.e., not to know how much, what or when will be received, and study their effects on humanitarian operations. Even though donations are the leading supplier when it comes to disaster relief operations, little research has been done to understand all their factors and implications.

For this literature review, exclusively academic papers were included, and to do the research, five databases were selected as search engines; EBSCO Academic Search Premier, ProQuest Science Journals, ProQuest ABI/INFORM, and Emerald Insight.

The keywords used were "donations & disaster relief," explored in the title, keywords, and abstract. The papers considered have been published between 2000 and 2019. With these criteria, a total of 24 papers were included in this review.

First, the papers are categorized according to the published date and the type of donations they study. After that, they are classified according to the focus of the study in two sections. The first focus analyzes the various factors that determine the quantity, timing, and nature of donations; the second focus is on different challenges presented in donations' management, including the nature and adverse effects of material convergence in disaster relief operations.

11.4.1 General Information

This section summarizes general information of the papers, such as the year of publication and the type of donations they address.

As presented in Fig. 11.2, the publications per year vary significantly since the year 2000, even though in recent years this topic has been studied more, the maximum number of publications rises to only four. Therefore, there is an essential opportunity for research on this topic.

Figure 11.3 shows that the type of donations studied relies importantly on the financial type. As stated before, organizations aim to have more financial aid for

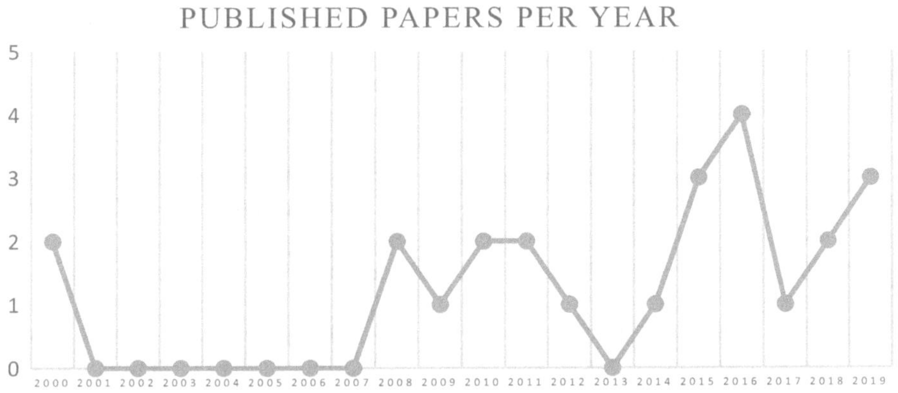

Fig. 11.2 Number of papers regarding donations per year

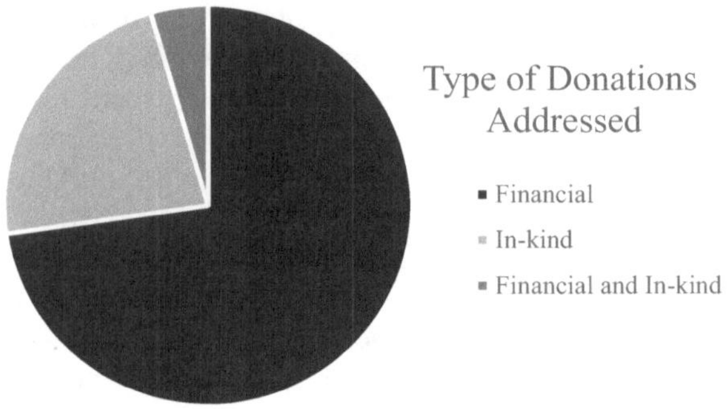

Fig. 11.3 Type of donation addressed in the research

their operations, but in many countries, this is still not the case, so it becomes very relevant also to study the nature and behavior of the in-kind donations.

11.4.2 Research Focus

The classification of the papers by research focus aims to identify what type of studies have been made to understand and better manage the received donations at the NGOs.

The first focus in the considered papers is the analysis and modeling of donations. In this classification, the papers attempt to understand the donations' behavior, which factors influence the donors to give and which are less critical; this significantly improves the quality of modeling and predicting donations for a particular disaster.

The second focus of research considered in this classification is the management of donations and material convergence. In this focus, the papers study different phenomena that happen when dealing with donations, especially material convergence that represents a significant challenge when managing material donations.

11.4.2.1 Donations' Analysis and Modeling

1. Bennett and Kottasz (2000) present a regression study that looks for the primary factor in encouraging donations. It presents that the significant triggers are influenced by the media presenting the unfortunate situation of the affected people, appealing to the emotional content. In contrast, some of the factors discouraging donations are also the media presenting unfair or inefficient relief operations.
2. Cheung and Chan (2000) developed a model that analyzes self-efficacy, outcome efficacy, moral obligation, need, and attribution as the determinants to donate to International Relief Organizations (IROs). They interviewed 277

subjects in Hong Kong, and the results revealed that the subjects were not strongly supportive of donating to charity, especially to IROs. On the contrary to what could be common belief, the study revealed as well that income, age, education, class, and gender exerted little effect on the intention to donate. On the other hand, social cognitive factors that proved to be crucial predictors of intention to donate were self-efficacy, trust in the charity organization, past donation, and expectations about the organization's use of donations, which led to conclude that IRO fund-raising programs can benefit from the consideration of social–cognitive factors that affect public donation to charity.

3. Brown and Minty (2008) analyzes via least squared fitting the impact of media coverage o n he specific case of the 2004 tsunami. They found that an additional minute or major newspaper article can raise the donations by 17–21%.
4. Chandler et al. (2008) question whether the name given to a storm affects the donors' contributions. They part from the fact that people tend to have a preference for the initial of their name. The study is made through seven significant hurricanes, and the results present that people seek to ameliorate the adverse effects of a disaster when they have shared characteristics.
5. Oosterhof et al. (2009) built a factorial experiment to research attitudes and the impact of media exposure. This research found out, as expected that the most significant predictor of the intention to donate is the past donations to disaster relief campaigns.
6. J. Kim et al. (2010) contrast through a multivariate analysis the willingness of an individual to pay more taxes so the government can sponsor relief operations and the willingness to make more voluntary donations for disaster recovery purposes. Their findings show that the opinion of the government is variable, so they rely more on non-profit organizations for disaster relief efforts.
7. Muller and Kräussl (2011) turns to corporate donations and use an OLS regression to identify how the reasons for donating affect the perception of the employees on corporate donations. The empirical investigation analyzed abnormal returns to announcements by U.S. Fortune 500 firms that documented their donations to Hurricane Katrina disaster relief in 2005. They proposed that announcements emphasizing employee involvement in the donation would send investors positive signals about the firm's ability to bounce back from the disaster's adverse effects. Nevertheless, their results indicated that the value of corporate philanthropy is contingent upon contextual factors and that significant value may be preserved by demonstrating employee involvement in corporate philanthropy.
8. Nogami (2014) analyzes the case of the Great East Japan Disaster and evaluates the causes that determine whether a person decides to donate or not. Factors such as the effectiveness of the monetary donation, the effectiveness of the disaster relief activities, and preparedness for disasters were considered. Around 300 Japanese subjects were interviewed, and a series of comparative analyses were conducted to examine differences between donors and non-donors. The study found that, although non-donors evaluated the effectiveness of the monetary donation more positively than the charitable donation,

donors evaluated the effectiveness of all the disaster relief activities more positively than non-donors. Moreover, donors were more prepared for disasters and more insightful into the current situation of disaster victims than non-donors.

9. Korolov et al. (2015) focus their study on the impact that social media has on donations. The level of chatter collected on Twitter about Hurricane Sandy was studied through an Erdos–Renyi Network Model. The study showed that the degree sequence of the social media "follower" network plays a crucial role in determining the scaling exponent. For random graphs and power-law graphs, the scaling exponent is at or near 2 (quadratic amplification). Understanding the scaling behavior that relates to social-media chatter to real physical actions is an essential step for estimating the extent of response and for determining social-media strategies to affect the response.
10. Ogawa and Ida (2015) explains donation behavior through dictator game experiments. The delay when the recipient obtains the donation and the individual characteristics of the donor are examined. The results revealed that donations decrease as the time delay rises and that gender, educational level, neuroticism, agreeableness, conscientiousness, and utilitarianism affect donations independent of the time delay.
11. Ülkü et al. (2015) address financial and in-kind donations and analyzes how the minimum amount of cash donation that is solicited impacts the donor's likelihood to donate cash instead of in-kind items. With this, they provide an optimal expression that maximizes this likelihood.
12. Korolov et al. (2016) study two donation scenarios, response to natural disasters, and regular donations. In these scenarios, the chatter on social media and the charitable donations are analyzed. Results show that in the emergency response case, there is a quadratic relation between those two variables, while in the regular donations, a near-linear relation is observed.
13. Y. Kim et al. (2016) research 31 US-based NGOs with and without operations in the four Asian countries most affected by the 2004 tsunami to compare their changes in revenues, explicitly studying the effect of going where there is the most considerable media attention on their private donations. The study principally found out weak indications that private donors systematically and strongly preferred NGOs with operations in the region.
14. Wei and Marinova (2016) focus on the global response to five significant earthquakes. They analyze the influence of gift-giving, geographical location, political regime, and trade openness on disaster donation decisions. The results show that the significant factors that influence the donations are the severity of the disaster and where the disaster is located influences this decision making the most.
15. Becchetti et al. (2017) test the impact of the tsunami shock on the generosity of a sample of Sri Lankan affected/unaffected microfinance borrowers, seven years after the event. The amount of damage experienced and the recovery aid received were the key factors to compare the participants. The results confirm

that the shock has effects in the long run since the tsunami negatively affects the generosity of those who suffered damage.

16. Eckel et al. (2018) compare the likelihood of donations depending on the operation level of the charities: national, state, local, or an individual affected by a fire. The experiment shows that the subjects prefer to donate to national or local organizations, but overall, they would choose to gift directly to the individual.
17. Boulianne et al. (2018) analyze the effects of caring and helping behaviors, along with the effects of social media on the response to the fire in McMurray, Alberta. The results show that the people that followed the events of the fire in social media would care more and led to help the affected people.
18. Manesi et al. (2019) explore the variables that predict disaster donations more strongly. They consider prosocial traits (social value orientation and social mindfulness), socio-demographic variables, and minimal social cues. The results show that social value orientation and social mindfulness were the most prominent predictors and were related to the donated amount.
19. Hickey et al. (2019) research the role of timing and salience of tax incentives on reported tax filer giving. This research finds that moving the timing of reporting of gifts on one's tax returns closer to the timing of giving increases average donations by approximately nine percentage points. Furthermore, the implications of these results are analyzed.

11.4.2.2 Donations Management and Material Convergence

1. Stapleton et al. (2010) present the advantages and disadvantages of different types of corporate donations to humanitarian response efforts. They mainly address the issue of matching corporate donations to humanitarian organizations' needs in response to disasters. In terms of cash donations, they found that in the inherent uncertainty context with information asymmetry, the appeals and donations process presents significant challenges to both the donor and the recipient organizations.
2. Destro and Holguín-Veras (2011) develop a model that explains material convergence in terms of socioeconomic characteristics using a database of donations from Hurricane Katrina.
3. Holguín-Veras et al. (2012) describe lessons learned and identify problems created by material convergence. They propose policies regarding material convergence management and control strategies.
4. Suárez-Moreno (2016) built a computational theoretical model to analyze and solve material convergence. Flows, actors, and their relationships are analyzed and measured in terms of their impact on the logistics performance of the system regarding speed and coverage. The author finally proposes a scenario that facilitates coordination between actors where logistics performance indicators show better results regarding response speed and coverage.

5. Sewordor et al. (2019) focus on the specific case of diasporas and diaspora NGOs. This research presents the different challenges these organizations face based on the case of the Haitian earthquake, which includes an overreliance on diaspora donors, competition among NGOs, and what is perceived as unfair funding practices towards diaspora NGOs.

11.5 Concluding Remarks

In this chapter, the importance of donations in disaster relief operations was highlighted. Humanitarian organizations all around the globe mostly rely on private donations for their operations. Therefore, their resources come from diverse and very uncertain sources.

Donations usually come in the form of financial or material aid. The former allows organizations to adjust their resources to the specific needs of the affected people and address fixed costs, such as staff, needed to sustain their operations. However, they can represent a challenge since there is a commitment not only with the affected people but with the donors to use those resources most efficiently, and using them for administrative costs is often seen as misused.

In many countries around the world; for different reasons, the majority of the donations are made as in-kind products. It leads the organizations to build an entire supply chain to take these products to the area affected by the disaster. All this logistical effort can be much more expensive than other alternatives that could have been held with financial donations. Another issue with receiving products is that the type of product is not always what is needed or wanted, and in the end, they end up taking valuable logistical resources of required items.

Finally, a literature review is presented with the papers that research the behavior and management of donations. There is little research about this topic despite its relevance in the humanitarian world; only 24 papers met the criteria presented in this review. These papers are classified according to the type of donations that they study; the majority addressed financial donations. The focus of their research also classified them, the first and largest category is focused on understanding the factors that influence donations, different authors analyze aspects like geographical location, socio-economic situation to media coverage or social media exposure.

The second category is focused on donations management and material convergence. As stated before, having an overflow of donations can become a logistical problem, and it is essential to understand and address this issue.

For future research, it is crucial to continue to understand and analyze the different implications that raising and managing donations for disaster relief have. It will lead to better decision making by the organizations, and in the end, more people will benefit from this.

References

Balcik, B., Beamon, B. (2008). Facility location in humanitarian relief. *International Journal of Logistics Research and Applications, 11*(2), 101–121.

Balcik, B., Beamon, B. M., & Smilowitz, K. (2008). Last mile distribution in humanitarian relief. *Journal of Intelligent Transportation Systems, 12*(2), 51–63.

Becchetti, L., Castriota, S., & Conzo, P. (2017). Disaster, aid, and preferences: The long-run impact of the tsunami on giving in Sri Lanka. *World Development, 94,* 157–173.

Bennett, R., & Kottasz, R. (2000). Emergency fund-raising for disaster relief. *Disaster Prevention and Management: An International Journal, 9*(5), 352–360.

Boulianne, S., Minaker, J., & Haney, T. J. (2018). Does compassion go viral? social media, caring, and the Fort McMurray wildfire. *Information, Communication & Society, 21*(5), 697–711.

Brown, P. H., & Minty, J. H. (2008). Media coverage and charitable giving after the 2004 tsunami. *Southern Economic Journal*, 9–25.

Chandler, J., Griffin, T. M., & Sorensen, N. (2008). In the "I" of the storm: Shared initials increase disaster donations. *Judgment and Decision Making, 3*(5), 404.

Cheung, C. -K., & Chan, C. -M. (2000). Social-cognitive factors of donating money to charity, with special attention to an international relief organization. *Evaluation and Program Planning, 23*(2), 241–253.

Destro, L., & Holguín-Veras, J. (2011). Material convergence and its determinants: case of hurricane Katrina. *Transportation Research Record, 2234*(1), 14–21.

Eckel, C., Priday, B., & Wilson, R. (2018). Charity begins at home: A lab-in-the-field experiment on charitable giving. *Games, 9*(4), 95.

Edwards, F. (2009). The Challenges of Donations Management. *The Public Manager, 38*(3), 72–75.

Fessler, P. (2013). Thanks, But No Thanks: When Post-Disaster Donations Overwhelm. N. P. Radio. https://www.npr.org/2013/01/09/168946170/thanks-but-no-thanks-when-post-disaster-donations-overwhelm.

Giving USA. (2018). The annual report on philanthropy for the year 2017. Available at https://lclsonline.org/wp-content/uploads/2018/12/Giving-USA-2018-Annual-Report.pdf.

Giving USA. (2019). The annual report on philanthropy for the year 2018. Available at https://givingusa.org/giving-usa-2019-americans-gave-427-71-billion-to-charity-in-2018-amid-complex-year-for-charitable-giving/.

Hickey, R. D., Minaker, B., & Payne, A. A. (2019). The sensitivity of charitable giving to the timing and salience of tax credits.

Holguín-Veras, J., Jaller, M., Van Wassenhove, L. N., Pérez, N., & Wachtendorf, T. (2012). Material convergence: Important and understudied disaster phenomenon. *Natural Hazards Review, 15*(1), 1–12.

IFRC (2018). What is a disaster?. http://www.ifrc.org/en/what-we-do/disaster-management/about-disasters/what-is-a-disaster/. Accessed 2018.

Information, C. f. I. D. (2017). In-kind can be unkind. United States Agency International Development. https://www.cidi.org/how-disaster-relief-works/in-kind-can-be-unkind/#.W47L1uhKjIU. Accessed 2018.

Jaller, M., & Holguín-Veras, J. (2011). Locating points of distribution in disasters with social costs considerations. Submitted to the Transportation Research Board (TRB) 90th Annual Meeting, vol. 15 Washington, DC Acessado em. p. 2012, 07.

Kim, J., Oh, S. S., & Jung, T. (2010). Funding for disaster recovery: Increased taxes or charitable donations to nonprofits? *International Journal of Public Administration, 33*(3), 151–159.

Kim, Y., Nunnenkamp, P., & Bagchi, C. (2016). The Indian Ocean tsunami and private donations to NGOs. *Disasters, 40*(4), 591–620.

Korolov, R., Peabody, J., Lavoie, A., Das, S., Magdon-Ismail, M., & Wallace, W. (2015). Actions are louder than words in social media. In *Proceedings of the 2015 IEEE/ACM International Conference on Advances in Social Networks Analysis and Mining 2015* (pp. 292–297). ACM.

Korolov, R., Peabody, J., Lavoie, A., Das, S., Magdon-Ismail, M., & Wallace, W. (2016). Predicting charitable donations using social media. *Social Network Analysis and Mining, 6*(1), 31.

MacLaughlin, S. (2019) The end of the beginning of online giving. npENGAGE. Available at https://npengage.com/nonprofit-fundraising/the-end-of-the-beginning-of-online-giving/.

Manesi, Z., Van Lange, P. A., Van Doesum, N. J., & Pollet, T. V. (2019). What are the most powerful predictors of charitable giving to victims of typhoon Haiyan: Prosocial traits, socio-demographic variables, or eye cues? *Personality and Individual Differences, 146*, 217–225.

McKenzie, M. (2011). The effects of natural disasters on donations to non-profits.

Muller, A., & Kräussl, R. (2011). The value of corporate philanthropy during times of crisis: The sense giving effect of employee involvement. *Journal of Business Ethics, 103*(2), 203.

National Civil Protection System. (2010). *Guía Para La Recepción, Organización Distribución Y Operación De Suministros Humanitarios Para La Asistencia De Poblaciones Afectadas Por Un Desastre*.

Nogami, T. (2014). What makes disaster donors different from non-donors. *Disaster Prevention and Management, 23*(4), 484–492.

Ogawa, K., & Ida, T. (2015). Investigating donating behavior using hypothetical dictator game experiments. *Review of Social Economy, 73*(2), 176–195.

Okten, C., & Weisbrod, B. A. (2000). Determinants of donations in private nonprofit markets. *Journal of Public Economics, 75*(2), 255–272.

Oosterhof, L., Heuvelman, A., & Peters, O. (2009). Donation to disaster relief campaigns: Underlying social cognitive factors exposed. *Evaluation and Program Planning, 32*(2), 148–157.

Oxenhaut, I. (2015). *National Director of the Disasters Division*. Mexican Red Cross/Interviewer: I. Mora-Ochomogo.

Pan-American Health Organization. (2001). *Logística y gestión de suministros humanitarios en el sector salud*. Pan American Health Org.

Pan-American Health Organization. (2008). *Saber donar*. Available at https://www.paho.org/uru/index.php?option=com_docman&view=download&category_slug=comunicacion-de-riesgo&alias=283-guia-saber-donar&Itemid=307.

Pearce, L. D. R. (2000). *An integrated approach for community hazard, impact, risk and vulnerability analysis: HIRV*. University of British Columbia.

Pérez Romero, L. A. (2004). *Marketing social: teoría y práctica*. Pearson Education.

ProPublica. (2018). Reporting on the Red Cross. Available at https://www.propublica.org/series/red-cross. Accessed June 2019.

Relief Web. (2008). Glossary of Humanitarian Terms. http://www.who.int/hac/about/reliefweb-aug2008.pdf. Accessed 2016.

Saunders, L. (2015). How to vet a charity. *The Wall Street Journal*. Retrieved from https://www.wsj.com/articles/how-to-vet-a-charity-1431093430.

Sewordor, E., Esnard, A. M., Sapat, A., & Schwartz, L. (2019). Challenges to mobilising resources for disaster recovery and reconstruction: Perspectives of the Haitian diaspora. *Disasters, 43*(2), 336–354.

Smith, T. (2019). Crowdfunding. https://www.investopedia.com/terms/c/crowdfunding.asp.

Stapleton, O., Van Wassenhove, L. N., & Tomasini, R. (2010). The challenges of matching corporate donations to humanitarian needs and the role of brokers. *Supply Chain Forum: An International Journal, 11*(3), 42–53. Taylor & Francis.

Suárez-Moreno, J. D., Osorio-Ramírez, C., & Adarme-Jaimes, W. (2016). Agent-based model for material convergence in humanitarian logistics. *Revista Facultad de Ingeniería Universidad de Antioquia, 81*, 24–34.

Van Wassenhove, L. N. (2006). Blackett memorial lecture—humanitarian aid logistics: Supply chain management in high gear. *Journal of the Operational Research Society, 57*(5), 475–489. https://doi.org/10.1057/palgrave.jors.2602125.

Wei, J., & Marinova, D. (2016). The orientation of disaster donations: Differences in the global response to five major earthquakes. *Disasters, 40*(3), 452–475.

Ülkü, M. A., Bell, K. M., & Wilson, S. G. (2015). Modeling the impact of donor behavior on humanitarian aid operations. *Annals of Operations Research, 230*(1), 153–168.

Chapter 12
Reliable Network Design: Case Study

Fabiola Regis-Hernández, Jaime Mora-Vargas, Angel Ruíz, and Diana Sánchez-Partida

Abstract During the last 20 years the climate-related disasters have dominated the picture accounting for 91% of all 7255 recorded events, being the floods the most frequent type of disaster. Several decisions, such as the allocation of shelters and relief distribution, are made to minimize the aftermath impact on the population. In this chapter, we present a model to develop a reliable network based on a hierarchical preferences multi-criteria framework. This work aims to minimize the distance between the affected populations and the available shelters as well as their exposition to the risk due to damaged routes integrating the several stakeholders' preferences. The proposed solution is tested with the hurricane Stan case, which impacted the Mexican Republic southeast in 2005, affecting Quintana Roo, Yucatan, Oaxaca, and Veracruz. The case study is based solely on the state of Veracruz's situation, which considers 27 available distribution centers, 109 affected populations, and 1,379 available shelters. The problem is solved in GAMS commercial software, and the results showed that a reduction of non-used capacity of the opened temporary shelters up to 90.33% could be obtained when the integration of stakeholders' preferences and adequate decision-making tool.

F. Regis-Hernández (✉)
Tecnologico de Monterrey, School of Engineering and Science, Av. Eugenio Garza Sada 300, Lomas del Tecnologico, 78211 San Luis Potosi, México
e-mail: fregisher@tec.mx

J. Mora-Vargas
Tecnologico de Monterrey, School of Engineering and Science, km 3.5 Carr. Lago de Guadalupe, Atizapan de Zaragoza, 52926 Estado de Mexico, México

A. Ruíz
Centre interuniversitaire de recherche sur les réseaux d'enterprise, la logistique et le transport (CIRRELT), Faculty of Business Administration, Université Laval, Quebec (Canada), 2325 Rue de l'Université, Québec G1V 0A6, Canada
e-mail: Angel.Ruiz@osd.ulaval.ca

D. Sánchez-Partida
Department of Logistics and Supply Chain Management, Universidad Popular Autónoma del Estado de Puebla (UPAEP University), 17 Sur 901, Barrio de Santiago, 72410 Puebla, Mexico

D. Sánchez-Partida et al. (eds.), *Disaster Risk Reduction in Mexico*,
https://doi.org/10.1007/978-3-030-67295-9_12

Keywords Allocation · Multi-criteria · Relief distribution · Optimization · Risk

12.1 Introduction

Several entities have developed disaster databases; one of them is the EM-DATA (since 1988), provided by the Centre for Research on the Epidemiology of Disasters (CRED). This database contains core data on the occurrence and effects of disasters from 1900 to the present. According to the EM-DATA database, a total of USD 656 billion losses, 2 billion affected people, and 142,088 fatalities have been reported since 1997 (EM-DATA, 2019). Also, Mexico is pointed as one of the countries with more hydrometeorological disasters occurrences per year, since its vulnerability of being struck from the Atlantic and Pacific Oceans sides. Only in Mexico, the hydrometeorological disasters represent 70.83% of the total natural disasters, resulting in 68.41% of total losses in infrastructure, and 65.76% of the total affected population (EM-DATA, 2019). One of the worst economic losses in Mexico caused by hydrometeorological disasters was in 2007, when the hurricanes Emily, Stan, and Wilma struck the country, resulting in the USD 7.91 billion in total damages. The Mexican southeast was the most affected, including Quintana Roo, Chiapas, Yucatan, Oaxaca, and Veracruz. This hurricane severely impacted the Sierra Region of the state of Chiapas, causing one of the most massive internal displacements of people, some of them suffering relocation to nearby places and also improvised actions by public agencies and limited implementations of international responds protocols.

Thomas and Kopczak (2005) estimates that over the next 50 years, natural disasters will increase by five times in number and severity. These overwhelming statistics show the need for the development of strategies to reduce the impact of natural disasters. As a result, researchers and practitioners have increased their interest in Humanitarian Logistics (HL), which, according to Natarajarathinam et al. (2009) is defined as:

> the process of planning, implementing and controlling the efficient, cost-effective flow and storage of goods and materials, as well as related information, from the point of origin to the point of consumption to alleviate the suffering of vulnerable people.

According to Altay and Green (2006), HL implies lifecycle operations, including mitigation, preparedness, response, and recovery. Thus, the ability to conduct efficient and effective humanitarian operations is a critical element of a disaster relief process (Leiras et al., 2014). According to Altay and Green (2006), the rapid response in the aftermath of a disaster "*depends on the ability of logisticians to procure, transport, and receive supplies at the site of the humanitarian relief effort.*"

In this chapter, we present a model to develop a reliable network based on a hierarchical preferences multi-criteria framework. The model aims to determine the assignment of affected people to temporary shelters considering the several stakeholders' preferences. The model is focused on attending the affected people in the aftermath of hydrometeorological disasters such as cyclones, hurricanes, storms, among others, considering the risk to which the evacuated people are exposed. In this case, the risk refers to the probability that an entity has to be damaged, e.g., the probability a landslide occurs while displaced people are walking through that site.

As noticed, the concepts "risk" and "reliability" are recurrent in this document; thus, we deepen on their definitions.

Risk: is often defined as the expected damage or loss of an active agent, resulting from the interaction between the vulnerability and the presence of a disturbing agent. The risk is usually represented as a percentage, cost, or people, depending on the context.
The Risk tightly integrates Danger (D), Exposition (E), and Vulnerability (V) concepts through the following expression:

$$Risk(R) = Vulnerability(V) * Danger(D) * Exposition(E) \tag{12.1}$$

where:
Danger (D) is presented as the occurrence probability of a disturbing phenomenon with a given magnitude within a time-lapse at a specific site. The potential danger is represented by a positive number within the range between 0 and 1.
Exposition (E) refers to the number of people, goods, infrastructure, and susceptible systems to be damaged, and it can be represented as economic losses or human lives. Finally,

Vulnerability (V) is defined as the susceptibility or propensity of the exposed systems to be affected. This concept is defined within the range of 0 and 1. It is worth mentioning that vulnerability is the only measure that can be controlled. Thus, Risk (R) can be controlled as a result.

Reliability: In literature, this concept has been defined in several ways. In order to find a definition focused on this research, first, we present several definitions already proposed by different researchers, and then we develop a final definition based on the conclusions from the previous research. According to Faturechi and Miller-Hooks (2014), a general definition for reliability is "*the probability that a system remains operative at a satisfactory post-disaster level.*" Leemis (2008) defined reliability as "*the probability that it will adequately perform its specified purpose for a specified lapse of time under specified environmental conditions.*" Rennemo et al. (2014) explained the concept as "the probability of executing an activity with success." In Edrissi et al. (2015), the concept of reliability represents a measure to travel time, capacity, and connectivity. "Where the travel time reliability definition is "*the probability that a trip between two nodes takes less time than some threshold value.*" The capacity reliability concept is defined as "*the* *probability that the network capacity can accommodate a certain volume of traffic demand at a required level-of-service*"; and finally, the concept of connectivity reliability is determined as "*the probability that the nodes of a network remain connected*" (Edrissi et al. 2015).

Once the reliability concept has been cleared, it is essential to define what does network reliability means. For this project, the network reliability concept is strongly related to the connectivity concept explained above, and a resulting definition is:

> The probability that the network remains completely available given a disaster occurrence, allowing the satisfaction of the on-time delivery and evacuation on-time requirements.

Some researchers have used the reliability as a performance measure. Vitoriano et al. (2009) proposed a Decision Support System (DSS) for building routes minimizing the risk over the routes. Indeed, Vitoriano et al. (2009) based the decisions on the minimization of the cost, ransack probability minimization, and reliability maximization.

Edrissi et al. (2015) presented a definition of a new emergency response reliability measure, evaluation, and selection of service level, and proposed a heuristic solution algorithm. In Vitoriano et al. (2011), a multi-criteria optimization model was proposed. This model simultaneously considered the time, cost, reliability, security, and equity as objective functions. Ukkusuri and Yushimito (2008) proposed an integer programming model for a location routing problem considering disruptions in the transportation network. The model pursued the prepositioning of facilities after evaluating the most reliable routes. Finally, Hamedi et al. (2012) tackled a routing and scheduling problem aiming to minimize the response time while unavailable paths are avoided.

The problem targeted in this work is focused on the network design decisions in the aftermath of a disaster. The decision-making in the aftermath of a disaster is defined by the high uncertainty level as well as the damages that infrastructure has suffered when a disaster strikes. More specifically, the reliability of the road is one of the main variables whenever the vehicle routing is defined. It is the main issue due to their impact on the departure of the vehicles and the delivery time, as well as the delivered quantity. It is worth to mention that these factors in humanitarian logistics cases become a death or live factor. Thus, to increase the reliability of the whole network due to the uncertainty scenario after a disaster is a challenge that must be considered in order to alleviate the suffering of the affected people.

This chapter is organized as follows: Sect. 12.2 shows the development of the network design optimization model. Section 12.3 presents the multi-criteria framework definition and its integration into the optimization model presented in Sect. 12.2. Section 12.4 presents the description of the Real-World instance, as well as the experiments and results. Finally, the conclusions are in Sect. 12.5.

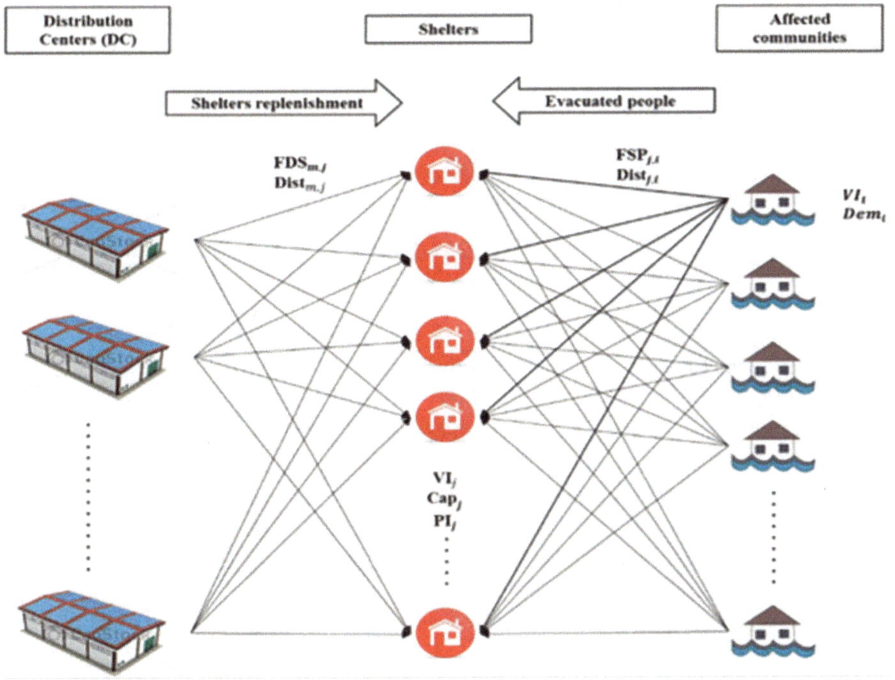

Fig. 12.1 Network design

12.2 Network Design

The problem considers the optimization of the network that will be used for the relief distribution. This network will be designed considering the affected zones, commonly known as the *hot* zones. The problem can be defined as to be a complete directed graph in which represents the vertices and refers to the arcs (Fig. 12.1). The vertices represent three types of entities: ***depots*** {1, ..., m} which location is known in advance; available ***shelters*** {1, ..., j} to be opened, which considers a vulnerability index (VI_j), a maximum number of people that can be sheltered (Cap_j); and an incident probability index (PI_j) which represents the probability of landfall or flood (also defined as *danger*). Finally, the ***affected communities*** {1, ..., i} are characterized by a number of people in need of shelter (Dem_i) and a vulnerability index (VI_j), which will be used to calculate the priorities for each shelter.

Each arc is defined by a distance between the affected community and shelter, as well as a risk factor, both based on the vulnerability index of the shelter and the maximum incident probability index between the shelter and the affected population.

The risk factor will be obtained with the expression (12.1). In this case, the risk will be represented as the number of people exposed to a particular danger (e.g., the people using a path with a high probability of a landfall), based on the vulnerability and danger of a specific path National Risk Atlas (2019). Finally, the exposition is represented by the affected population. Displaced people must be assigned to the available shelters allowing the distribution decisions. Alike the arcs among the affected zones and shelters, the ones between shelters and shelters are given as (j, k), as well as a risk factor (FSS_{jk}), and a distance ($Dist_{jk}$).

After the problem statement, the assumptions and main constraints are stated below:

Assumptions

1. The location of an affected community is given by its gravity center.
2. The physical location and the capacity of the shelters are known in advance.
3. The physical location of the distribution centers is known in advance.
4. The population coming from the same affected community can be assigned to different shelters.

Main constraints

1. The capacity of the shelter cannot be exceeded.
2. All the people in need must be assigned to one shelter.
3. Only one distribution center can be opened.

12.2.1 Optimization Model

The optimization model proposed is a multi-objective model. The sets, parameters, and decision variables are shown in Table 12.1.

The optimization model is given by the following expressions. It considers four objective functions, which refers to the minimization of: (1) non-used capacity, (2) total traveled distance, (3) cumulative risk through the entire network, and (4) the risk to which the evacuated people are exposed.

$$Min \sum_{j=1}^{J} nuc_j \tag{12.2}$$

$$Min \sum_{i=1}^{I} \sum_{j=1}^{J} dist_{ij} * z_{ij} \tag{12.3}$$

$$Min \sum_{j=1}^{J} \sum_{m=1}^{M} \left(dist_{mj} * FS_j * v_m\right) + \sum_{j=1}^{J} \sum_{k=1}^{K} \left(dist_{jk} * FSS_{jk} * x_j\right) \tag{12.4}$$

Table 12.1 Sets, parameters, and decision variables of the network design optimization model

Sets	
i	Affected population
j, k	Available shelter
m	Available distribution center
Parameters	
Dem_i	Number of persons to be evacuated from affected zone i
Cap_j	The capacity of shelter j in number of people
DSC_{mj}	Distance between distribution center m and shelter j
DSS_{jk}	Distance between shelter j and shelter k
IM_i	Marginal index of the affected community i
FS_j	Risk factor when shelter j is opened
FSS_{jk}	Risk factor when the arc (j, k) is active
FSP_{ij}	Risk factor when the arc (i, j) is active
Decision variables	
x_j	Binary variable, 1 if a shelter j is opened; 0 otherwise
v_m	Binary variable 1 if the distribution center m is opened; 0 otherwise
z_{ij}	Number of evacuated people from the affected zone i to shelter j
nuc_j	The non-used capacity of shelter j
PR_i	Auxiliary variable to obtain the priority of shelter j for the next stage

$$Min \sum_{i=1}^{I} \sum_{j=1}^{J} \left(z_{ij} * \left(FS_j + FSP_{ij} \right) \right) \tag{12.5}$$

The constraints of the model are given for the demand, capacity, shelters' operation, and shelters' coverage.

Demand

$$\sum_{j=1}^{J} z_{ij} = dem_i \quad \forall i \in I \tag{12.6}$$

Capacity

$$\sum_{i=1}^{I} z_{ij} \leq Cap_j * x_j \quad \forall j \in J \tag{12.7}$$

Equations (12.6) ensure that the demand for the affected populations will be satisfied. In the same way, Eq. (12.7) ensures that each shelter's capacity is not exceeded.

Shelters' operation

$$4 * x_j \geq \sum_{i=1}^{I} z_{ij} \quad \forall j \in J \tag{12.8}$$

$$\sum_{m=1}^{M} v_m = 1 \tag{12.9}$$

Equations (12.8) make sure that affected people will be assigned only to those shelters that are opened. In order to justify the opening of a shelter, a minimum assignment of 4 people is required. Currently, the decision of 4 people relies on the fact that a relief kit can satisfy four people. Equations (12.9) ensures that only one distribution center will be opened.

Shelters' coverage

$$\sum_{i=1}^{I} IM_i * z_{ij} = PR_j \quad \forall j \in J \tag{12.10}$$

Equations (12.10) determine the priority of each shelter j for in advance decision whenever the fairness in the delivery is considered (e.g., to decide how many relief kits will be assigned to a specific temporary shelter based on the average priority of the population sheltered in that site). The value of PR_i is based on the vulnerability index (VI_i) and the number of persons within each shelter j.

Finally, the no negativity constraints are given by Eqs. (12.11), (12.12), and (12.13).

No negativity

$$x_{ij}, v_m \in \{0, 1\} \quad \forall i \in I, j \in J, k \in K, m \in M \tag{12.11}$$

$$z_{ij} \in N^+ \quad \forall i \in I, j \in J \tag{12.12}$$

$$PR_j \in R^+ \quad \forall j \in J \tag{12.13}$$

12.3 Multi-criteria Framework

People face daily decisions, where the right decision depends on multiple criteria. To solve multiple criteria decision problems represents a hard task for the decision-makers. Most of the time, there is no perfect option to suit all the criteria, but decision-makers may deal with an excellent option that considers all criteria (Kahraman, 2008).

In the case of humanitarian logistics, decision-makers deal with this kind of problem knowing that the final decision can jeopardize affected peoples' lives.

According to literature, the increasing motivation to measure the performance of the humanitarian supply chain has led to amplify the number of goals to pursue in humanitarian logistics, and with this, the need of multi-objective contributions.

Considering the relief distribution problem, several multi-objective contributions have been proposed during the last decade. However, the impact that each goal has in the final decision must not be considered lightly in such a way that the minimization of the costs has not the same importance as the minimization of human life losses or any other objective functions. In order to assign the right weight to each of the pursued goals, multi-criteria decision analysis has been developed to support the decision-making, providing techniques for finding the right solution that satisfies all the pursued goals simultaneously.

As far as we know, only one multi-criteria framework for horizontal coordination has been developed. Gralla et al. (2014) developed a methodology to set priorities to different objectives, based on a horizontal coordination humanitarian logistics experts' preference.

The methodology presented in Fig. 12.2 (Regis-Hernández et al. 2017) is applied to integrate the objective functions into a single multi-criteria objective function.

Phase 1. Identification of goals, criteria, and performance indicators

The first phase aims to identify the stakeholders' goals and preferences. Indeed, all the actors in a humanitarian situation would undoubtedly agree that their goal is to mitigate the needs of the affected people. However, each stakeholder will be more or less sensible to the different aspects, including efficacy, efficiency, and service

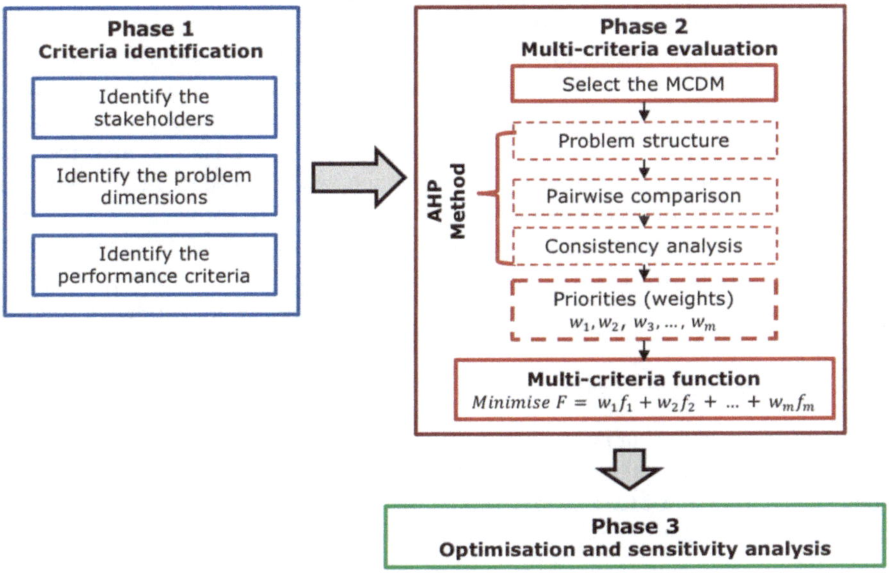

Fig. 12.2 Proposed framework

level. Therefore, this will promote different lines of action. Consequently, a structured process encompassing generic and specific data collection is required to select pertinent and satisfactory criteria. Generic data collection aims at identifying potential and consistent criteria. To this end, international standards and reports from government agencies, non-governmental organizations, and scientific literature are revealed as the most useful sources.

On the other hand, specific data collection aims at reaching stakeholders' agreement on the goals and criteria to be pursued. The literature proposes various techniques to select the main criteria in a consensual or agreed manner. Among them, the Delphi survey, the Nominal Group technique, consensus conferences, and the TRIAGE technique (Gervais and Pépin 2002; Lamontagne et al. 2010) have been successfully used in different contexts to help a team of experts or stakeholders resolve differing views or opinions. Finally, during a structured discussion process, the experts select a given number of criteria to be used in Phase 2.

Phase 2. Multi-criteria evaluation

The purpose of Phase 2 is to assess and quantify the relative importance of the selected criteria. To do so, the Analytic Hierarchy Process (AHP), a well-known multi-criteria Decision Making (MCDM) method, was proposed because of its flexibility and ability to handle the imprecise judgment of experts in order to prioritize the stakeholders' preferences. This method assesses the relative importance of the criteria and alternatives but also allows for translating subjective judgments into relative weights of importance (Saaty, 1980).

The AHP decomposes the problem hierarchically. The overall decision objective is at the top, and the criteria, sub-criteria, and decision alternatives are at each descending level of the hierarchy (Dolan et al. 1989; Partovi et al. 1990). Decision-makers compare each factor with all the other factors (at the same level of the hierarchy) using a pairwise comparison matrix to determine each factor's weight or relative importance. The AHP contains three main stages, as shown in Fig. 12.2. The first stage refers to the "structuring the problem." In this stage, a hierarchy is designed based on the characteristics of the problem in such a way that the first level represents the goal to be achieved, the second level represents the performance criteria, and the third level presents the alternatives to the problem. The second stage consists of the "pairwise comparison," in which experts compare the relative importance of each pair of considered criteria according to their preferences. In the third and last stage, a "consistency analysis" is performed, and the relative weights are generated.

Phase 3. Formulation of the multi-criteria objective function and model

The third step of the multi-criteria framework aims at formulating a model (for example, a mathematical model or a simulation model) that represents the situation under study. All the problem constraints, including limitations on the available resources, will be considered, and a set of decision variables is proposed. The goal of the formulation is to identify the values for the decision variables that maximize (or minimize) one or several objectives or utility functions that should quantify the stakeholder's wishes. So, a crucial part of this process is to ensure a durable consistency between what stakeholders expect from the distribution network and how the proposed decision variables influence the network to achieve the expected results. In other words, coherency analysis aims at validating how alternative decisions impact the performance of the supply chain under study.

After applying the methodology, the objective functions for each stage (network design and relief distribution) can be showed as a single multi-criteria objective function.

$$Min\ w_1 * \alpha + w_2 * \beta + w_3 * \gamma + w_4 * \delta \tag{12.14}$$

where w_1, w_2, w_3, w_4 represent the weights for each objective, and the sum must be 1: $w_1 + w_2 + w_3 + w_4 = 1$.

12.3.1 Network Design Multicriteria Model

Phase 1. Identification of goals, criteria, and performance indicators

In Table 12.2, the dimensions, performance criteria, and their description are presented for the network design. The dimensions refer to the classification of the objective functions, while the performance criteria are those that will be evaluated in order to achieve a dimension. In particular, we decided to have one economic indicator related to the infrastructure costs (C1_I1), one indicator related to social impacts of the distribution (C2_I1 = displacement distance), and two more indicators related to the reliability of the network (C3_I1 = network length and C3_I2 = overall risk), as described in Table 12.2, where the relationship between the performance criteria and the performance indicators is presented.

Table 12.2 Dimensions and performance criteria considered for the network design optimization model

Dimensions	Performance criteria	Description
C1. Economic	Infrastructure (C1_I1)	The used infrastructure represents the costs; in this case, the shelters to be opened. The criterion is based on total non-used capacity
C2. Social	Displacement distance (C2_I1)	It measures the distance that affected people must travel to reach their assigned shelter
C3. Reliability	Network length (C3_I1)	Measure the total distance between the opened shelters and the distribution center. Assumes a direct relationship between the length of the network and the risk to which the displaced people are exposed
	Overall risk (C3_I2)	Represents the measure of the risk that displaced people are exposed. The vulnerability, exposition, and occurrence probability of flood or landslide are considered to obtain a risk value

Phase 2. Multi-criteria evaluation

This step seeks to translate stakeholders' qualitative preferences into quantitative weights that can be used in a mathematical optimization model. To illustrate this phase, and to consider that we did not have access to stakeholders involved in real situations, we determine the preference of each stakeholder over a performance criterion using the expertise of the humanitarian logistics researchers. To do so, an extensive literature review was performed as shown in Tables 12.3, 12.4, 12.5 and 12.6. We proposed the set of objectives selected in phase 1 to all the stakeholders, and we required each of them to compare the importance of each objective concerning the others. To this end, and based on the AHP (Saaty, 1987), we structured the problem, as shown in Fig. 12.3.

Since we did not have access to real stakeholders in the real situation, we determine each stakeholders' preferences based on the perspective of the researchers. An exhaustive literature review was performed as shown in Tables 12.3—12.6 to determine the preferences that each stakeholder (Government (G), Non-Governmental Organization (NGO), Private Sector/Donor (PS/D), and Military (M)) has.

Based on the literature review (Tables 12.3–12.6), a pairwise matrix containing the relative importance RI (i,j) of each objective over all the others can be obtained. For example, the Government (Gov) believes that objective C3_I2 is twice more important than C1_I1. Therefore, the cell (C3_I2, C1_I1) in his pairwise matrix takes the value 2, and, inversely, the cell (C1_I1, C3_I2) takes the value 1/2. Tables 12.7, 12.8, 12.9 and 12.10 report the pairwise matrix for the Government, NGO's, Private Sector/Donors, and Military.

Finally, the preferences of all the stakeholders must be integrated into a single set of weights. However, because in most of the real cases, not all the stakeholders

Table 12.3 Network design. The relationship between the performance criteria C1_I1 and stakeholders based on the number of contributions in literature

Gov	NGO	PS/D	M
Akhtar et al. (2012), McLachlin et al. (2011)	Tomasini and Van Wassenhove (2009), Tatham and Kovacs (2010)	Balland (2013), Hovhanessian (2012), Cozzolino (2012), Huang et al. (2011), Moshtari and Goncalves (2011), Heaslip (2012), Tomasinim and Van Wassenhove (2009), Egan (2010), Heaslip (2013), Scholten et al. (2010)	Barbarosolu et al. (2002)

Table 12.4 Network design. The relationship between the performance criteria C2_I1 and stakeholders based on the number of contributions in literature

Gov	NGO	PS/D	M
	Heaslip (2012, 2013), Hovhanessian (2012), Cozzolino (2012), McLachlin and Larson (2011), Pettit et al. (2007), Kovács and Spens (2007), Heaslip and Barber (2014)	Balland (2013)	Barbarosolu et al. (2002)

Table 12.5 Network design. The relationship between the performance criteria C3_I1 and stakeholders based on the number of contributions in literature

Gov	NGO	PS/D	M
Balland (2013), McLachlin and Larson (2011)	McLachlin and Larson (2011), Schulz and Blecken (2010), Balland (2013), Hovhanessian (2012), Cozzolino (2012), Tzeng et al. (2007), Moshtari and Goncalves (2011), Pettit et al. (2007), Kovács and Spens (2007), Tomasini and Van Wassenhove (2009), Egan (2010), Tatham and Kovacs (2010), Heaslip (2013), Dolinskaya et al. (2011), Scholten et al. (2010)	Balland (2013), McLachlin and Larson (2011), Kovács and Spens (2007), Tomasini and Van Wassenhove, (2009), Heaslip (2013)	

Table 12.6 Network design. The relationship between the performance criteria C3_I2 and stakeholders based on the number of contributions in literature

Gov	NGO	PS/D	M
Balland (2013)	Balland (2013), Huang et al. (2011)		Heaslip (2012), McLachlin and Larson (2011), Barbarosolu et al. (2002), Pettit et al. (2007), Kovács and Spens (2007), Heaslip and Barber (2014), Heaslip (2013)

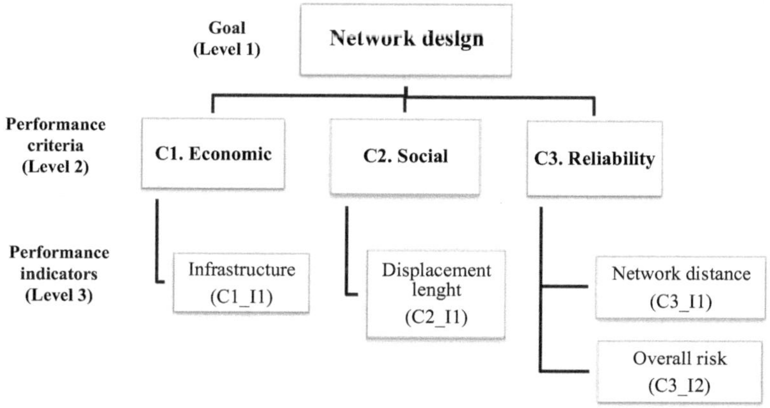

Fig. 12.3 The proposed hierarchy for a reliable network design

have the same experience, expertise, or credibility, we assign a specific numerical value that indicates the weight or the importance of their opinions on the group decision. A pairwise matrix based on the literature review (Tables 12.3–12.6) is

Table 12.7 Government's performance criteria pairwise comparison matrix

	C1_I1	C2_I1	C3_I1	C3_I2
C1_I1	1.00 (2/2)	9.00 (2/0)	1.00 (2/2)	2.00 (2/1)
C2_I1	0.11 (0/2)	1.00 (0/0)	0.11 (0/2)	0.11 (0/1)
C3_I1	1.00 (2/2)	9.00 (2/0)	1.00 (0/0)	2.00 (2/1)
C3_I2	0.50 (1/2)	9.00 (1/0)	0.50 (1/2)	1.00 (1/1)

Table 12.8 NGOs' performance criteria pairwise comparison matrix

	C1_I1	C2_I1	C3_I1	C3_I2
C1_I1	1.00 (2/2)	0.25 (2/8)	0.13 (2/15)	1.00 (2/2)
C2_I1	4.00 (8/2)	1.00 (8/8)	0.53 (8/15)	4.00 (8/2)
C3_I1	7.50 (15/2)	1.88 (15/8)	1.00 (15/15)	7.50 (15/2)
C3_I2	1.00 (2/2)	0.25 (2/8)	0.13 (2/15)	1.00 (2/2)

Table 12.9 Private sector/donors' performance criteria pairwise comparison matrix

	C1_I1	C2_I1	C3_I1	C3_I2
C1_I1	1.00 (10/10)	10.00 (10/1)	2.00 (10/5)	9.00 (10/0)
C2_I1	0.10 (1/10)	1.00 (1/1)	0.20 (1/5)	9.00 (1/0)
C3_I1	0.50 (5/10)	5.00 (5/1)	1.00 (5/5)	9.00 (5/0)
C3_I2	0.11 (0/10)	0.11 (0/1)	0.11 (0/5)	1.00 (0/0)

Table 12.10 Military's performance criteria pairwise comparison matrix

	C1_I1	C2_I1	C3_I1	C3_I2
C1_I1	1.00 (1/1)	1.00 (1/1)	9.00 (1/0)	0.14 (1/7)
C2_I1	1.00 (1/1)	1.00 (1/1)	9.00 (1/0)	0.14 (1/7)
C3_I1	0.11 (0/1)	0.11 (0/1)	1.00 (0/0)	0.11 (0/7)
C3_I2	7.00 (7/1)	7.00 (7/1)	9.00 (7/0)	1.00 (7/7)

Table 12.11 Stakeholders' pairwise comparison matrix

	GOV	NGO	PS/D	M
GOV	1.00 (3/3)	0.17 (3/18)	0.25 (3/12)	0.43 (3/7)
NGO	6.00 (18/3)	1.00 (18/18)	1.50 (18/12)	2.57 (18/7)
PS/D	4.00 (12/3)	0.67 (12/18)	1.00 (12/12)	1.71 (12/7)
M	2.33 (7/3)	0.39 (7/18)	0.58 (7/12)	1.00 (7/7)

provided in order to determine the importance that each stakeholder has on the group decisions to do so. The pairwise matrix is presented in Table 12.11.

Based on the AHP (Saaty, 1987), the following weights were obtained:

$$w_1 = 0.3653$$

$$w_2 = 0.1402$$

$$w_3 = 0.1722$$

$$w_4 = 0.3223$$

where, the infrastructure (C1_I1) is the most important with the 36.53% importance; the overall risk is the second important with 32.23%; the network length (C3_I1) has 17.22% importance; and finally, the less important is the displacement distance (C2_I1) with 14.02% importance.

The integration of the objective functions (12.2)–(12.5) into one multi-criteria objective function (12.15) has the following structure:

$$\begin{aligned} & Min w_1 \sum_{j=1}^{J} nuc_j \\ & + w_2 \sum_{i=1}^{I} \sum_{j=1}^{J} dist_{ij} * z_{ij} \\ & + w_3 \sum_{j=1}^{J} \sum_{m=1}^{M} \left(dist_{mj} * FS_j * v_m\right) + \sum_{j=1}^{J} \sum_{k=1}^{K} \left(dist_{jk} * FSS_{jk} * x_j\right) \\ & + w_4 \sum_{i=1}^{I} \sum_{j=1}^{J} \left(z_{ij} * (FS_j + FSP_{ij})\right) \end{aligned} \tag{12.15}$$

Leading to the following update to the single multi-criteria objective function (12.16) for the reliable network design model as follows:

$$\begin{aligned} & Min 0.3653 * \sum_{j=1}^{J} nuc_j \\ & + 0.1402 * \sum_{i=1}^{I} \sum_{j=1}^{J} dist_{ij} * z_{ij} \\ & + 0.1722 * \sum_{j=1}^{J} \sum_{m=1}^{M} \left(dist_{mj} * FS_j * v_m\right) + \sum_{j=1}^{J} \sum_{k=1}^{K} \left(dist_{jk} * FSS_{jk} * x_j\right) \\ & + 0.3223 * \sum_{i=1}^{I} \sum_{j=1}^{J} \left(z_{ij} * (FS_j + FSP_{ij})\right) \end{aligned} \tag{12.16}$$

Once the stakeholders' preferences are integrated within the multi-criteria optimization model, the model can be applied to evaluate the solution of the case study.

12.4 Case Study: Hurricane Stan

The hurricane Stan struck Mexico in 2005, resulting in several deaths and infrastructure damages. The implementation of the network design based on a multi-criteria framework is presented within this section. The analysis of the results that compare the solution proposed by CENAPRED and the one obtained with the two-staged model is presented.

12.4.1 Background

According to the EM-DATA database, between 2000 and 2016, a total of USD 756,000 million world economic losses, 543 million affected people, and 191,189 thousand million affected people have been reported. It induced by 1,635 hydrometeorological disasters in 64 countries (EM-DATA, 2016). Struck by the Atlantic and the Pacific, Mexico is one of the most affected countries by hydrometeorological disasters per year. One of the worst economic losses in Mexico was caused by hydrometeorological disasters in 2005, when hurricanes Emily, Stan, and Wilma strike, leaving USD 7.91 billion in total damages. As a result, the Mexican Republic southeast was the most impacted, affecting Quintana Roo, Yucatán, Oaxaca, and Veracruz.

This case study considers only the state of Veracruz analysis, which reported 2,755, 218 people in 109 different communities, where the most crowded are Veracruz, Xalapa, Coatzacoalcos, Cordoba, and Papantla (CENAPRED, 2005).

Veracruz is one of the biggest states in Mexico, plains, and low high hills define and its geographical characteristics, as well as it is joined with Puebla and Hidalgo states. Although, due to its coastal extension, Veracruz is susceptible to several storms and only in the last 136 years, 36 tropical storms struck. In 2005, three tropical storms and one hurricane category one in the scale Saffir-Simpson struck Veracruz. At that moment, Veracruz had 6,908,975 people, which represents 7.09% of the total country population. The more populated cities are Veracruz, Xalapa, Coatzacoalcos, Cordoba, and Papantla. Veracruz is tagged as one of the most vulnerable states in the country (see Fig. 12.4).

The damage and the resilience capacity of each population rely on their vulnerability index. Considering the formula to determine the risk (12.1) based on the exposition, the probability of occurrence, and the vulnerability, Veracruz has a high-risk index.

12.4.2 Experiments and Results

The case study considers 27 available distribution centers, 109 affected populations, and 1,379 available shelters with a specific capacity. For this case study, and for the

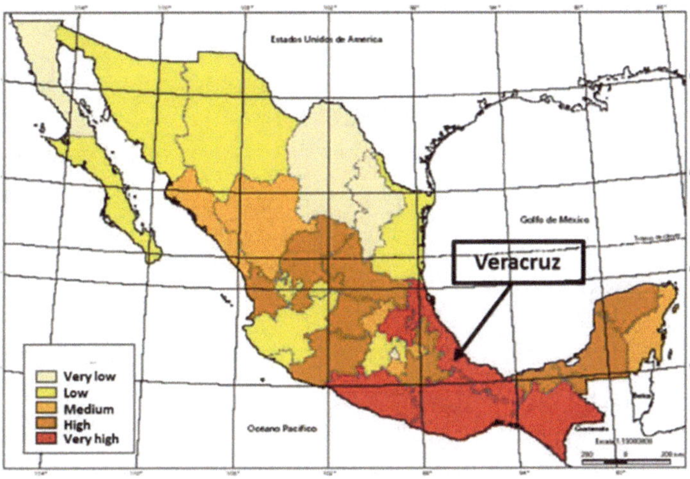

Fig. 12.4 Vulnerability index

network design stage, the demand is considered as the number of people in need per shelter. The risk between the affected populations and their assigned shelter is determined using Eq. (12.1).

To determine the assignment of affected people to each shelter and the relief distribution, we applied the multi-criteria network design, optimization model. The information related to the case study is the one reported by CENAPRED (2005). Table 12.12 shows the solution presented by CENAPRED in 2005 in the aftermath of Hurricane Stan specifically for Veracruz. Also, the percentage of shelters' non-used capacity, opened shelters, and the risk to which people are exposed when traveling through a path is presented in Table 12.12.

Due to the availability of information related to hurricane Stan, the parameters to be evaluated are based mainly on the network design (risk exposition, distance, and opened shelters). In order to compare the solution given by the Mexican government and the solution obtained applying the proposed model, the vulnerability, danger, and exposition of each shelter are obtained from the Mexican National Risk Atlas (Atlas Nacional de Riesgos, 2019).

For this case study, the risk is measured by the number of people exposed to landfall or flood whenever they use the temporary shelters or the evacuation routes. The obtained solution applying the multi-criteria optimization model is used given in Tables 12.13, 12.14, 12.15 and 12.16. Finally, Table 12.17 shows the comparison among the solution given by CENAPRED and the solution resulting from the implementation of the proposed multi-criteria framework decision-making process.

CENAPRED proposed to open 50 temporary shelters with room for 42,313 people, while the solution obtained considering the design of a reliable network lead to open only four temporary shelters with room for 9,000 people. Currently, the non-used capacity resulting from the government decision and the one obtained

Table 12.12 CENAPRED solution analysis flood

Location	Op. sh.	Ass. to sh.	S Cap.	NUC	Vul.	FD	LD	TD	ER
Hueyapan de Ocampo	3	500	1,044	544	0.25	1	1	1	125
Catemaco	3	105	2,360	2,255	0.17	1	1	1	17.85
San Andrés Tuxtla	5	150	150	0	0.21	1	1	1	31.5
Santiago Tuxtla	1	115	2,692	2,577	0.15	1	0.75	1	17.25
Acayucan	3	162	2,800	2,638	0.24	1	0.75	1	38.88
San Juan Evangelista	2	420	420	0	0.23	1	1	1	96.6
Soteapan	1	5	451	446	0.35	0.75	1	1	1.75
Angel R. Cabada	3	745	745	0	0.15	1	1	1	111.75
Agua Dulce	1	29	1,155	1,126	0.16	1	0.75	1	4.64
Las Choapas	1	65	1,300	1,235	0.18	1	1	1	11.7
Juan R. Clara	2	300	3,588	3,288	0.27	1	0.75	1	81
Isla	1	50	500	450	0.14	1	0.75	1	7
José Azueta	1	10	2,920	2,910	0.22	1	0.75	1	2.2
Tlacojalpan	1	250	2,150	1,900	0.21	1	0.75	1	52.5
Pajapan	1	20	20	0	0.39	1	1	1	7.8
Texistepec	2	240	240	0	0.26	1	0.75	1	62.4
Coatzacoalcos	1	25	2,065	2,040	0.14	1	1	1	3.5
Tierra Blanca	1	18	18	0	0.16	1	1	1	2.88
Ursulo Galván	1	30	2,000	1,970	0.19	1	1	1	5.7
Ignacio de Llave	1	200	1,200	1,000	0.33	1	1	1	66
Veracruz	3	850	850	0	0.16	1	1	1	136
Atlahuilco	1	50	50	0	0.43	1	0.5	1	21.5
Xalapa	2	650	650	0	0.16	0.75	1	1	104
Nautla	1	60	3,715	3,655	0.25	1	1	1	15
Cazones Herrera	3	115	8,960	8,845	0.28	1	1	1	32.2
Tampico Alto	1	20	20	0	0.2	1	1	1	4
Tempoal	4	250	250	0	0.2	1	1	1	50
Total shelters	50	5,434	42,313	36,879					1,110.6

Op. Sh. Opened shelters, *Ass. To sh.* assignment to shelters, *S. Cap.* Shelter capacity, *NUC* non-used capacity, *%NUC* percentage of non-used capacity, *Vul.* Vulnerability, *F.D.* Flood danger, *L.D.* Landfall danger, *T.D.* Total danger, *E.R.* Exposition risk

with the proposed methodology is the 36,879 people, and 3,566 respectively. These results were expected, such as the implementation of the hierarchical multi-criteria framework with the infrastructure as the most critical criterion (36.53% importance), followed by the overall risk (32.23% importance), the network length (17.22% importance); and finally, the displacement distance (14.02% importance). Nevertheless, the solution does not favor the displacement distance due to the given

Table 12.13 Proposed assignment for the shelter located in Jilotepec

Affected population	Demand (people)	Distance (Sm) shelter—aff. pop.	Trav. dist. (km)
Acayucan	57	10.738	612.07
Acula	9	10.321	92.89
Acultzingo	37	13.692	506.59
Alpatlahuac	18	15.424	277.64
Amatitlan	14	15.424	215.93
Amatitlan de los Reyes	75	11.869	890.20
Apazapan	7	11.395	79.77
Aquila	4	16.731	66.92
Astacinga	10	17.610	176.10
Atlahuilco	18	15.908	286.34
Atoyac	42	12.131	509.49
Atzacan	35	10.959	383.56
Calcahualco	25	10.603	265.07
Camaron de Tejeda	11	10.997	120.97
Cordona	100	50.414	5,041.41
Cosoleacaque	6	45.762	274.57
Playa de Vicente	74	37.830	2,799.45
Zentla	24	35.044	841.06
Zongolica	77	35.160	2,707.29
Zaragoza	19	46.302	879.73
Yanga	31	33.358	1,034.11
Rafael Delgado	34	37.660	1,280.43
Xoxocotla	9	38.401	345.61
Orizaba	100	31.695	3,169.53
Puente Nacional	39	37.783	1,473.54
Otatitlan	11	48.263	530.89
Paso del Macho	54	49.001	2,646.08
Omealca	43	32.688	1,405.58
Rio Blanco	79	37.125	2,932.86
Uxpanapa	49	38.644	1,893.56
Olutla	26	37.961	986.99
Oteapan	22	47.555	1,046.20

weight to this criterion. Indeed, only 22 out of 127 affected populations were less than 10 km far from their assigned temporary shelter. Considering a displacement by feet, this solution could be improved.

It is worth to mention that further analysis must be done based on interviews with each of the involved stakeholders instead of based on the literature review. Nevertheless, the case study demonstrates the usefulness of the multi-criteria framework methodology.

Table 12.14 CENAPRED assignment for the shelter located in Emiliano Zapata

Affected population	Demand (people)	Distance (Sm) shelter—aff. pop.	Trav. dist. (km)
Chinameca	27	30.143	813.86
Chocaman	33	30.027	990.90
Coetzala	4	22.835	91.34
Comapa	33	22.718	749.68
Cordoba	100	25.535	2,553.47
Cosamaloapan	100	25.406	2,540.56
Coscomatepec	100	21.887	2,188.73
Cosoleacaque	100	21.884	2,188.44
Cotaxtla	36	22.063	794.26
Soteapan	55	4.878	268.30
Sochiapa	6	4.993	29.96
Tierra Blanca	68	89.829	6,108.34
Nogales	63	45.847	2,888.34
Soledad Atzompa	38	5.114	194.33
Tehuipango	40	4.805	192.22
Minatitlan	99	277.260	27,448.77
Orizaba	100	39.891	3,989.08
Soconusco	24	3.183	76.40
Sayula de Aleman	55	3.359	184.77
Paso de Ovejas	59	47.725	2,815.78
Tenampa	11	3.097	34.07
Tepatlaxco	15	3.039	45.59
Cuitlahuac	45	14.849	668.21
Cuichapa	22	14.751	324.51
Emiliano Zapata	97	19.913	1,931.53
Santiago Tuxtla	7	7.650	53.55
Soledad de Doblado	53	9.694	513.78
Oteapan	3	45.456	136.37

Table 12.15 CENAPRED assignment for the shelter located in Cosamaloapan

Effected population	Demand (people)	Distance (Sm) shelter—aff. pop.	Trav. dist. (km)
Camerino Z. Mendoza	77	96.159	7,404.25
Carlos A. Carrillo	43	98.799	4,248.37
Carrillo Puerto	28	97.264	2,723.38
Chacaltianguis	22	97.924	2,154.33
Cordoba	68	125.849	8,557.73
Huatusco	97	13.880	1,346.32
Las Choapas	100	165.787	16,578.68
Medellin de Bravo	76	167.960	12,764.99
Fortin	100	13.885	1,388.45
Hueyapan de Ocampo	75	12.701	952.57
Mariano Escobedo	60	167.521	10,051.24
Ixtaczoquitlan	100	13.850	1,384.96
La Perla	37	166.229	6,150.48
Juan Rodriguez Clara	67	166.955	11,186.01
Minatitlan	100	166.432	16,643.19
Jose Azueta	45	166.489	7,491.99
Isla	74	13.785	1,020.12
Ixhuatlan del café	38	13.771	523.30
San Juan Evangelista	55	40.307	2,216.91
Los Reyes	10	165.112	1,651.12
Ixmatlahuacan	11	14.046	154.51
Ixhuatlan del sureste	27	14.116	381.14
Ignacio de la llave	34	13.882	471.98
Manilo Fabio Altamirano	40	167.372	6,694.88
Huiloapan de Cuahutemoc	12	13.922	167.06
Ixhuatancillo	31	13.920	431.53
Hidalgotitlan	34	14.092	479.12
Magdalena	5	164.632	823.16
Maltrata	29	166.137	4,817.98
San Andres Tenejapan	5	40.150	200.75

Table 12.16 CENAPRED assignment for the shelter located in Juchique de Ferrer

Affected population	Demand (people)	Distance (Sm) shelter—aff. pop.	Trav. dist. (km)
Cordoba	100	32.297	3,229.73
Cosamaloapan	16	32.211	515.38
Coscomatepec	40	29.518	1,180.72
Cosoleacaque	100	29.565	2,956.55
Jalcomulco	9	77.452	697.07
Las Choapas	34	276.812	9,411.61
Fortin	5	117.962	589.81
Texistepec	35	80.932	2,832.63
Tierra Blanca	100	81.882	8,188.23
Ixtaczoquitlan	19	118.076	2,243.44
Tlilapan	9	39.122	352.10
Tomatlan	12	38.625	463.50
Tlcojalpan	9	78.480	706.32
Tlaquilpan	13	35.975	467.67
Texonapa	94	81.023	7,616.18
Minatitlan	100	274.989	27,498.94
Tlalixcloyan	69	78.530	5,418.56
San Juan Evangelista	5	104.214	521.07
Jesus Carranza	46	82.993	3,817.66
Orizaba	31	50.162	1,555.03
Tlacoltapan	27	78.708	2,125.11
Texhuacan	9	82.017	738.15
Tequila	24	6.945	166.67
Totutla	30	54.134	1,624.02
Moloacan	31	44.309	1,373.59
Naranjal	4	50.975	203.90
Tuxtlla	4	50.545	202.18
Tlcotepec de Mejia	7	78.623	550.36
Jamapa	19	82.933	1,575.72
Tres Valles	83	47.399	3,934.09
Santiago Sochiapan	15	104.240	1,563.60
Nanchital	53	46.470	2,462.93
Santiago Tuxtla	100	6.886	688.56
Tlatetela	27	35.926	970.01
Jaltipan	73	81.960	5,983.06
Saltabarranca	11	103.974	1,143.71
Mixtla de Altamirano	19	48.255	916.84

Table 12.17 Comparison of the applied solution (CENAPRED) and the proposed solution

Sol.	Op. Sh.	Cap. (people)	NUC	Network distance (km)	Risk	Disp.Risk
Gov.	50	42,313	36,879	218,081.40	1,110.60	N/A
Prop. sol.	4	9,000	3,566	442.75	544.69	405,835.71

Op. Sh. Opened shelters, *Cap.* Capacity, *NUC* Non-used capacity, *Disp. Risk* Displacement risk

12.5 Conclusions

Aftermath disasters decisions are challenging due to the number of stakeholders with diverse goals, but also to the urgency of the decisions to be made. This paper proposes a reliable network design based on a multi-criteria optimization model. Throughout the proposed decision support tool, decision-makers are guided in the decisional process, thus ensuring the balanced integration of the stakeholders' preferences (humanitarian, governmental, donors, and beneficiaries). The proposed multi-criteria optimization model allows adequate translation, from quantitative to quantitative, of different stakeholders' preferences for their integration into a multi-criteria optimization model. The applicability of the integration of the stakeholders' preference was demonstrated with the Hurricane Stan case study. Performance indicators, such as costs, coverage, reliability, and risk, were considered to formulate an objective function.

However, this study presents various limitations. Firstly, it assumes that all the required information, including the needs of the affected people or the real situation of routes and infrastructures, is available and accurate. Indeed, the integration of technologies for gathering information about the consequences of a disaster in an accurate and timely manner, mostly the situation of routes and infrastructures as well as the real needs of the population, is a promising topic of future research. Secondly, we had to rely on the literature review to identify and evaluate decision-makers' preferences and goals to validate how the solutions provided by the model satisfy their needs. Additional research is required to obtain information from different organizations involved in the humanitarian supply chain identifying their real drivers and goals as well as other constraints not considered by our model but which impact their day-to-day operations.

References

Akhtar, P., Marr, N. E., & Garnevsca, E. V. (2012). Coordination in humanitarian relief chains: Chain coordinators. *Journal of Humanitarian Logistics and Supply Chain Management, 2*(1), 85–103. https://doi.org/10.1108/20426741211226019.

Altay, N., & Green, W. G. (2006). OR/MS research in disaster operations management. *European Journal of Operational Research, 175*(1), 475–493. https://doi.org/10.1016/j.ejor.2005.05.016.

Atlas Nacional de Riesgos. (2019). Available from: http://atlasnacionalderiesgos.gob.mx.

Balland, J. (2013). Humanitarian relief organizations and its relationship with logistics service providers. Jönköping University.

Barbarosolu, G., Özdamar, L., & Çelvik, A. (2002). An interactive approach for hierarchical analysis of helicopter logistics in disaster relief operations. *European Journal of Operational Research, 140*(1), 118–133. https://doi.org/10.1016/S0377-2217(01)00222-3.

CENAPRED. (2005). Características e impacto socioeconómico de los principales desastres ocurridos en la República Mexicana en el año 2005. México (In Spanish).

Cozzolino, A. (2012). Humanitarian logistics: cross sector cooperation in disaster relief management. ISBN 978-3-642-30186-5.

Dolan, J. G., Isselhardt, B. J., & Cappuccio, J. D. (1989). The analytic hierarchy process in medical decision making: A tutorial. *Medical Decision Making, 9*(1), 40–50.

Dolinskaya, I. S., Shi, Z. E., Smilowitz, K. R. (2011). *Decentralized approaches to logistics coordination in humanitarian relief.* In Proceeding of the 2011 industrial engineering research conference.

Edrissi, A., Nourinejad, M., & Roorda, M. J. (2015). Transportation network reliability in emergency response. *Transportation Research Part E: Logistics and Transportation Review, 2015*(80), 56–73. https://doi.org/10.1016/j.tre.2015.05.005.

Egan, M. J. (2010). Private goods and services contracts: Increased emergency response capacity or increased vulnerability? *International Journal of Production Economics, 126*(1), 46–56. https://doi.org/10.1016/j.ijpe.2009.10.005.

EM-DATA. (2016). The international disaster database. Centre for Research on the Epidemiology of Disasters (CRED). Retrieved 20, September 2016 from http://www.emdat.be/country_profile/index.html.

EM-DATA. (2019). The international disaster database. Centre for Research on the Epidemiology of Disasters (CRED). Retrieved 15, July 2019 from http://www.emdat.be/country_profile/index.html.

Faturechi, R., & Miller-Hooks, E. (2014). Measuring the performance of transportation infrastructure systems in disasters: A comprehensive review. *ASCE Journal of Infrastructure Systems, 21*(1), 1–15. https://doi.org/10.1061/(ASCE)IS.1943-555x.0000212.

Gralla, E., Goentzel, J., Fine, C. (2014). Assessing trade-offs among multiple objectives for humanitarian aid delivery using expert preferences. *Production and Operations Management, 23*(6), 978–989. https://doi.org/10.1111/poms.12110.

Leiras, A., Irineu De Brito, J. R., Peres, E. Q., Bertazzo, T. R., Yoshizaki, H. T. Y. (2014). Literature review of humanitarian logistics research: Trends and challenges. ISBN 0960003071073. http://www.emeraldinsight.com/10.1108/JHLSCM-04-2012-0008.

Gervais, M., & Pépin, G. (2002). TRIAGE: A new group technique gaining recognition in evaluation. *Evaluation Journal of Australasia, 2*(2), 45–49.

Hamedi, M., Haghanim, A., Yang, S. (2012). Reliable transportation of humanitarian supplies in disaster response: Model and heuristic. *Procedia—Social and Behavioural Sciences, 54*, 1205–1219. https://doi.org/10.1016/j.sbspro.2012.09.835.

Heaslip, G. E. (2012). The logistical challenges of coordinating military and civilian agencies in humanitarian operations. *Irish Academy of Management Conference, 353,* 1–31.

Heaslip, G. (2013). Services operations management and humanitarian logistics. *Journal of Humanitarian Logistics and Supply Chain Management, 5*(1), 35–60. https://doi.org/10.1108/20426741311328501.

Heaslip, G., & Barber, E. (2014). Using the military in disaster relief: Systemising challenges and opportunities. *Journal of Humanitarian Logistics and Supply Chain Management, 4*(1), 60–81. https://doi.org/10.1108/JHLSCM-03-2013-0013.

Hovhanessian, M. (2012). Coordination barriers between humanitarian organizations and commercial agencies in times of disaster. Jönköping International Business School.

Huang, Y., Li, X., Omitaomu, O. (2011). Conceptual supernetwork model for coordination mechanisms in humanitarian relief chain. In *Proceedings of the 2011 industrial engineering research conference.*

Kahraman, C. (2008). Fuzzy multi-criteria decision making. *Theory and Applications with Recent Developments, 16*(1), 1–19.

Kovács, G., & Spens, K. M. (2007). Humanitarian logistics in disaster relief operations. *International Journal of Physical Distribution & Logistics Management, 37*(2), 99–114. https://doi.org/10.1108/09600030710734820.

Lamontagne, M. E., Swaine, B. R., Lavoie, A., Champagne, F., Marcotte, A. C. (2010). Consensus group sessions: A useful method to reconcile stakeholders 2019 perspectives about network performance evaluation. *International Journal of Integrated Care, 10.*

Leemis, L. M. (2008). Reliability. In *Operations research and management science handbook.* Taylor &Francis Group, LLC.

McLachlin, R., & Larson, P. D. (2011). Building humanitarian supply chain relationships: Lessons from leading practitioners. *Journal of Humanitarian Logistics and Supply Chain Mangement, 1* (1), 32–49.

Moshtari, M., Goncalves, P. (2011). Understanding the drivers and barriers of coordination among humanitarian organizations. In *POMS 23rd annual conference*. Chicago, Illinois, U.S.A.

Natarajarathinam, M., Capar, I., & Narayanan, A. (2009). Managing supply chains in times of crisis: A review of literature and insights. *International Journal of Physical Distribution & Logistics Management, 39*(7), 535–573. https://doi.org/10.1108/09600030910996251.

Partovi, F. Y., Burton, J., & Banerjee, A. (1990). Application of analytical hierarchy process in operations management. *International Journal of Operations & Production Management, 10* (3), 5–19.

Pettit, S. J., Beresford, A. K. C., & Drive, C. (2007). Emergency relief logistics: an evaluation of military, non-military, and composite response models. *International Journal of Logistics Research and Applications: A Leading Journal of Supply Chain Management, 8*(4), 313–331. https://doi.org/10.1080/13675560500407325.

Regis-Hernández, F., Mora-Vargas, J., Ruiz, A. (2017). A multi-criteria vertical coordination framework for a reliable aid distribution. Special issue: Logistics systems design in latin America. *Journal of Industrial Engineering and Management, 10*(4), 789–815.

Rennemo, S. J., RØ, K. F., Hvattum, L., & Tirado, G. (2014). A three-stage stochastic facility routing model for disaster response planning. *Transportation Research Part E: Logistics and Transportation Review, 62,* 116–135. https://doi.org/10.1016/j.tre.2013.12.006.

Saaty, T. L. (1980). *The analytic hierarchy process.* McGraw-Hill.

Saaty, R. W. (1987). The analytic hierarchy process—What it is and how it is used. *Mathematical Modelling, 9*(3–5), 161–176.

Scholten, K., Scott, P. S., & Fynes, B. (2010). (Le) agility in humanitarian aid (NGO) supply chains. *International Journal of Physical Distribution Management, 40*(8/9), 623–635.

Schulz, S. F., & Blecken, A. (2010). Horizontal cooperation in disaster relief logistics: Benefits and impediments. *International Journal of Physical Distribution & Logistics Management, 40* (8/9), 636–656. https://doi.org/10.1108/09600031011079300.

Tatham, P., & Kovacs, G. (2010). The application of "swift trust" to humanitarian logistics. *International Journal of Production Economics, 126*(1), 35–45. https://doi.org/10.1016/j.ijpe.2009.10.006.

Thomas, A. S., & Kopczak, L. R. (2005). From logistics to supply chain management: the path forward in the humanitarian sector. *Fritz Institute, 15,* 1–15.

Tomasini, R. M., & Van Wassenhove, N. (2009). From preparedness to partnerships: case study research on humanitarian logistics. *International Transactions in Operational Research, 16*(5), 549–559. https://doi.org/10.1111/j.1475-3995.2009.00697.x.

Tzeng, G.-H., Cheng, H.-J., & Huang, T. D. (2007). Multi-objective optimal planning for designing relief delivery systems. *Transportation Research Part E: Logistics and Transportation Review, 43*(6), 673–686. https://doi.org/10.1016/j.tre.2006.10.012.

Ukkusuri, S. V., & Yushimito, W. F. (2008). Location routing approach for the humanitarian prepositioning problem. *Transportation Research Record: Journal of the Transportation Research Board, 2008*(2089), 18–25. https://doi.org/10.3141/2089-03.

Vitoriano, B., Ortuño, M. T., & Tirado, G. (2009). HADS, a goal programming-based humanitarian aid distribution system. *Journal of Multi-Criteria Decision Analysis, 16,* 55–64. https://doi.org/10.1002/mcda.

Vitoriano, B., Ortuño, M. T., Tirado, G., Montero, J. (2011). A multi-criteria optimization model for humanitarian aid dis-tribution. *Journal of Global Optimization*.

Part III
After the Disaster

Chapter 13
The Design of a Humanitarian Aid Assignment Mechanism in the Post-disaster Stage

Damián-Emilio Gibaja-Romero and Diana Sánchez-Partida

Abstract The humanitarian logistics literature points out that Humanitarian Aid (HA) allocations are unfair and inefficient at the post-disaster stage. Specifically, such problems arise during the allocation of indivisible goods due to instability and the short life of humanitarian supply chains. In this research, we focus on designing an assignment procedure to deal with the previous problem through a mechanism design approach since it is possible to elicit victims' preferences and institutions' priorities; then, we model the HA allocation problem as a three-sided market under victims get HA only if some member of the humanitarian supply chain can provide her. So, we propose the Nested Deferred Acceptance mechanism as an assignment procedure to allocate HA where HSC' members do not have a (minimum or maximum) quota of goods to allocate among the disaster victims. It is possible to demonstrate that this mechanism generates the best possible assignment for the agents that make the requisition, and such mechanism is rational, fair, and strategy proofness (no victim can manipulate it to get a better allocation).

Keywords Assignment problem · Humanitarian logistics · Nested Deferred Acceptance

D.-E. Gibaja-Romero (✉)
Department of Mathematics, Graduate School of Engineering, Universidad Popular Autónoma del Estado de Puebla (UPAEP University), 17 Sur 901, Barrio de Santiago, 72410 Puebla, Mexico
e-mail: damianemilio.gibaja@upaep.mx

D. Sánchez-Partida
Department of Logistics and Supply Chain Management, Universidad Popular Autónoma del Estado de Puebla (UPAEP University), 17 Sur 901, Barrio de Santiago, 72410 Puebla, Mexico

D. Sánchez-Partida et al. (eds.), *Disaster Risk Reduction in Mexico*,
https://doi.org/10.1007/978-3-030-67295-9_13

13.1 Introduction

Given the socioeconomic damages caused by a natural disaster, a Humanitarian Supply Chain (HSC) does not only assist people in the first moments after a catastrophe happens (Adivar et al., 2010), it also develops activities before and during the disaster to ameliorate the phenomenon's impact (Khan et al., 2008). The literature suggests splitting such phenomena into phases to facilitate disaster management. Whybark (2015) considers four primary phases: mitigation and preparedness, both develop in a pre-disaster stage; meanwhile, response and recovery occur in during-disaster and post-disaster stages, respectively. However, other authors consider different phases depending on how important each stage is for them. For example, Ilhan (2011) considers a management process of five phases (prevention, mitigation, preparedness, recovery, and reconstruction) since he emphasizes on the importance of defining pre-disaster strategies and points out that response capacity is almost null at the during-disaster stage. In opposition, Lin Moe and Pathranarakul (2006) highlight the post-disaster stage by considering the phases of prediction, warning, emergency relief, rehabilitation, and reconstruction; and Brown et al. (2017) include a sixth phase, resolution, to evaluate the consequences of the catastrophe.

Although there is no general agreement on the phases of disaster management, HSC's issues arise from the pre-disaster, during-disaster, and post-disaster stages (Nagurney et al., 2011). The post-disaster stage represents an opportunity area for HL's practitioners, authorities and academicians (Oloruntoba & Gray, 2006; Chandraprakaikul, 2010) given the lack of coordination among the HSC's members to allocate humanitarian aid (HA) that leads to non-desirable allocations (Tatham et al., 2017). In this chapter, we analyze the issues that arise during the allocation of humanitarian aid (HA) following a mechanism design approach to propose an assignment procedure that allocates HA in the best possible way during the post-disaster stage.

Since HL pursue recovery and reconstruction of affected communities at the post-disaster stage, the HSC assigns indivisible goods, like housing, among victims. Our analysis focuses on indivisible HA resources since the concerning literature identifies two issues on their allocation: (i) they are not assigned to disaster's victims that need them with urgency (de Montclos, 2009), and (ii) there are population sectors that do not get any HA while such resources are wasted in other sectors (Kwon & Kim, 2018). In terms of the mechanism design literature, the previous empirical evidence points out that HA's assignment procedures are unfair and inefficient at the post-disaster stage.

The designing of an assignment procedure is relevant since HL deals with affectations in the supply chain to restore the welfare of disaster victims; this establishes a critical difference between Commercial Logistics (CL) and HL. While CL pursues products' improvement, or benefit maximization, through a bargaining process between decision-makers (Kovács & Spens, 2007a, 2007b), HL oversees the assignment of resources that come from donors. Even more, HSC's members are volunteers, public organizations, and donor institutions that pretend to alleviate the impact of a catastrophe (Jahre et al. 2007).

Summarizing, the HSC works with a shared pool of resources that any member can allocate, and their intervention is necessary to complete the assignment procedure. So, the allocation of indivisible HA casts similarities with the matching through the institution's problem (Bloch et al., 2020), where a shared pool of apartments is allocated among households through the intermediation of institutions. In this sense, we analyze the allocation of indivisible HA as an application of the model proposed by Bloch et al. (2020), where HSC's members, disaster victims, and indivisible HA mimic the role of institutions, households, and apartments, respectively. Thus, we use the Nested Deferred Acceptance (NDA) mechanism to allocate indivisible HA among disaster victims.

The chapter is organized as follows. In Sect. 13.2, we briefly discuss the post-disaster stage literature. Later, Sect. 13.3 models the HA allocation problem as a matching with institutions problem where there are no capacity constraints, i.e., any member of the humanitarian supply chain can allocate any quantity of HA resources. In Sect. 13.4, we discuss the designing of an allocation mechanism that satisfies desirable properties like fairness and strategy-proofness; hence, we explain how property rights impact the allocation of indivisible resources since humanitarian aid comes from donations. We first establish a framework only under the allocation of HA resources is exclusive for certain HSCs' members; then, we apply the deferred acceptance algorithm to deal with such a market. Later, we introduce the nested deferred acceptance algorithm to deal with the fact that any HSC's member can allocate, if it is possible for them, all HA resources. Section 13.5 closes the chapter with the conclusions.

13.2 Literature Review

As we mentioned in the introduction, we deal with the allocation of humanitarian aid during the post-disaster stage. Specifically, we focus on the reconstruction face since the affected community has restored their communication means; i.e., in such a phase, humanitarian organizations and the government are aware of the needs of the disaster victims (Holguín-Veras et al., 2012). Hence, the reconstruction phase requires the establishment of a procedure that allocates HA resources, which is not always possible since the HSC is instable at this stage (Scholten et al., 2010). So, our chapter is closely related to the matching in practice literature that analyses the building of assignment procedures in environments under money generates disgusting transactions, i.e., selling HA is not ethical (Büthe et al., 2012).

Kidney transplantation "markets" is a classic example where money generates disgusting transactions (Roth et al., 2004). However, the mechanism design approach proposes an assignment procedure under patients' preferences are considered; even more, the application of the deferred acceptance and the top trading cycle procedures generate assignments that satisfy desirable properties. like fairness and efficiency, for such a market. Freedman et al. (2018) emphasize that mechanism design is a useful tool to align engineering algorithms with human values, which is also necessary for the HA allocation problem.

In our case, fairness and strategy-proofness are basic requirements that any assignment should satisfy to guarantee the effectiveness and credibility of the humanitarian supply chain. Thus, we determine the most appealing assignment procedure by following the experience of practitioners in the residency matching program (Roth, 2008), housing (Andersson et al., 2016), and school choice (Abdulkadiroğlu & Sönmez, 2003; Hatfield et al. 2016), to mention the most representative and successful cases of this literature.

Therefore, we model the HA allocation as a matching through institutions problem (Bloch et al., 2020) where members of the HSC do not have a (minimum or maximum) quota of goods to allocate among the disaster victims. Although the NDA applies directly, we observe that HL's objectives are not fully captured by desirable properties of the mechanism design literature. Specifically, the preference-priority structure leads with scenarios under victims and resources belong to pairs that are not acceptable by any member of the HSC. So, we introduce the notion of the last HA resource provider, to name a member of the HSC that can allocate any good to any victim. In other words, it is essential the existence of an HSC's member with similar features to the government and international organizations. Emphasizing on this last feature contributes to enhancing the HSC since no framework establishes that government collaboration is mandatory.

13.3 The Model

13.3.1 Basic Elements

Humanitarian logistics, at the post-disaster stage, are necessary to distribute and allocate those goods that are necessary for the recovery and reconstruction of the affected communities. Since these goods come from donations, or public institutions, and the ultimate goal of HL is welfare restoration, we have that HA are common resources; this means that any member of the HSC can allocate them to guarantee the effectiveness of HL. Even more, only members of the HSC can allocate HA during the post-disaster stage to avoid the arising of non-ethical exchanges such as grey markets (Büthe et al., 2012). Thus, we have a matching through institutions problem, as in Bloch et al. (2020), and we model the HA allocation problem as a three-sided market.

We consider a market composed of a set of disaster victims $V = \{v_1, v_2, \ldots, v_\eta\}$, HSC's members $M = \{m_1, m_2, \ldots, m_\mu\}$, and indivisible HA resources $R = \{r_1, r_2, \ldots, r_\rho\}$.

In opposition to HL developed at pre-disaster and during-disaster stages, in the reconstruction phase is possible to establish a functional communication structure between the HSC and the victims, which is not possible in the first moments after the catastrophe happens (Kusumasari et al., 2010). Thus, HSC's members can identify the victims' preference over the set of indivisible HA, i.e., at the post-disaster stage, it is possible to elicit victims' preferences. So, each victim v has a strict preference list P_v over the set $R \cup \{\emptyset\}$, where the empty set represents the outside option. Thus, $rP_v r'$ means that v strictly prefers good r to r'. Recalling that HSC's members have a complete communication structure, it is possible to classify HA according to victims' features; so, the outside option indicates that not all goods in R are useful (acceptable) for all victims. Formally, we say that r is acceptable for v if and only if $rP_v\emptyset$. Let $P = (P_{v_1}, P_{v_2}, \ldots, P_{v_\eta})$ be the profile of victims' preferences, where P_{-v} is the profile of all victims' preferences different from victim v.

We use m to denote a generic member of the HSC. We recall that a member m represents a volunteer, donor organization, or public institution that collaborates to restore the welfare of disaster victims, i.e., all $m \in M$ can allocate humanitarian aid.

In our case, the first way to cope with welfare restoration is by easing the allocation of HA; second, members also can contribute with the acquisition of HA through funding or by donating the good that victims require. So, let γ be a function from R to $M \cup \{\emptyset\}$ such that $\gamma(r) \in M$ means that the donator of r actively participate in the allocation of it, while $\gamma(r) = \emptyset$ means the opposite. Thus, $\gamma^{-1}(m)$ is the set of all goods that m donates, and $\gamma^{-1}(m) = \emptyset$ means that m does not provide any indivisible good. If γ is a bijective function, all donators of HA actively participate in the activities of post-disaster HL.

Concerning the allocation of HA among victims, the existence of a communication structure during the post-disaster stage allows us to assume that HSC's members evaluate the pertinence of assigning a good r to some victim v; i.e., each member m has a priority $\rhd_m$ over sets of pairs (r, v). In other words, m can attend sets of victims in different regions, but its intermediation depends on factors like distance. We denote by $\rhd = (\rhd_{m_1}, \rhd_{m_2}, \ldots, \rhd_{m_\mu})$ the profile of all members' priorities, where $\rhd_{-m}$ is the profile of all members' priorities different to m.

We assume that $\rhd_m$ is responsive, which means that, for all subsets $U \in 2^{R\times V}$ and all pairs $(r, v), (r', v') \in (R \times V)\backslash U$, we have that

i. $U \cup \{(r, v)\} \rhd_m U \cup \{(r', v')\}$ if and only if $\{(r, v)\} \rhd_m \{(r', v')\}$, and
ii. $U \cup \{(r, v)\} \rhd_m U$ if and only if $\{(r, v)\} \rhd_m \emptyset$.

Point (i) indicates that m attends a broader set of victims instead of another set of victims whenever the first has a pair resource-victim with a higher priority than the second. Similarly, Point (ii) establishes that m attends an additional pair resource-victim if and only if such pair is acceptable.

In the previous paragraphs, we discussed the lack of coordination in the assignment of HA during the post-disaster stage. Among the reasons behind the coordination problem, the absence of property rights stands; this means that no agent in M owns any resource in R because this is a shared pool of indivisible goods whose elements are donations that will become the victims' property when the assignment procedure finishes. To overcome the common use of HA, we assume that each resource r has a priority π_r over the set M; so, $m\pi_r m'$ means that m has a strict priority over m' to assign r.

The priority π_r summarizes the participation level of HSC members to get r. For example, if r is an apartment, priority π_r reflects the level of funding that each HSC member provided for building/rehabilitating the apartment; so, the member with the higher priority is the one that provides the more considerable amount of funding to build the apartment. Thus, donor members are the ones with top priority at their donations. Let $\pi = \left(\pi_{r_1}, \pi_{r_2}, \ldots, \pi_{r_\rho}\right)$ be the profile of all resources' priorities, where π_{-r} is the profile of resources' priorities different to r.

Finally, a humanitarian market is a tuple $(V, M, \gamma, R, P, \rhd, \pi)$.

13.3.2 The Solution Concept

We know that the ultimate goal of an HSC is the allocation of resources in the "best possible" way. At the post-disaster stage, specifically, it is possible to elicit the preferences to disaster victims because there exists a complete communication structure. So, we have a matching market where humanitarian resources only can be assigned through the intermediation of an HSC's member, i.e., the solution to our problem cannot be expressed by only considering victims and resources, as it happens in school choice (Abdulkadiroğlu & Sönmez, 2003; Hatfield et al., 2016) or kidney exchange (Roth et al., 2004; Ashlagi et al., 2015), to mention a few representative examples of two-sided matching markets.

Consequently, an assignment of HA relates victims, resources, and members of the HSC. We assume, for this chapter, that victims only can get one indivisible good, while members do not have constraints on the number of resources that can allocate. Hence, an assignment in our problems is composed by (i) a many-to-one matching between resources and members, and (ii) a one-to-one matching between victims and pairs resources-members. Below, we present the mathematical formulation of an assignment in a humanitarian market $(V, M, \gamma, R, P, \rhd, \pi)$.

An assignment is a pair $\Gamma = (\theta, \phi)$ such that

i. $\theta : R \cup M \rightarrow 2^R \cup M \cup \{\emptyset\}$, where

   a. $\theta(r) \in M \cup \{\emptyset\}$,
   b. $\theta(m) \in R \cup \{\emptyset\}$,
   c. $r \in \theta(m)$ if and only if $\theta(r) = m$;

ii. $\phi : R \times M \cup V \rightarrow R \times M \cup V \cup \{\emptyset\}$, where

   a. $\phi(r, m) \in V \cup \{\emptyset\}$,
   b. $\phi(v) \in R \times M \cup \{\emptyset\}$,
   c. $\phi(v) = (r, m)$ if and only if $\phi(r, m) = v$.

iii. $r \in \theta(m)$ if and only if $(r, m) = \phi(v)$ for some $v \in V$.

In the previous definition, point (iii) establishes the intermediation made by members of the HSC. In words, a victim gets a good if and only if there is a member that assigns it.

Although the assignment of humanitarian aid is the goal of HL at the post-disaster stage, this allocation also must satisfy desirable properties. As we mentioned in the introduction, the literature recognizes the importance of assigning humanitarian but also emphasizes on the necessity to improve such assignment procedures. Mainly, the assignments of HA are unfair and inefficient (de Montclos, 2009; Kwon & Kim, 2018).

The designing of fair and efficient mechanisms requires the specification of how members in the HSC make decisions, i.e., in which pairs, composed by a HA resource and a victim, a member intermediate. We assume that all members are rational, i.e., they only intervene in those pairs that are acceptable for him. Formally, we consider that each member m follows a choice function $Ch_m : 2^{R\times V} \rightarrow 2^{R\times V}$ such that for all sets $U \in 2^{R\times V}$, we have that $Ch_m(U) = \{(r, v) \in U | (r, v) \triangleright \emptyset\}$. Given this choice function, we can establish the desirable properties that assignments of HA must satisfy.

An assignment Γ is **individually rational (IR)** if every agent in the market (victims and members) get an allocation that is weakly better to the outside option, i.e., all agents get an acceptable assignment. Formally, an IR assignment Γ satisfies

i. For all $v \in V$, either $\phi(v)P_v\emptyset$ or $\phi(v) = \emptyset$, and
ii. For all $m \in M$, $\Gamma(m) = Ch_m(\Gamma(m))$.

Given that HL suffers from coordination problems at the post-disaster stage, it is common that some HA is wasted, i.e., resources remain in distribution centers while there still exist victims that require them. In other words, the HA allocation must satisfy the non-wastefulness property. An assignment Γ is **non-wasteful** if no victim-member pair (v, m) can claim a resource r, i.e., there is no v, m, r such that

i. $rP_v\phi(v)$,
ii. $(r, v) \in Ch_m(\Gamma(m) \cup (r, v))$, and
iii. $\Gamma(r) = \emptyset$.

Also, we observe in the literature that post-disaster allocation of HA suffers from coping with "fairness," i.e., victims do not get what they need even if they are the victims with the higher priority. In other words, it is possible the existence of justified envy among the beneficiaries of the HSC. We say that (v, m) has **justified envy** over (v', m') at an IR assignment Γ if

i. $\phi(v')P_v\phi(v)$,
ii. $(\phi(v'), v) \in Ch_m(\Gamma(m) \cup (\phi(v'), v))$, and
iii. $m\pi_{\Gamma(v')}m'$.

Therefore, justified envy arises when an SHC' member has a higher priority to assign a good that a victim prefers to her allocation. So, an assignment Γ is **fair** if it is individually rational, non-wasteful, and there is no justified envy.

Also, efficiency is a desirable property for HA assignments. Concerning such property, we say that an assignment is efficient if no victim can improve without harming others. In other words, Γ is **efficient** if there is no Γ' such that all victims weakly prefer their allocation on Γ' to their allocation on Γ.

Finally, it is worth mentioning that disaster's victims have incentives to misreport their preferences in order to get the best possible HA for them; sometimes, we say that victims manipulate the game when they report a false preference list that provides them with a better good than the one assigned in the opposite case. So, the designing of an assignment procedure must ensure that victims do not manipulate the allocation procedure; in the mechanism design literature, this means that victims must truthfully report their preferences. Before to formalize this concept, we introduce additional concepts.

An assignment procedure Λ is a map from the set of all triads $(P, \triangleright, \pi)$ into the set of all possible assignments Γ, i.e., $\Lambda(P, \triangleright, \pi)$ is an assignment where $\Lambda(P, \triangleright, \pi)(e)$ is the assignment of e for all $e \in V \cup M \cup R$. Suppose that P is the profile of victims' true preferences, and P' is a profile where at least some victim misreports her valid preferences, we say that Λ is **strategy-proof for victims** if and only if

$$\Lambda(P, \triangleright, \pi)(e) P_v \Lambda(P', \triangleright, \pi)(e)$$

for all $v \in V$. In terms of game theoretical concepts, reporting valid preferences is a dominant strategy under a strategy-proof mechanism.

13.4 Designing a Mechanism for the Allocation of Indivisible HA

The allocation of indivisible goods represents a significant problem in economics since indivisibility drives to the non-existence of a competitive equilibrium, i.e., there is no price vector that induces an allocation that clear the market (Roth & Postlewaite, 1977). In the previous observation, it is worth noticing that market allocation mechanisms (typically) rely on price generation. Given that HA, in general, comes from donations, pricing mechanisms are not suitable for the allocation of indivisible goods (Abidi et al., 2013). In other words, it is not ethical selling HA goods among disaster victims. Therefore, it is necessary to design a centralized mechanism that considers the preferences and priorities of victims and HSC's members; such a procedure should allocate HA without considering money exchange.

In this section, we first present the Deferred Acceptance (DA) (Gale & Shapley, 1962) algorithm as a procedure to generate a fair allocation. However, the implementation of such a mechanism requires to pre-allocate HA among HSC's members. In other words, the DA cannot deal with the fact that multiple members can allocate an HA good. So, we later discuss the implementation of the Nested Deferred Acceptance algorithm (Bloch et al., 2020) to deal with the flexible allocation that characterizes HA.

13.4.1 The Deferred Acceptance Algorithm

Introduced by Gale and Shapley in 1962, the DA algorithm generates an allocation in a two-sided market of men and women (the marriage problem) or students and universities (the college admission problem). So, this mechanism generates a match between two types of agents, each agent with a preference relation over the agents of the other type; or allocation between people and objects, where objects typically prioritize how people get them while people have preferences over objects.

Although our modeling approach represents the allocation of indivisible HA as a three-sided market, this problem is a two-sided market whenever there exists a partition $\mathbb{P} = \{R_1, \ldots, R_\mu\}$ of R and an injective function $\wp : \mathbb{P} \to M$, i.e., we have that

i. $\cup_{j=1}^{\mu} R_j = R$,
ii. $R_i \cap R_j = \emptyset$ for all $i \neq j$, and
iii. If $\wp(R_i) = \wp(R_j)$, then $R_i = R_j$.

In words, the pair $(\mathbb{P}, \wp)$ indicates that each $R_j \in \mathbb{P}$ represents a set of HA resources that are exclusively assigned by HSC's member $\wp(R_i)$. So, it is natural to assume that each member m of the humanitarian supply chain prioritize which disaster victim should get a good in R_m; we denote this priority as $\rhd_m^r$ for all $r \in R_m$. To avoid extra notation, we denote by $\wp(\mathrm{r})$ the HSC member that can assign the resource r, i.e., $\wp(\mathrm{r}) = \mathrm{m}$ if and only if $r \in R_m$. So, $v \rhd_m^r v'$ means that v has priority over v' to get the aid r when m is the unique member of the HSC that can allocate r. Consequently, the allocation of indivisible HA is a two-sided market when we consider that $\rhd_m = \left(\rhd_m^r\right)_{r \in R_m}$.

Note that, by considering $\rhd_m$ as a vector of priorities, the pair (r, m) can be understood as an agent of the HSC since it has a priority $\rhd_m^r$. The previous discussion indicates how each resource must be assigned among the disaster's victims. So, the allocation of HA casts similarities with the college admission problem; in such a situation, the DA procedure is recognized as a proper assignment procedure to generate fair allocations due to the empirical evidence (Gale & Shapley, 1962; Roth, 2008).

By considering the market $(V, M, \gamma, R, P, \rhd, \mathbb{P}, \wp)$, where $\rhd_m = \left(\rhd_m^r\right)_{r \in R_m}$, the DA proceeds as follows.

Step 1. Each disaster victim v asks for her most preferred acceptable HA resource, i.e., the top resource at preference P_v. Each pair $(r, \wp(\mathrm{r}))$ observes all victims' requisitions, and $\wp(\mathrm{r})$ tentatively allocates r to the victim with the highest priority at $\rhd_{\wp(\mathrm{r})}^r$ while other victims are rejected.

Step t. Every rejected victim requests the most preferred resource among those resource that has not previously rejected them. Resource-member pairs observe all requisitions, and the member allocates the resource to the victim with the highest priority; other victims are rejected.

The algorithm stops when no further requisitions can be made, i.e., no victim can request a resource either all her acceptable resources rejected her, or she tentatively is assigned to some good. When the algorithm stops, all victims get the resource tentatively assigned to them.

In the following example, we show how the DA algorithm applies for a market $(V, M, \gamma, R, P, \rhd, \mathbb{P}, \wp)$.

Example 1 *Consider* $V = \{v_1, v_2, v_3\}$, $M = \{m_1, m_2\}$ *and* $R = \{r_1, r_2, r_3\}$, *where* $\wp$ *is such that we have the following resource-member pairs* $(r_1, m_1), (r_2, m_1)$ *and* (r_3, m_2). *Preferences and priorities are given as follow*

$$P = \begin{pmatrix} P_{v_1} & P_{v_2} & P_{v_3} \\ r_1 & r_1 & r_3 \\ r_3 & r_3 & r_2 \\ r_2 & r_2 & r_1 \end{pmatrix} \text{ and } \rhd = \begin{pmatrix} \rhd_{m_1}^{r_1} & \rhd_{m_1}^{r_2} & \rhd_{m_2}^{r_3} \\ v_3 & v_1 & v_1 \\ v_2 & v_3 & v_2 \\ v_1 & d & v_3 \end{pmatrix}.$$

Finally, we consider that no member of the HSC in M is a donor of any of the resources in R.

During the first step of the DA algorithm, victims v_1 *and* v_2 *request resource* r_1, *while victim* v_3 *requests* r_3. *Since* $v_2 \rhd_{m_1}^{r_1} v_1$, *we have that member* m_1 *tentatively allocates resource* r_1 *to victim* v_2. *Moreover,* m_2 *tentatively allocates HA* r_3 *to member* v_3 *because of no other victim requests* r_3. *Step 1 of the DA finishes with the following tentative allocation*

$$\Gamma^1 = \begin{pmatrix} v_1 & v_2 & v_3 \\ \emptyset & m_1 & m_2 \\ \emptyset & r_1 & r_3 \end{pmatrix}$$

where v_1 *is rejected from resource* r_1.

At the beginning of the second step, victims v_2 *and* v_3 *retains their allocation, while* v_1 *requests an acceptable HA from those that have not rejected her. Note that* v_1 *has priority over* v_3 *to get the resource* r_3 *from the member* m_2. *Thus,* v_3 *is rejected from* r_3, *at the second finishes with the following tentative allocation.*

$$\Gamma^2 = \begin{pmatrix} v_1 & v_2 & v_3 \\ m_2 & m_1 & \emptyset \\ r_3 & r_1 & \emptyset \end{pmatrix}.$$

In the third step, v_1 *and* v_2 *retain their allocation, while* v_3 *requests an acceptable good from those that have not previously reject her. So,* v_3 *requests* r_2.

Since v_1 *and* v_2 *are tentatively assigned to* r_3 *and* r_1*, we have that* m_1 *tentatively assigns* r_2 *to* v_3*. The step finishes with the following tentative assignment*

$$\Gamma^3 = \begin{pmatrix} v_1 & v_2 & v_3 \\ m_2 & m_1 & m_1 \\ r_3 & r_1 & r_2 \end{pmatrix}.$$

Note that all disaster victims get a resource at the end of step 3. Thus, the algorithm finishes in step 3, and the final assignment is Γ^3. □

The Deferred Acceptance algorithm is widely used in the matching literature given the desirable properties that it satisfies. Gale and Shapley (1962) demonstrate that the DA is a strategy-proof mechanism that generates a stable matching in the marriage problem. In such a context, stability means that there are no blocking pairs, i.e., there is no man-woman pair such that they prefer to be matched instead of the match that they get in the procedure.

If we consider that a side of the marriage market is a set of indivisible objects, the existence of a blocking pair implies the existence of justified envy. Hence, whenever we apply the DA algorithm to the market $(V, M, \gamma, R, P, \rhd, \mathbb{P}, \wp)$, the assignment is fair.

It is worth mentioning that the DA is not efficient, as also Gale and Shapley point out (Kesten, 2010). However, it is possible to demonstrate that this mechanism generates the best possible assignment for the agents that make the requisition, while it generates the worst possible assignment for the agents that receive the requisition. The following theorem summarizes the previous discussion for the allocation of humanitarian aid when we model it as a two-sided market.

Theorem 1 *Whenever the DA procedure is applied to allocate HA resources in a market* $(V, M, \gamma, R, P, \rhd, \mathbb{P}, \wp)$*, the DA mechanism is strategy-proof for disaster victims and generates a fair assignment. Even more, every disaster victim is at least as well off under* Γ^{DA} *as she would be under any other fair assignment.*

Proof *In the paragraphs above, we point out that the HA allocation problem casts similarities with a two-sided market when*

$\rhd_m = \left(\rhd_m^r\right)_{r\in R_m}$ *for all* $m \in M$.

Specifically, the market $(V, M, \gamma, R, P, \rhd, \mathbb{P}, \wp)$ *can be seen as a college admission problem where pairs* (r, m) *are universities that admit students. In this case, an HA resource, among students (disaster victims). Thus, the proof follows the reasoning in Gale and Shapley* 1962. □

13.4.2 The Intervention of the HSC's Members

By Theorem 1, the DA algorithm is an appealing assignment procedure for the allocation of HA when it is clear the HSC's member that **should** assign each HA resource. However, the last is not always evident in the post-disaster stage since the HSC characterizes by its lack of agility (Oloruntoba & Gray, 2006), which means that members of the humanitarian supply chain do not have the same responsive level than those that belong to a commercial supply chain. Hence, the two-sided framework $(V, M, \gamma, R, P, \triangleright, \mathbb{P}, \wp)$ cannot deal with the fact that any HSC member can assign any HA resource since it assumes a pre-allocation of goods γ. Thus, any member m_j has the ability to allocate any resource in R_{m_j}.

Consequently, the post-disaster stage entitles each HSC member with different abilities to allocate HA resources, and such abilities depend on the community affectations caused by the disaster. For example, not all aid donors actively collaborate in the allocation of HA since they do not have the capacity and knowledge to distribute HA in the affected communities, as we explain in the description of the function γ. So, the inclusion of the resource priority vector π captures the fact that any HSC members do not own HA resources; even more, such priority vector contributes identifying which HSC's member is the most appropriated to allocate an aid resource among disaster's victim. In other words, the HA allocation has a three-sided behavior that is not directly managed by the DA algorithm.

Given the limitations of the two-sided framework $(V, M, \gamma, R, P, \triangleright, \mathbb{P}, \wp)$, the literature on mechanism design discusses DA's modifications to deal with market structures that are not two-sided markets in nature (Kamada & Kojima, 2018). In such literature, we observe the introduction of priorities that generates "artificial" property rights over indivisible goods, which is the theoretical basis of function $\wp$. However, the previous approach ignores that any member in M can assign HA resource like it happens with housing.

Another stream of the literature splits three-sided assignment procedures into two stages in which property rights over resources are determined in the first stage, while the second stage focuses on allocating goods among the ones that need the good. Nevertheless, this second approach raises criticism since two-step assignment procedures are not strategy-proof (Goto et al. 2014; Diebold and Bichler, 2017).

By the previous discussion, the NDA mechanism deals with the fact that HA allocation is flexible; also, such a mechanism endogenizes the establishment of rights to use common resources. Notably, this mechanism uses the DA as its basis, but it also interlinks two deferred acceptance algorithms. In the first one, the outer DA, disaster victims request HA, but the assignment is not completed since it is necessary for the intervention of the HSC's members. Thus, in the second DA, the nested one or the inner DA, HSCs members compete to get the resources that they can assign. So, the inner loop indicates which HSC's member should assign each good, and the algorithm goes back to the outer loop where HSC's member assigns HA among the victims that request it.

Below, we formally describe how the NDA applies to a market $(V, M, \gamma, R, P, \triangleright, \pi)$.

Step 1. This step is divided into the following phases:

Phase A1 (the outer loop). Each disaster victim v requests her most preferred indivisible good at the preference list P_v, which means that each victim elicits their demand at this phase. Let $\mathfrak{M}^1$ be the set of all pairs (r, v) such that r is the HA demanded by v. The algorithm goes to phase B.

Phase B1 (the inner loop). Each member m of the HSC observes the set $\mathfrak{M}^1$. Let $R_m^{\mathfrak{M}^1} = \{v \in R | (r, v) \rhd \emptyset \text{ for some } (r, v) \in \mathfrak{M}^1\}$ be the set of all resources in R such that m can intervene in its allocation. Then, we start a deferred acceptance algorithm under agents in M request an indivisible object in $R_m^{\mathfrak{M}^1}$. Note that this nested DA solves an allocation problem that is like a school choice problem; when the nested DA finishes, the Phase B of the NDA stops, and the algorithm goes to Phase C.

Phase C1 (the tentative allocation). Let $\theta^{B^1}(m)$ the tentative allocation of goods that each member of the HSC gets at the end of Phase B. So, each member m tentatively assigns an HA resource $r \in \theta^{B^1}(m)$ to the victim v if and only if $(r, m) \in Ch_m(\theta^{B^1}(m))$. Other disaster victims are rejected, and the algorithm goes to step t.

Step t. This step is divided into the following phases:

Phase At (the outer loop). Each disaster victim v requests her most preferred indivisible good, among the acceptable goods that have not rejected her, at preference list P_v. Let $\mathfrak{M}^t$ be the set of all pairs (r, v) such that r is the HA demanded by v. The algorithm goes to phase B.

Phase Bt (the inner loop). Each member m of the HSC observes the set $\mathfrak{M}^t$. Let $R_m^{\mathfrak{M}^t} = \{v \in R | (r, v) \rhd \emptyset \text{ for some } (r, v) \in \mathfrak{M}^t\}$ be the set of all resources in R such that m can intervene in its allocation. Then, we start a deferred acceptance algorithm in which agents in M request an indivisible object in $R_m^{\mathfrak{M}^t}$. Note that this nested DA solves an allocation problem that is like a school choice problem; when the nested DA finishes, the Phase B of the NDA stops, and the algorithm goes to Phase C.

Phase Ct (the tentative allocation). Let $\theta^{B^t}(m)$ the tentative allocation of goods that each member of the HSC gets at the end of Phase B. So, each member m tentatively assigns an HA resource $r \in \theta^{B^t}(m)$ to the victim v if and only if $(r, m) \in Ch_m(\theta^{B^t}(m))$. Other disaster victims are rejected, and the algorithm goes to step t.

The NDA algorithm finishes when all disaster victims are assigned to some HA resource or all their acceptable resources have rejected them. We denote by Γ^{NDA} the assignment produced by the NDA algorithm.

In the example below, we illustrate how the NDA algorithm works in a market $(V, M, \gamma, R, P, \rhd, \pi)$.

Example 2 *Consider* $V = \{v_1, v_2, v_3\}$, $M = \{m_1, m_2\}$ *and* $R = \{r_1, r_2, r_3\}$. *Victims' preferences and members' priorities are given as follow*

$$P = \begin{pmatrix} P_{v_1} & P_{v_2} & P_{v_3} \\ r_1 & r_1 & r_3 \\ r_3 & r_3 & r_2 \\ r_2 & r_2 & r_1 \end{pmatrix} \quad \text{and} \quad \rhd = \begin{pmatrix} \rhd_{m_1} & \rhd_{m_2} \\ (r_1, v_1) & (r_1, v_1) \\ (r_2, v_1) & (r_1, v_2) \\ (r_1, v_3) & (r_3, v_2) \\ (r_2, v_1) & (r_2, v_2) \end{pmatrix}.$$

Also, we consider that no member of the HSC in M is a donor of any of the resources in R. Finally, the profile of resources priorities is

$$\pi = \begin{pmatrix} \pi_{r_1} & \pi_{r_2} & \pi_{r_3} \\ m_1 & m_2 & m_1 \\ m_2 & m_1 & \end{pmatrix}.$$

Step 1

Phase A1. At this phase, all disaster victims request their most preferred HA resource, which implies that $\mathfrak{M}^1 = \{(r_1, v_1), (r_1, v_2), (r_3, v_3)\}$. Although the previous set indicates the demand of victims, the allocation of resources requires the intervention of the HSC's members. Thus, the NDA goes to Phase B1.

Phase B1. During this phase, members in M compete to allocate the HA resources only for acceptable pairs in $\mathfrak{M}^1$; so, we have that $R_{m_1}^{\mathfrak{M}^1} = \{(r_1, v_1)\}$ and $R_{m_2}^{\mathfrak{M}^1} = \{(r_1, v_1), (r_1, v_2)\}$. Then, we run a DA between members and resources where both members ask for resource r_1, at step 1, because (r_1, v_1) is the top pair for m_1 and m_2. Given that $m_1 \pi_{r_1} m_2$, we have that m_1 is the member that can allocate resource r_1.. This DA algorithm goes to step 2 where m_1 retains r_1, while m_2 demands r_1 since (r_1, v_2) is the second pair with the highest priority at $\rhd_{m_2}$. However, the tentative allocation (m_1, r_1) remains since m_1 has a higher priority than m_2 under priority π_{r_1}.

Phase C1. By the previous phase, we have that v_2 is rejected from r_1 and v_3 is rejected from r_3, while m_1 tentatively allocates good r_1 to the victim v_1. In other words, the tentative assignment is

$$\Gamma^{NDA^1} = \begin{pmatrix} v_1 & v_2 & v_3 \\ r_1 & \emptyset & \emptyset \\ m_1 & \emptyset & \emptyset \end{pmatrix}.$$

Then, Step 1 of the NDA algorithm finishes.

Step 2

Phase A2. Disaster victim v_1 retains resource r_1, while victims v_2 and v_3 request resources r_3 and r_2, respectively. In other words, we have that

$\mathfrak{M}^2 = \{(r_1, v_1), (r_3, v_2), (r_2, v_3)\}$.

Phase B2. At the beginning of this phase, each member observes the set $\mathfrak{M}^2$, and chooses those pairs that are acceptable for them. So, we have that $R_{m_1}^{\mathfrak{M}^2} = \{(r_1, v_1)\}$ and $R_{m_2}^{\mathfrak{M}^2} = \{(r_3, v_2)\}$. So, the deferred acceptance between M and R starts. So, m_1 and m_2 demand resources r_1 and r_3, respectively. Note that members of the HSC do not compete to allocate an HA resource, the deferred acceptance of the inner loop finishes.

Phase C2. By the previous phase, we have that m_1 tentatively allocates r_1 to victim v_1, while m_2 tentatively allocates r_3 to v_2 since HSC's members do not get requisitions from other victims to these HA resources. Then, v_3 is rejected from r_2, and the tentative assignment is

$$\Gamma^{NDA^2} = \begin{pmatrix} v_1 & v_2 & v_3 \\ r_1 & r_3 & \emptyset \\ m_1 & m_2 & \emptyset \end{pmatrix}.$$

Step 3

Phase A3. Victim v_1 and v_2 retain resource r_1 and r_3, respectively. Since v_3 was rejected from r_2, she requests her last acceptable HA resource, which is r_1. So, we have that $\mathfrak{M}^3 = \{(r_1, v_1), (r_3, v_2), (r_1, v_3)\}$.

Phase B3. Since each member of the humanitarian supply chain observes the requisitions made by the victims, they choose those requisitions that can attend, i.e., those that are acceptable. Thus, the set of allocation under each HSC's member can intervene are $R_{m_1}^{\mathfrak{M}^3} = \{(r_1, v_1), (r_1, v_3)\}$ and $R_{m_2}^{\mathfrak{M}^2} = \{(r_3, v_2)\}$. Analogously to Phase B in Step 2, members m_1 and m_2 can allocate resources r_1 and r_3, respectively, since they do not compete for the allocation of the same resources.

Phase C3. Note that member m_2 only receives requisitions from the victim v_2, which implies that m_2 tentatively allocates r_3 to v_2. On the opposite, member m_1 receives requisitions from v_1 and v_3 to get the resource r_1. Given that $(r_1, v_1) \rhd_{m_1} (r_1, v_3)$, we conclude that member m_1 tentatively assigns resource r_1 to v_1. Hence, v_3 is rejected from resource r_1, and the tentative assignment is

$$\Gamma^{NDA^3} = \begin{pmatrix} v_1 & v_2 & v_3 \\ r_1 & r_3 & \emptyset \\ m_1 & m_2 & \emptyset \end{pmatrix}.$$

At the end of step 3, victims that are not assigned to some resources have been rejected from all their acceptable HA. Thus, the NDA finishes, and the final assignment is

$$\Gamma^{NDA} = \begin{pmatrix} v_1 & v_2 & v_3 \\ r_1 & r_3 & \emptyset \\ m_1 & m_2 & \emptyset \end{pmatrix}.$$

Example 2 illustrates how the NDA mechanism deals with the fact that HA resources are donations that can be allocated by any HSC's member. Specifically, during the inner loop, the members of the humanitarian supply chain compete to determine which agents are the most capable of allocating HA resources by following the same reasoning of the DA algorithm, i.e., tentative assignments that arise from proposals made in a descendent manner concerning victims' preferences and rejections based on members' priorities.

Also, it is worth mentioning that the HA allocation problem is a particular case of matching through institutions' problems. The last considers the existence of distributional constraints, i.e., institutions should allocate a fixed number of apartments.

In our case, HSC's members, that play the role of institutions, do not have a capacity constraint since, in the post-disaster stage, it is essential to allocate the most considerable amount of resources. In other words, the imposition of distributional constraints limits members to allocate resources. So, we have a problem under distributional constraints not useful in the discussion performed by Bloch, Cantala, and Gibaja (2020). Even more, the application of the NDA to the allocation of HA inherits all the properties that the NDA satisfies in the matching through institutions problem. The following theorem summarizes the previous discussion.

Theorem 2 *Consider the market of HA resources* $(V, M, \gamma, R, P, \triangleright, \pi)$. *When we apply the NDA mechanism to allocate HA resources, the output of this assignment is fair. Even more, every disaster victim is at least as well off under* Γ^{DA} *as she would be under any other fair assignment, and they cannot manipulate the algorithm.*

Proof *See the ineffective quotas case of Bloch, Cantala, and Gibaja* (2020).

13.5 Discussion

From a theoretical point of view, Theorem 2 establishes that the NDA is an appealing assignment procedure for the HA allocation problem due to it satisfies desirable properties in a post-disaster stage and deals with the fact that HA is composed mainly of donations (common resources). However, the NDA's application relies on considering that disaster victims and members report their preferences and priorities to the mechanism, assumption that may lead to undesirable consequences like the one illustrated in Example 2. In such an example, victim v_3 is rejected from all HA resources, i.e., she does not get any aid at the end of the procedures. Also, we observe that resource r_2 remains unassigned; in other words, the resource is "wasted" in the sense that no HSC member can allocate the pair (r_2, v_3).

The previous scenario is not an unexpected result since the majority of the members of an HSC are volunteers (Jahre et al., 2007). Nevertheless, the overall performance of an HSC may be considered as inefficient due to the existence of a wasted resource or a victim that does not get a resource. Since HL pursue welfare restoration in the affected communities during the post-disaster stage, any of the previous situations do not contribute to coping with the HSC's objectives.

Note that victim v_3 does not get resource r_2 since the pair (r_2, v_3) is not acceptable for any member of the HSC, despite that r_2 is acceptable for v_3. In this case, we say that the structure of preferences and priorities is incomplete whenever the NDA is applied to the market $(V, M, \gamma, R, P, \rhd, \pi)$. So, a **complete market whenever the NDA is applied** is a market $(V, M, \gamma, R, P, \rhd, \pi)$ such that no pair (r, v) remains unassigned in the assignment whenever the good is acceptable for the disaster victim v.

Although a **complete** market $(V, M, \gamma, R, P, \rhd, \pi)$ guarantees non-wasted resources; it is a strong assumption in a post-disaster scenario since it considers that the market is over-demanded and implies that any HA resource can be applied for at least some member of the HSC. As we mentioned before, this is not always possible (Kwon & Kim, 2018).

To weaken the complete market property, we can consider that the government actively participates in HL during the post-disaster stage. For instance, the government excels as a member of HSC since it has abilities/resources that no other member, volunteer or not, in M has. Specifically, the government has access to all transportation and distribution structure, which allow it to allocate any HA resource among the disaster victims (Kunz and Gold, 2017). By following the idea of Heredia-Roldán et al. (2019), the government may participate in the HSC as a last resource provider of HA resources to disaster victims. In our humanitarian context, the last resource provider means that all pairs in $R \times V$ are acceptable for the government. Below, we formalize the previous concept.

We use g to denote the government; which means that $\rhd_g$ is the government's priority. We say that the government is the **last HA provider** of a market $(V, M, \gamma, R, P, \rhd, \pi)$ if

i. $g \in M$, (and)
ii. $(r, v) \rhd_g \emptyset$ for all $(r, v) \in R \times V$.

Proposition 1 *If the government is the last resource provider of the market* $(V, M, \gamma, R, P, \rhd, \pi)$, *then the market is complete.*

Proof *We proceed by contradiction, i.e., we consider the existence of a pair* (r, v) *that remains unassigned, even if* $rP_v\emptyset$, *when the NDA algorithm is applied to allocate HA goods.*

Since the HA resource r is acceptable for victim v, this implies that the victim requests the resource at some step t of the NDA algorithm. Moreover, the government is the last resource provider of this market, which means that g competes to allocate r during phase Bt. However, r and v remain unassigned when the algorithm finishes. So, we have the following cases

i. *The government rejects v because there exists v′ such that* $(r, v') \triangleright_g (r, v)$. *In words, the government allocates r to the victim v′, which is a contradiction since the HA good remains unassigned at the end of the mechanism.*
ii. *The government cannot allocate the resource r since there exists member m such that* $m\pi_r g$, *i.e.,* $r \in \theta^{NDA}(m)$. *It is a contradiction with the fact that r remains unassigned.*

In any case, a contradiction arises by assuming the existence of an unassigned pair (r, v). *Therefore, the market is complete.* □

The veracity of Proposition 1 relies on the features of the last result provider, an argument that is independent of the fact that the government belongs or not to the HSC.

Although the government satisfies the definition of the last resource provider, the government is not a unique agent with such features. Even more, no framework guarantees the participation of the government in the post-disaster stage. In humanitarian crisis derived from political issues, governments disappear or lost some of the properties of a last HA resource provider (Kunz and Gold, 2017). However, international organizations, such as the United Nations, also have the abilities to allocate any resource and their governance structure imply the acceptability of all possible pairs (r, v).

Remark 1 *The existence of an HA resource provider guarantees that no resource is wasted and implies that the most significant number of disaster victims get an HA resource (whenever the HA demand overpasses HA supply).*

13.6 Conclusions

In this chapter, we analyze the allocation of HA through a mechanism design perspective. Such theoretical framework allows us to deal with the problems that HA allocation faces during the post-disaster stage. For instance, we focus on the lack of fairness that characterizes the assignment procedures in such stage and the fact that disaster victims have incentives to manipulate the assignment procedure given the affectations caused by the catastrophe.

To overcome the previous issues, we propose a variation of the DA algorithm that considers pre-defined property rights, i.e., there exists a pre-allocation of HA resources among members of the HSC that determines which members should allocate each resource. Although such DA's variation copes with fairness and strategy-proofness, the pre-allocation of resources is not a realistic assumption in the post-disaster stage since any member may allocate any HA resource. Even more, from a theoretical point of view, the pre-allocation of HA goods induces a two-stage assignment procedure under victims have incentives to misreport their preferences. Hence, we apply the NDA algorithm for the allocation of HA resources, which satisfies the same properties as the DA and deals with the possibility that any HSC's member allocates HA.

Although the post-disaster stage allows that victims and HSC's members report their preferences and priorities since communication technologies are available for the reconstruction process, the building of a fair assignment may be too restricted for reaching with HL's objectives. Specifically, not all allocations are possible, given the features of the volunteers that integrate the HSC. Thus, the presence of the last resort provider is necessary for the effectiveness of the HSC; in other words, the humanitarian supply chain needs a member that can attend all victims regardless of the good that they need. The participation of the last provider deals with the instability and short life that characterize HSCs, i.e., the last resort provider remains in the HSC even if other members come back to their usual activities.

References

Abdulkadiroğlu, A., & Sönmez, T. (2003). School choice: A mechanism design approach. *American Economic Review, 93*(3), 729–747.

Abidi, H., de Leeuw, S., & Klumpp, M. (2013). Measuring success in humanitarian supply chains. *International Journal of Business and Management Invention, 2*(8), 31–39.

Adivar, B., Atan, T., Sevil Oflaç, B., & Örten, T. (2010). Improving social welfare chain using optimal planning model. *Supply Chain Management: An International Journal, 15*(4), 290–305.

Andersson, T., Ehlers, L., & Svensson, L. G. (2016). Transferring ownership of public housing to existing tenants: A market design approach. *Journal of Economic Theory, 165,* 643–671.

Ashlagi, I., Fischer, F., Kash, I. A., & Procaccia, A. D. (2015). Mix and match: A strategyproof mechanism for multi-hospital kidney exchange. *Games and Economic Behavior, 91,* 284–296.

Bloch, F., Cantala, D., & Gibaja, D. (2020). Matching through institutions. Games and Economic Behavior, *121*, 204-231.

Brown, N. A., Rovins, J. E., Feldmann-Jensen, S., Orchiston, C., & Johnston, D. (2017). Exploring disaster resilience within the hotel sector: A systematic review of literature. *International Journal of Disaster Risk Reduction, 22,* 362–370.

Büthe, T., Major, S., & e Souza, A. D. M. (2012). The politics of private foreign aid: humanitarian principles, economic development objectives, and organizational interests in NGO private aid allocation. *International Organization, 66*(4), 571–607.

Chandraprakaikul, W. (2010). Humanitarian supply chain management: Literature review and future research. In *The 2nd international conference on logistics and transport, Queenstown* (Vol. 18).

De Montclos, M. P. (2009). Humanitarian aid and the Biafra war: Lessons not learned. *Africa Development, 34*(1).

Diebold, F., & Bichler, M. (2017). Matching with indifferences: A comparison of algorithms in the context of course allocation. *European Journal of Operational Research, 260*(1), 268–282.

Freedman, R., Borg, J. S., Sinnott-Armstrong, W., Dickerson, J. P., & Conitzer, V. (2018, April). Adapting a kidney exchange algorithm to align with human values. In *Thirty-Second AAAI Conference on Artificial Intelligence*.

Gale, D., & Shapley, L. S. (1962). College admissions and the stability of marriage. *The American Mathematical Monthly, 69*(1), 9–15.

Goto, M., Hashimoto, N., Iwasaki, A., Kawasaki, Y., Ueda, S., Yasuda, Y., & Yokoo, M. (2014, May). Strategy-proof matching with regional minimum quotas. In *Proceedings of the 2014 international conference on Autonomous agents and multi-agent systems* (pp. 1225–1232). International Foundation for Autonomous Agents and Multiagent Systems.

Hatfield, J. W., Kojima, F., & Narita, Y. (2016). Improving schools through school choice: A market design approach. *Journal of Economic Theory, 166,* 186–211.

Heredia-Roldán, M. J., Gibaja-Romero, D. E., Martínez-Flores, J. L., & Caballero-Morales, S. O. (2019). The impact of trust in the strategic decisions of a decentralized supply chain. *OPSEARCH*, 1–23.

Holguín-Veras, J., Jaller, M., Van Wassenhove, L. N., Pérez, N., & Wachtendorf, T. (2012). On the unique features of post-disaster humanitarian logistics. *Journal of Operations Management, 30*(7–8), 494–506.

Ilhan, A. M. (2011). The humanitarian relief chain. *South East European Journal of Economics and Business*. https://doi.org/10.2478/v10033-011-0015-x.

Jahre, M., Persson, G., Kovács, G., & Spens, K. M. (2007). Humanitarian logistics in disaster relief operations. *International Journal of Physical Distribution & Logistics Management.*

Kamada, Y., & Kojima, F. (2018). Stability and strategy-proofness for matching with constraints: A necessary and sufficient condition. *Theoretical Economics, 13*(2), 761–793.

Kesten, O. (2010). School choice with consent. *The Quarterly Journal of Economics, 125*(3), 1297–1348.

Khan, H., Vasilescu, L. G., & Khan, A. (2008). Disaster management cycle-a theoretical approach. *Journal of Management and Marketing, 6*(1), 43–50.

Kovács, G., & Spens, K. (2007a). Humanitarian logistics in disaster relief operations. *International Journal of Physical Distribution & Logistics Management, 37*(2), 99–114.

Kovács, G., & Spens, K. M. (2007b). Logistics theory building. *The Icfai Journal of Supply Chain Management, 4*(4), 7–27.

Kunz, N., & Gold, S. (2017). Sustainable humanitarian supply chain management–exploring new theory. *International Journal of Logistics Research and Applications, 20*(2), 85–104.

Kusumasari, B., Alam, Q., & Siddiqui, K. (2010). Resource capability for local government in managing disaster. *Disaster Prevention and Management: An International Journal, 19*(4), 438–451.

Kwon, I. W., & Kim, S. H. (2018). Humanitarian supply chain/logistics: Roadmap to effective relief effort. *Journal of International & Interdisciplinary Business Research, 5*(1), 95–109.

Lin Moe, T., & Pathranarakul, P. (2006). An integrated approach to natural disaster management: public project management and its critical success factors. *Disaster Prevention and Management: An International Journal, 15*(3), 396–413.

Nagurney, A., Yu, M., & Qiang, Q. (2011). Supply chain network design for critical needs with outsourcing. *Papers in Regional Science, 90*(1), 123–142.

Oloruntoba, R., & Gray, R. (2006). Humanitarian aid: an agile supply chain? *Supply Chain Management: an International Journal, 11*(2), 115–120.

Roth, A. E. (2008). Deferred acceptance algorithms: History, theory, practice, and open questions. *International Journal of Game Theory, 36*(3–4), 537–569.

Roth, A. E., & Postlewaite, A. (1977). Weak versus strong domination in a market with indivisible goods. *Journal of Mathematical Economics, 4*(2), 131–137.

Roth, A. E., Sönmez, T., & Ünver, M. U. (2004). Kidney exchange. *The Quarterly Journal of Economics, 119*(2), 457–488.

Scholten, K., Sharkey, Scott, P., & Fynes, B. (2010). (Le) agility in humanitarian aid (NGO) supply chains. *International Journal of Physical Distribution & Logistics Management, 40*(8/9), 623–635.

Tatham, P., Spens, K., & Kovács, G. (2017). The humanitarian common logistic operating picture: A solution to the inter-agency coordination challenge. *Disasters, 41*(1), 77–100.

Whybark, D. C. (2015). Co-creation of improved quality in disaster response and recovery. *International Journal of Quality Innovation, 1*(1), 3.

Part IV
Ways to Create Resilience in the Economic Activities

Chapter 14
Disaster Resilience Index in the Agricultural Sector in the State of Mexico

Diana Sánchez-Partida, Alejandro Monterroso-Rivas, and María-del-Carmen Ferruzca-Albarrán

Abstract This paper quantifies the resilience of the agricultural sector in the State of Mexico when the frost-disrupting agent threatens it. This sector represents that 66.9% of the municipalities are engaged in agriculture, and there are very few works develop on the subject. This analysis is essential because with climate change, the number of frost events has increased, and it has become more severe. In 2015, one municipality was affected by frosts, while in 2016, 24 municipalities were affected, which mainly affects the farmer sector being able to reach 100% of the loss of the crops. Given this, it is crucial to managing measures that measure and help to strengthen the sector. For that, it is necessary to know the current level of resilience, considering indicators that help the first development of the sector. So far it has not been taken into account that a large number of farmers have or not adaptive experience when these disturbing agents are present, so a methodological approach is proposed for the development of a disaster resilience index taking into account this experience and other indicators such as infrastructure capacity, local economic capacity, socio demographic capacity, and agricultural production capacity.

Keywords Farmer sector · Frosts · State of mexico · Disaster resilience index · Climate change

Electronic supplementary material The online version of this chapter (https://doi.org/10.1007/978-3-030-67295-9_14) contains supplementary material, which is available to authorized users.

D. Sánchez-Partida · M. C. Ferruzca-Albarrán
Department of Logistics and Supply Chain Management, Universidad Popular Autónoma del Estado de Puebla, (UPAEP University), 17 Sur 901, Barrio de Santiago, 72410 Puebla, CP, Mexico

A. Monterroso-Rivas (✉)
Departamento de Suelos, Universidad Autónoma Chapingo, Km 38.5 Carretera México-Texcoco, 56230 Chapingo, CP, estado de México, Mexico
e-mail: aimrivas@correo.chapingo.mx

D. Sánchez-Partida et al. (eds.), *Disaster Risk Reduction in Mexico*,
https://doi.org/10.1007/978-3-030-67295-9_14

14.1 Introduction and Literature Review

Resilience is defined as the ability of a system and its components to anticipate, absorb, adapt or recover from the effects of a dangerous event in a timely and efficient manner, including ensuring the preservation, restoration or improvement of its essential basic structures and functions United Nations Intergovernmental Panel on Climate Change (IPCC, 2007). It should be mentioned that a resilient community is less vulnerable to danger (Cutter, 2018).

So we can see that climate change has radically changed temperature regimes (Altieri and Nicholls, 2008). Since the twentieth century, there have been at least 253 events in Mexico considered disasters, of which 80% are related to water, that is, they have a hydro-meteorological origin (EM-DAT, 2017). Moreover, because of its location, the State of Mexico is highly susceptible to hydrometeorological events, and as has been observed over the last few years, they have been increasing due to global warming (World Bank, 2012). Agricultural production is susceptible to climate change since crop yields are diminished, estimates indicate that in 2050 there will be losses in the value of production, in addition to a pronounced instability, which will have negative impacts on income agricultural sector, which will generate vulnerability in food consumption (SAGARPA, 2015).

The most vulnerable elements in this type of event are communication channels and agriculture. The effects of climate change on agricultural yields may vary from region to region (Easterling et al., 2007). Small changes in the climate can have disastrous impacts because the reduction of the half to one ton of production can mean for the farmer the difference between life and death (Rosenzweig and Hillel, 2008).

Research on the agricultural sector is crucial because it is a branch of the national economy, with several functions, as the primary source of business activity that uses the workforce. This technical and cultural tradition represents civilization in itself, as well as food security. It is why it is relevant to develop resilience measurement indices that help policymakers create initiatives to support the agriculture sector, and the farmers themselves to know what level they are in and what strategies they should take to strengthen their resilience.

Altieri and Nicholls (2008) point out through official statistics that the most impoverished farmers in developing countries are considered less resilient to these impacts of climate change due to their geographical exposure, low income, greater dependence on agriculture for their survival and their limited ability to find another alternative to live.

Resilience measurement continues to be a meeting ground between policymakers and academics. However, there are inherent limitations in measuring disaster resilience. For example, resilience indicators produced by Federal Emergency Management Agency (FEMA) and one produced by an independent academic group measure community resilience by defining and quantifying community resilience at a national level, but they each have a different conceptual model of the resilience concept. Users of resilience measurement tools need to be keenly aware of the conceptual framing, input data, and geographic scale of any schema before implementation as these parameters can and do make a difference in the outcome even when they claim to be measuring the same concept (Cutter and Derakhshan, 2019).

Until now, several works about resilience index have been reported in the literature; for example, Orencio and Fujii (2013) proposed an index for a disaster-resilient coastal community at the local level. The composites of the index were determined through a process of prioritizing national-level components of a risk-management and vulnerability-reduction system.

Cutter et al. (2014) create an empirically-based resilience metric called the Baseline Resilience Indicators for Communities (BRIC) that is both conceptually and theoretically sound yet, easy enough to compute for use in a policy context. A common set of variables was used to measure the inherent resilience of counties in the United States according to six different domains or capitals, as identified in the extant literature—social, economic, housing and infrastructure, institutional, community, and environmental.

Cimellaro et al. (2016) proposing a framework for measuring community resilience at different spatial and temporal scales. Seven dimensions are identified for measuring community resilience: population and demographics, environmental and ecosystem, organized governmental services, physical infrastructures, lifestyle and community competence, economic development, and social-cultural capital. Each dimension is characterized by a corresponding performance metric that is combined with the other dimensions using a multilayered approach.

Arouri et al. (2015) use commune fixed-effects regressions to estimate the effect of natural disasters on welfare and poverty of rural households in Vietnam, and subsequently examines household and community characteristics that can strengthen the resilience of households to natural disasters. They find that all three disaster types considered in the study, including storms, floods, and droughts, have adverse effects on household income and expenditure.

Balica et al. (2012) develop methodologies and tools to assess vulnerability. One of the most important goals of assessing coastal flood vulnerability, in particular, is to create a readily understandable link between the theoretical concepts of flood vulnerability and the day-to-day decision-making process.

Measure resilience in a community is very important because it can help to develop programs that can support it, but it is very general the point of view. For that reason, it is suggested to develop sectorized research in agriculture because that is preponderant.

Rhiney (2017) draws attention to the varying ways underlying forces of economic globalization and global environmental change have been threatening the livelihood security of farmers throughout the Caribbean. The paper also sheds light on some of the local-scale implications of these broader changes. It highlights the fact that the impacts are likely to produce uneven vulnerability outcomes mediated mainly around differences in the social and economic landscapes in which individual farmers operate.

Mahoo et al. (2013) review and assess the vulnerability and risks to Ethiopian agriculture as a result of climate change; explore the threats faced by Ethiopian agriculture as a result of climate change; identify gaps and opportunities in addressing the challenges of climate change; and create a plan for integrating adaptation and mitigation actions and policies into the national framework.

In similar studies, many indicators have been determined that affect vulnerability such as the frequency of events, environmental problems, climate, population, health, agriculture, human capital, financial capital, and natural capital (Monterroso and Conde, 2015). However, some other indicators can be added, such as those taken into consideration for this work, such as infrastructure capacity and adaptive capacity, which have not been taken into consideration in other investigations. They are essential to determine an index of more strong resilience.

Many traditional farmers in many rural areas have adapted to environmental changes, developing various resilient systems in response to the constraints they have faced over time. Many of these farming systems around the world serve as sustainability models that offer examples of adaptation measures that can help many other villagers to increase their resilience to the impact of climate change (Altieri and Nicholls, 2008).

This research intends to create a Disaster Resilience Index (DRI) of the Agricultural Sector in the State of Mexico that can be applied to the farmers considering infrastructure, economic-municipal, sociodemographic, adaptive, and agricultural production capacities that can measure the capacities of the farmer and increase their resilience.

14.2 Problem Description

14.2.1 Characteristics of the State of Mexico

The State of Mexico represents 1.14% of the surface of the country, has an area of 22,346 km^2, and by 2015 it was divided into 125 municipalities according to (INEGI, 2015a). It adjoins in the North with the states of Querétaro and Hidalgo; to the south with Guerrero, Morelos, and the Federal District; to the East with the states of Puebla and Tlaxcala; and to the west with Michoacán and Guerrero (IGECEM, 2016).

The number of inhabitants of this state is 16 million 187 thousand 608 people, where 8 million 353 thousand 540 are women, and 7 million 834 thousand 068 are men. 73% of the state has a temperate sub-humid climate, located in the high valleys of the north, center, and east; 21% is warm subhumid and is southwest, 6% dry and semi-dry, present in the northeast, and 0.16% cold weather, located in the upper parts of volcanoes. 66.9% of the state is dedicated to irrigated and seasonal agriculture, with the main crops being corn, green peas, barley, beans, potatoes, alfalfa, wheat, avocado, and guava, among others (INEGI, 2015b). This State is one of the states of the Mexican Republic, with the most significant number of municipalities dedicated to agriculture.

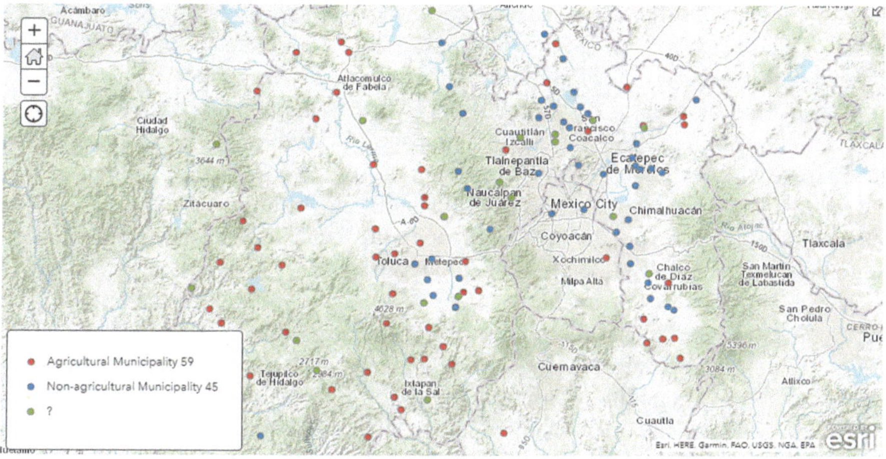

Fig. 14.1 Farmers and non-farmers municipalities of the State of Mexico

In Fig. 14.1, we can visualize that 47.2% of the total of the state is dedicated mainly to agriculture, which is represented with green dots. While 36% of this is not dedicated mainly to agriculture, nevertheless, they have places of sowing and are represented with a blue dot, while the municipalities that are represented with green points, which are 16.8% is unknown if they are engaged in agriculture or have planting places.

The State of Mexico has 125 municipalities and is located in the central part of the country, belonging to two major physiographic provinces, the Transversal Volcanic System and the Balsas River Basin. It is subject to the heterogeneity of its geographical characteristics such as climate, soil, and vegetation so that most of the territory of the State of Mexico is permanently exposed to the action of phenomena of a natural origin of different kinds such as processes of landslides, volcanic manifestations, areas subject to flooding or the presence of frost (INEGI, 2018).

14.2.2 Frosts and Declarations of Emergency in the State of Mexico

Frost is a meteorological phenomenon characteristic of regions where temperature fluctuations are considerable, especially in arid and semi-arid areas or in high-altitude areas. The difference between frost and snow is that, during a frost, precipitation does not occur because the water vapor contained in the air instead of rising freezes and is deposited on the floor. While, in the snowfall, there is precipitation (CENAPRED, 2008).

The amount of atmospheric humidity at the surface level will be outstanding since its presence will determine its occurrence and type of frost, whether white or black. The white frost is easy to recognize by the presence of frost on the surface, either of the soil or vegetation, in which case it is the humidity of the air that suffers the freezing process. As for the black frost can be more harmful, depending on the place and time of year of occurrence, since, in this case, the same moisture contained in the plants freezes, obstructing the passage of the sap through the tissues of vegetables, with fatal consequences for them. In rural areas if the phenomenon is of low intensity, it does not cause damage to agriculture, however, if the snowfall is an intense or black frost, the damage can reach 100%, depending on the crop and the stage of growth in the one that is (National Risk Atlas, 2018).

The occurrence of frosts in the State of Mexico is very associated with its geographical location, as already mentioned, its atmospheric dynamics where the eastern winds predominate; and also the displacement of cold air masses of Arctic and polar origins, in addition to its rugged relief, a factor that ultimately favors the presence of intense radiation (National Risk Atlas, 2018).

In this sense, the most affected areas are those that comprise the mountain systems, as shown in Fig. 14.2, which represents the frequency of frosts in the State of Mexico from the years 2000 to 2018. Data obtained from the database of declarations of Emergency, Disasters, and Climatological Contingency issued by the Center National Disaster Prevention (CENAPRED, 2000–2018, and FONDEN database from 2013–2016. For example, the Nevado de Toluca region presents some of the highest rates of occurrence, varying from one to two thirds a year; the same thing that happens in the zone of the Sierra Nevada; encompassing portions of the municipalities of Toluca, Calimaya, San Mateo Atenco, Tenango del Valle, Temascaltepec, Texcoco, Ixtapaluca, Tlalmanalco, Amecameca, Atlautla, and Ecatzingo.

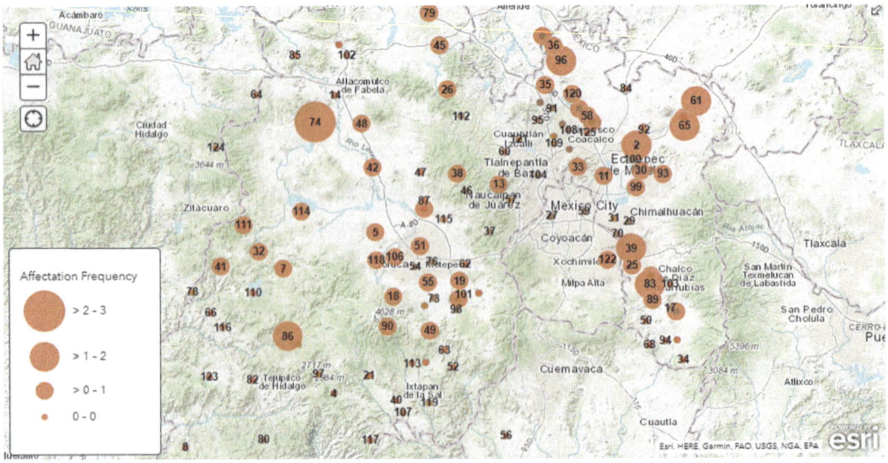

Fig. 14.2 Frequency of affectations against hydrometeorological events (Frosts) of the State of Mexico

Then there are the areas of mountains and mountains that circumscribe the Lerma River Basin, which reach 3,000 m. of altitude, where they are recorded from 10 to 160 days with the presence of frost per year. In the valleys of the Basin, the frequency fluctuates in the same range, except for some small elevations of hills and hills and the Jocotitlán Volcano. The lower incidence is observed in the portions located towards the North of the State, in part of what belongs to the ancient Lake of Texcoco and, to a lesser degree, the portions of the South of the State, where they are practically uncommon, with less than 20 days Annual (INEGI, 2018).

The database of emergency declarations issued by the Natural Disasters Fund (FONDEN, 2017) is a financial instrument through which it is oriented to provide resources to the Federal Entities and Municipalities that have exceeded their financial capacity to address the effects. In 2016, the State of Mexico registered 24 municipalities in emergencies due to the effects caused by the presence of a frost and one municipality in 2015, while in the database of declarations of Emergency, Disasters and Climatological Contingency issued by the Center National Disaster Prevention (CENAPRED, 2000–2018) who support the National System of Civil Protection (SINAPROC) in the technical requirements that its operational demands. In 2016 the state of Mexico registered 24 municipalities in emergency and 13 in 2015 under the presence of 3 touches of frost.

14.3 Methodology

Taking into account the definition of the resilience of the IPCC, the following five fundamental indicators are proposed for the calculation of an empirical DRI; all they are capabilities that refer to the ability to recover from damage.

1. Infrastructure Capacity.
2. Economic-Municipal Capacity.
3. Sociodemographic Capacity.
4. Adaptive Capacity
5. Agricultural Production Capacity.

Table 14.1 shows the Dimensions. These indicators were selected to help us determine an indicator of real resilience. Since many indicators that are involved in the resilience of the agricultural sector are estimates, and an indicator was considered, which is not had previously considered which is the adaptive capacity which its sub-indicators are the simulacra and emergency plans which are preventive actions that many of the farmers make based on the experience obtained by previous events. We can also find the sub-indicators or variables of the other Dimensions with their description, their measurement base, the unit of measurement, and the source from which the information was obtained, which is possible see in Appendix 14.A.

Table 14.1 Dimensions and indicators used for the creation of the DRI

Dimension	Indicator	Description	Unit	Source
Infrastructure capacity	Public services	Percentage of households with access to public services such as electricity, gas, waste collection, water, transportation	%	INEGI (2016)
	Housing situation	Percentage of population in inhabited dwellings	%	INEGI (2016)
Economic-municipal capacity	Human capital	Percentage of professionalism of the population and the percentage of active workers	%	INEGI (2010)
	Economic dynamism	Percentage of the specific cup of economic participation	%	SECRETARY OF ECONOMY
Sociodemographic capacity	Education	Percentage of population with studies	%	INEGI (2016)
	Health coverage	Percentage of the population that has health services	%	INEGI (2016)
Adaptive capacity	Drills	Percentage of the number of drills performed in 6 months	%	MUNICIPAL GOVERNMENT
	Emergency plans	Percentage of emergency plans, if the farmer has them or does not have them	%	MUNICIPAL GOVERNMENT
Production capacity	Implanted surface	Percentage of surface im-planted	%	SAGARPA

Once the data were obtained, a weighting of equal value was given to all the indicators, and the following Eq. 14.1 is taken from (Monterroso and Conde 2015) to obtain our standardized values of the Dimensions.

$$Z = (X_i - X)/DS \tag{14.1}$$

Where Z represents the standardized values, X_i is the observed value, X is the average of the values i, and DS is the standard deviation of the values i. The value of each Indicator studied was calculated. Subsequently, all the Dimensions are added, obtaining a DRI. The weights used for the five dimensions of capacity used a weight of 0.2, that is, each of the capacities has the same importance value among them. Subsequently, a weight of 0.6 was assigned for the result of these capacities to emphasize the resilience and 0.4 weight to frost disrupting agents. The data must be normalized between (0–100) to be interpreted and facilitate an explanation of what will be the municipality with much resilience and which one with less.

14.4 Results

As indicated above, five Dimensions were integrated, each with a pair of Indicators for each municipality of the State of Mexico, and according to the equation, and to the sum of these standardized values of each Indicator, a DRI was obtained. Therefore, in this section, the results obtained from the DRIs of the 125 municipalities of the State of Mexico with equal weighting for each Dimension will be shown.

Figure 14.3 summarizes the data obtained from the total DRIs of each municipality.

Below is Table 14.2 of the results of the five most resilient municipalities and the five least resilient municipalities. In this table, it is possible to observe that Amatepec Lerma, San Antonio la Isla, Coatepec Harina, and Villa del Carbón are the first five resilient municipalities. Conversely, San Felipe del Progreso, Xalatlaco, Tequixquiac, Ixtapaluca, and Ecatepec de Morelos are the less. The interpretation for Amatepec, for example, is the most resilient because even though it does not have economic dynamism and human capital, it has an adaptative capacity that can mitigate the impact of the frost. The interpretation for Ecatepec de Morelos, for example, is the municipality less resilient due to the impact of the frost in this area and the few capacities to face it. In Appendix 14.B, there is the complete list of the indices of all the municipalities.

Moreover, in Fig. 14.4 is possible to see a map with all the municipalities with its resilient level. On the one hand, here it is more notorious that the municipality San Felipe del Progreso with code 074, is very affected by snowfall and is not very resilient to cope with the situation. On the other hand, the municipality of Lerma, with code 051, has greater resilience in the event of frost.

The results of the most resilient municipalities yielded to two municipalities that do not dedicate more than 50% to the farmer sector. However, they are resilient because, in the economic part, they can obtain more benefits than even municipalities that if they dedicate more than 80% to agriculture.

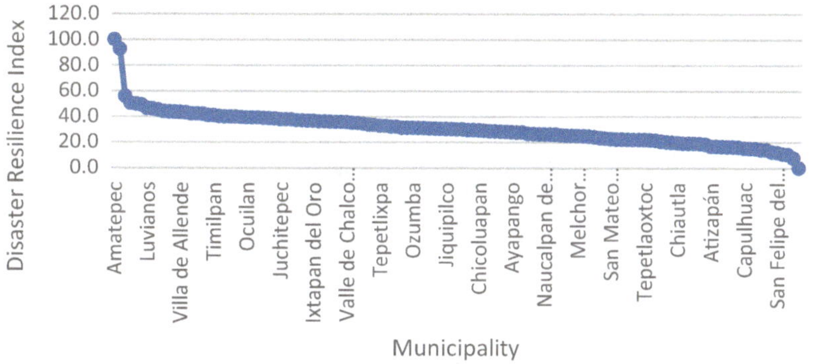

Fig. 14.3 Total indices in percentages by the municipality

Table 14.2 Table of results of the five most resilient municipalities and the five least resilient ones

Municipality	Infrastructure capacity	Economic-municipal capacity	Socio-demographic capacity	Adaptive capacity	Production capacity	Frost-disruptive agent	Disaster resilience index %
Amatepec	0.272024	−0.841592	1.252536	11.133355	0.990938	1.039457	100.000000
Lerma	1.073431	8.417741	1.716212	0.018861	0.883406	−0.777776	92.766375
San Antonio la Isla	0.874647	−0.199993	4.323696	−0.093407	0.162310	−0.777776	56.092445
Coatepec Harinas	0.984323	0.635378	0.973873	−0.093407	1.610827	1.039457	50.667773
Villa del Carbón	1.073716	0.577075	2.405795	−0.093407	0.048453	−0.777776	50.104105
San Felipe del Progreso	0.799455	−0.640086	0.722049	−0.093407	1.104795	−0.777776	12.056684
Xalatlaco	0.136824	−0.766433	−0.750717	−0.093407	−1.396902	1.039457	11.062353
Tequixquiac	1.083557	0.485720	−1.336268	−0.093407	−0.520834	−0.777776	10.532271
Ixtapaluca	0.958924	−0.905345	−0.593354	−0.093407	−0.220377	1.039457	7.852852
Ecatepec de Morelos	1.052444	−0.635705	−3.169603	−0.093407	−1.396902	1.039457	0.000000

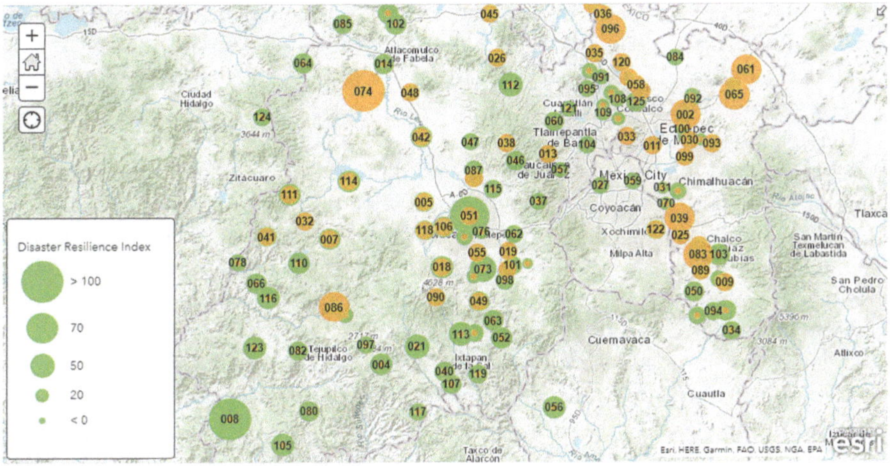

Fig. 14.4 Map with more resilient and less resilient municipalities

Moreover, as we can see, the less resilient municipalities, only one of the municipalities, is the one that devotes a more significant percentage to agriculture. Moreover, therefore, we can conclude that this municipality requires more attention, economic, infrastructure, production to be resilient. The other four less resilient municipalities are not engaged in agriculture, so it is logical that in the production and socio-demographic are meager rates in addition to that, most of these municipalities are overcrowded. Hence, the income distribution is unequal, and this results in a lower value in the subscripts that make up the Dimension.

14.5 Discussion

The agricultural sector should be ready for any eventuality such as frost; In this case, it is sought that the municipalities dedicated to agriculture in the State of Mexico present a high level of resilience in front of this type of event. However, as we can see in the graphs, the State of Mexico cannot be considered a resilient State, since none of the 125 municipalities that make it exceed 40% of the DRI.

It should be mentioned that the primary indicators that were taken into consideration are the minimum necessary to determine a useful DRI. However, there are many more base indexes, such as the technological capacity in which water harvest systems can be placed as an Indicator of rainfall, mulching, water management, diversification strategies, natural indicators for climate forecasting another may be the capacity of community organization whose Indicators can be residential stability, civic infrastructure, and civic participation. Even with these fundamental indicators can be strengthened with the addition of Indicators for example in the infrastructure the Indicator of roads and means of mobility in the local economic capacity, the diversity of businesses, the distribution of income, among others, which can contribute to the increase or decrease of the percentage of the resilience or for each municipality.

The indicator whose relevance has gained importance in importance is the adaptive capacity of the inhabitants of the municipality since the frequent occurrence of this type of event is the experience of people knowing how to react from what they should or can do to replace or otherwise recover. The whole field is beneficial to be able to say that a municipality is resilient (>50%), but not all municipalities have simulations or emergency plans that can help them to have an adaptive capacity.

Other actions that can be taken to increase resilience in the indicators are the economic investment, resources, and time to improve or implement the Indicators proposed in this work.

14.6 Conclusions

The State of Mexico is one of the States of the Republic that has the most significant agricultural sector, so it is imperative to determine what type of events affect its frequency. That is why a method was used to determine the DRI of the agricultural sector of the State of Mexico against the appearance of frosts, which is classified as a hydrometeorological event. It based on five Dimensions which in turn were divided into doubles, giving a negative percentage for all 125 municipalities that comprise it, meaning that more work and effort is needed in these states to be able to say that they are resilient since no municipality presented more than 40% of resilience.

The Dimensions with their pairs of Indicators are the minimum necessary to obtain a useful DRI; however, more indicators and Indicators can be considered and thus have a more holistic index.

For the allocation of the weighting, the same is taken for each Dimension, since among scientists, it is possible to defer in which indicator it is the most relevant than another so that each one is weighted equitably.

With this work, it helps determine the municipalities with the least resilience, and then design a development strategy to strengthen the farmer system and make it more sustainable.

As future work, it is recommended to review and add more Dimensions with their Indicators in order to obtain a more accurate resilience indicator, in addition to focusing not only on a hydrometeorological event but also on various events to be able to extrapolate to various states of the Mexican Republic with high occurrence.

References

Altieri, M. A., & Nicholls, C. (2008). Impactos del cambio climatico sobre las comunidades campesinas y de agricultores tradicionales y sus respuestas adaptativas. *Agroecología, 3,* 7–24.

Arouri, M., Nguyen, C., & Youssef, A. B. (2015). Natural disasters, household welfare, and resilience: Evidence from rural Vietnam. *World Development, 70,* 59–77.

Balica, S. F., Wright, N. G., & Van der Meulen, F. (2012). A flood vulnerability index for coastal cities and its use in assessing climate change impacts. *Natural Hazards, 64*(1), 73–105.

CENAPRED (2000–2018). Declaratorias sobre emergencia, desastre y contingencia climatológi-ca a nivel municipal entre 2000 y 2017. National Center for Disaster Prevention. Retrieved March 10th, 2018, from https://datos.gob.mx/busca/dataset/declaratorias-sobre-emergencia-desastre-y-contingencia-climatologica/resource/41444ebe-6a35-4631-8f91-9237d5114488 [in Spanish].

CENAPRED (2008). Serie Fascículos. Secretaría de Gobernación. Retrieved January 10th, 2020, from http://proteccioncivil.gob.mx/work/models/ProteccionCivil/Resource/372/1/images/fasciculo_heladas.pdf [in Spanish].

Cimellaro, G. P., Renschler, C., Reinhorn, A. M., & Arendt, L. (2016). PEOPLES: A framework for evaluating resilience. *Journal of Structural Engineering, 142*(10), 04016063.

Cutter, S. L., Ash, K. D., & Emrich, C. T. (2014). The geographies of community disaster resilience. *Global Environmental Change, 29,* 65–77.

Cutter, S. L. (2018). Linkages between vulnerability and resilienc. In S. Fuchs & Thaler, T. (Eds.), *Vulnerability and Resilience to Natural Hazards* (Chapter 12, pp. 257–270). Cambridge University Press.

Cutter, S. L., & Derakhshan, S. (2019). Implementing disaster policy: Exploring scale and measurement schemes for disaster resilience. *Journal of Homeland Security and Emergency Management, 16*(3), 1–14. https://doi.org/10.1515/jhsem-2018-0029.

Easterling, W., Aggarwal, P., Batima, P., Brander, K, Erda, L., Howden, M., Kirilenko, A., Morton, J., Soussana, J. F., Schmidhuber, S., Tubiello, F. (2007). Food, fibre and forest products. In M. L. Parry, Canziani, O. F., Palutikof, J. P., van der Linden, P. J., Hanson, C. E. (Eds.) *Climate change 2007: Impacts, adaptation and vulnerability. contribution of working group II to the fourth assessment report of the intergovernmental panel on climate change* (pp. 273–313). Cambridge University Press.

EM-DAT: The emergency events database, Brussels, Belgium. Retrieved June 5th, 2017, from www.emdat.be.

FONDEN. (2017). Recursos autorizados por declaratoria de desastre. National System of Civil Protection. Retrieved March 15th, 2018, from http://www.proteccioncivil.gob.mx/es/ProteccionCivil/Recursos_Autorizados_por_Declaratoria_de_Desastre [in Spanish].

IGECEM. (2016). Estadísticas Básicas Municipales del Estado de México, Estado de México. Retrieved May 20th, 2018, from http://igecem.edomex.gob.mx/agenda%20basica_edomex [in Spanish].

INEGI. (2010). Banco de Información Económica—Estado de México. Retrieved May 20th, 2018, from https://www.inegi.org.mx/sistemas/bie/ [in Spanish].

INEGI. (2015a). Resultados del Conteo de población y vivienda 2015. Instituto Nacional de Estadística y Geografía. Retrieved May 20th, 2018, from https://www.inegi.org.mx/programas/intercensal/2015/ [in Spanish].

INEGI. (2015b). Censo Agrícola, Ganadero y Forestal 2015 Retrieved May 20th, 2018, from https://www.inegi.org.mx/programas/cagf/2007/ [in Spanish].

INEGI. (2016). Encuesta Nacional de Ingresos y Gastos de los Hogares 2016. Estado de México, México. Retrieved May 20th, 2018, from https://www.inegi.org.mx/programas/enigh/nc/2016/ [in Spanish].

INEGI. Estado de México. Retrieved May 20th, 2018, from https://www.inegi.org.mx/ [in Spanish].

IPCC. (2007). Summary for policymakers. In M. L. Parry, Canziani, O. F., Palutikof, J. P., Linden, P. J., & Hanson, C. E. (Eds.) *Climate change 2007: Impacts, adaptation and vulnerability. Contribution of working group II to the fourth assessment report of the intergovernmental panel on climate change*. Cambridge: Cambridge University Press.

Mahoo, H., Radeny, M. A., Kinyangi, J., Cramer, L. (2013). Climate change vulnerability and risk assessment of agriculture and food security in Ethiopia: Which way forward? Research programme on climate change, agriculture and food security. Working paper #59.

Monterroso, A., & Conde, C. (2015). Exposure to climate and climate change in Mexico. *Geomatics. Natural Hazards and Risk, 6*(4), 272–288.

National Risk Atlas. Retrieved August 31st, 2018, from http://www.atlasnacionalderiesgos.gob.mx.

Orencio, P. M., & Fujii, M. (2013). A localized disaster-resilience index to assess coastal communities based on an analytic hierarchy process (AHP). *International Journal of Disaster Risk Reduction, 3,* 62–75.

Rhiney, K. (2017). Livelihood in/securities, vulnerability and resilience to global change in the Caribbean agriculture sector. In *Caribbean in/securities: Creativity and negotiation in the Caribbean (CARISCC)*. Working Papers Series, 1–3.

Rosenzweig, C., & Hillel, D. (2008). *Climate change and the global harvest: impacts of El Nino and other oscillations on agroecosystems*. New York: Oxford University Press.

SAGARPA. (2015). SIAP, Sistema de Información Agropecuaria y Pesquera de México. Secretaría de Agricultura, Ganadería y Pesca. Retrieved March 15th, 2018, from https://www.gob.mx/siap [in Spanish].

World Bank. (2012). Improving the assessment of disaster risks to strengthen financial resilience: A special joint G20 publication by the Government of Mexico and the World Bank. Washington, DC. Retrieved April 22nd, 2018, from http://documents.worldbank.org/curated/en/606131468149390170/Improving-the-assessment-of-disaster-risks-to-strengthen-financial-resilience.

Chapter 15
Development of a Resilience Strategy for a Supply Chain of a Tool Manufacturer

Ricardo Hernandez-Zitlalpopoca, Diana Sánchez-Partida, Patricia Cano-Olivos, and Santiago-Omar Caballero-Morales

Abstract This paper focuses its study in the generation of a Supply Chain Resilience strategy for a metal-transformation company located in the City of Puebla, Mexico. The study tends to strengthen the capacity of the company, in terms of resilience in case of any logistic or operational disruption caused by the negative impacts of a disaster. It was suggested to start with a Risk Management Analysis (RMA) following by a Business Continuity Plan implementation. Using the (Define, Measure, Analyze, Improve, and Control) DMAIC methodology, disturbing agents from a national federal agency were analyzed to detect potential risks on the complete Mapping Production Process of the company, to sort those risks per weighted damage impact later. The strategy set up would help to the Tool Manufacturer to control risks better and improve the resilience culture of the company. The risk cost impact was estimated to be reduced from 1.2 M USD to USD 500 k USD. In the second scenario, an AHP was used, but considering other aspects like infrastructure, roads, and so forth, the safety sites were found in the northwest, center-west, center-east, southwest, southeastern, and central areas of the state of Puebla.

Keywords Supply chain resilience · Risk management analysis · Business continuity plan · Disaster · Disturbing agent

15.1 Introduction

It is of utmost importance for companies to protect their production sites in case of any disruption due to natural disasters. Sites can be affected in direct and indirect physical damages, disturbing them from business interruptions, harm to different

R. Hernandez-Zitlalpopoca (✉) · D. Sánchez-Partida · P. Cano-Olivos · S.-O. Caballero-Morales
Department of Logistics and Supply Chain Management, Universidad Popular Autónoma del Estado de Puebla (UPAEP University), 17 Sur 901, Barrio de Santiago, CP 72410 Puebla, Mexico
e-mail: ricardo.hernandez02@upaep.edu.mx

D. Sánchez-Partida et al. (eds.), *Disaster Risk Reduction in Mexico*,
https://doi.org/10.1007/978-3-030-67295-9_15

sectors such as infrastructure, information, raw material, and in their complete Supply Chain process. The Industry, as an essential mean of production and economy for society, can be critically affected if they do not count with a robust resilience framework included in the Risk Management Plan of the company, these set of rules or ideas, can support the company to take decisions in case of a disruption or disaster. This project tends to have a good impact since the company is located in Puebla City. The city is found in the Fire Belt per expert geologist, where the string of volcanos and sites of seismic activity is. It is said that 75% of active volcanos of all around the world are positioned in this area, not to mention that Puebla City is one of the top ten states in the country to provide Gross Domestic Product (GDP) due to their infrastructure and industrial development. As it is true that breakdowns do not only depend on hazards and natural exposures, this paper focuses on this kind of disaster as there is not a Risk Mitigation Plan in place for these kinds of matters Kreimer (1999).

This chapter will have delimited its study in the Supply Chain because it plays a vital role in companies, with its proper management, it is possible to increase competitiveness and customer satisfaction, creating value and competitive advantage against other companies. Though supply chain resilience is an essential subject to study, this scope has been growing a lot in the literature in the last years because it has turned more frequent to hear about several numbers of natural disasters that have threatened the operations of the supply chain in different companies; internally and in a global level. Indeed, the scale of natural disasters has also increased; some authors from the UNISDR, like Pascaline (2018), state that climate change as one of the leading causes. These disasters have endangered life, well-being, and the complete environment. Every single department in a company is essential. However, the supply chain is a crucial part of the operations of the company, the efficient performance of the supply chain will reflect the accurate delivery regarding quantity, location, time, and cost to the end customer in the right way. Begin this said, it is essential to work on a Supply Chain Resilience Strategy for this Tool Manufacturer. This strategy will support the company to react against disturbances and continue operations with a quick recovery. Indeed, the company has an RMA; however, there is not a consideration when it comes to disasters of nature. With this, a Business Continuity Plan will be proposed so it can be added to the Quality System of the company.

15.2 Literature Review

15.2.1 Disasters

According to the Pascaline (2018), the world has witnessed several kinds of disasters during the last 30 years, different kind of disturbing agents such as earthquakes, wildfires, and floods, have affected economically to several countries along with disruption and breakdowns in the Industry and its Supply Chain. Statistics have shown that from the 80s' till the end of the first decade of 2000, the number of events has increased on average, from 150 to 450 events, and each event can cost up to US 35 billion for the global reinsurance Industry. However, a relevant data to consider per the United Nations International Strategy for Disaster Reduction (UNISDR), is that the severity of the disasters has dramatically increased over the last two decades, there has been a great support and encouragement by the international cooperation to reduce disaster risk, for instance, the Sendai risk mitigation framework. UNISDR states a framework of 15-year, voluntary, where the State, along with stakeholders and local government, share the responsibility to reduce risk in natural disasters. Nonetheless, natural disasters do not stop from happening. This same study by the UNISDR shows that in the last 20 years, between 1998 and 2017, the accumulative total of disaster events registered was around 7255, on which per category, 80% of the disasters were related to floods, storms, and earthquakes. The rest of the categories impacting disasters are extreme temperature, landslide, drought, wildfire, volcanic activity, and mass movement (dry). It is fundamental to consider the UNISDR only consider a disaster; those were people are harmed, internal or external, not natural events presented during a year, Pascaline (2018).

15.2.2 Resilience for the Industry

The statistics showed how natural disasters are increasing for companies, so it is crucial to enhance their resilience performance as the odds of having disruption increases. Sertyesilisik (2017) defines resilience as the way of improving societies and their strength against disasters/hazards by reducing their adverse consequences, as well as the probability of a human-made hazard. Also, the Department for International Development in the (2011) defines resilience as the ability of a system, community or society which is exposed to hazards to resist, absorb, accommodate to and recover from the effects of a hazard in a timely and efficient manner. Resilience is essential to identify the areas of opportunity to mitigate disaster consequences. In the case of the industry, it is essential before analyzing its resilience capability to identify four critical elements, first is the context, second the disturbance, third the capacity to deal with the disturbance, and lastly, the reaction to the disturbance. Figure 15.1 shows this information about the four critical elements.

This framework will somehow show us a general understanding of the current situation on any system or process, in the case of this study, the Industry. Most of the companies would indeed like to bounce back quickly when recovering from a disturbance. However, the fact is that sometimes companies are not prepared from such disasters that they were not able to overcome such devastating effects. Norrman (2004), in the analysis made for Ericsson regarding disruption in their supply chain due to natural disasters, gives several examples of different types of firms shutting down. These disruptions are propitiated due to different types of disturbances, to mention some, the Hurricane Floyd that affected Daimler-Chrysler plant producing suspension parts along with seven plants across the Toyota fire causing a shutdown of around 18 plants for more than two weeks. Hoa (2015) also

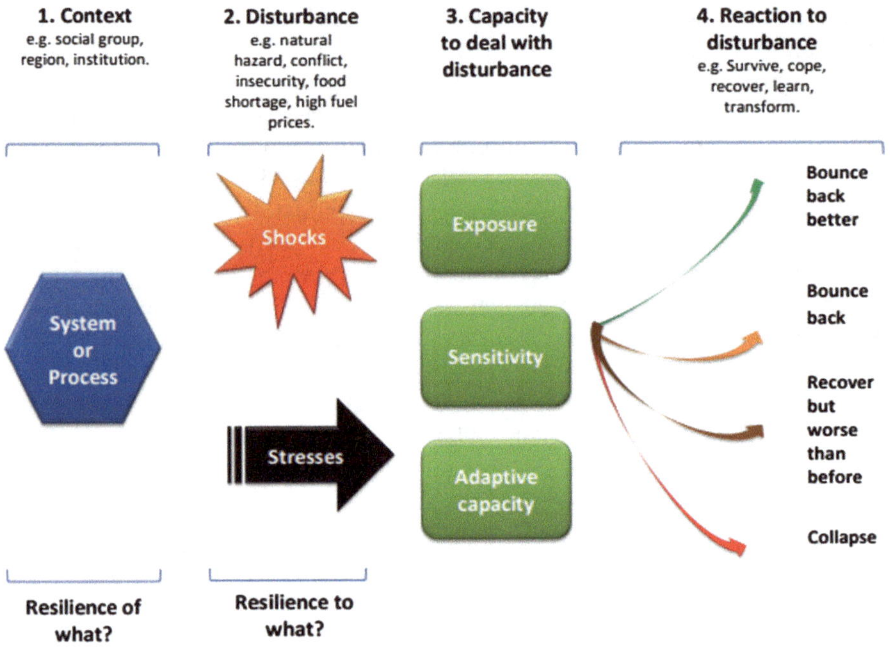

Fig. 15.1 Four Elements of Resilient Framework, DFIF (2011)

gives other disruption examples like the fire in Phillips that interrupted deliveries to Ericsson or the Tsunami in Japan shutting down production in different plants, including Toyota, in around 40,000 vehicles. In this direction, it is vital to understand the complete framework of the resilience of a company, to more in-depth study it, and to set a strategy to strengthen actions for when disturbances come up.

15.2.3 Where Industry Can Be Affected

Natural disasters can hit Industry potentially, as reviewed previously, there are different types of natural disasters or other man hazard risk that can jeopardize the production system of the industry. Merza (2013), divides the impacts affecting the industry in direct and indirect ways. The first type of impact, the direct, Merza states in Cruz (2015) and will have to do with damages on buildings, materials, and production equipment that will obstruct labors to continue working on the production. Besides, Merz refers to Van den Ver, with the other direct impact, the Natech hits or Natural-Disaster Triggered Technological Accidents, as they can have a devastating impact and consequence to the society after a Natural Disaster, an example of a Natech can be the risk of the nuclear plants, radiating rays to people in Japan, after the tsunami in 2011. This second type, the indirect, will have to do with the inability to get material or information, Van den Ver (2017). The indirect damages Messner (2007) state that can incur on a domino effect since all the supply chain industry is integrated, the damage can propagate even to companies that did not have the natural disaster event. The indirect damages can have more effects financially speaking since the complete supply chain can be disrupted, and more than one company can be affected indirectly.

Getting to know that natural disasters can affect the Industry, directly and indirectly, Chung-Hung (2010) works on the risk weight that natural disasters can harm the Industry. He states that four essential variables need to be considered when measuring the hazard level or disaster risk: exposure, location, vulnerability, and hazard. These variables will help to understand the level of catastrophe that any risk can incur in the Industry. Secondly, Chiang (2003), reduces the vulnerability and location variables into the ratio of occurrence disaster, leaving the analysis of the disaster risk as to the product of the ratio of occurrence disaster times the damage times exposure on a person or property. This equation is a helpful indicator when reviewing the impact of a natural disaster, which can also be used in the Framework Resilience for a Company.

15.2.4 Supply Chain Resilience

There are several studies focused on the Supply Chain Risk to set strategies to reduce vulnerability or ensure continuity in case of any disturbance or natural disaster, whenever a company works on a Risk Mitigation program, they are working on their resilience improvement. Biggs (2011), states that the importance to understand the resilience in the industry is vital if the management is looking for a resilience improvement. For example, essential points to identify are the independent variables, enterprise experience, size and age, the current condition, how attentive is the staff the working regarding resilience, and how the motivation is being implemented regarding resilience.

Companies due to different types of disturbances and natural disasters are being required to investigate their vulnerable sectors. Business environments have diversified, the operation should be useful, the production process including suppliers, warehouses, consumers, and the complete supply chain network should be taken care of to assure besides competitiveness and continuity in case of any disruption, Simba (2017). Supply Chain Resilience and Supply Chain Risk Management (SCRM) has received increasing attention from researches because of the complexity to predict and uncertainty within the supply chain. According to Leat and Revoredo-Giha (2013), the Supply Chain Risks will be events that will affect in a negative way the supply chain operations, so companies are concentrated to understand its vulnerability and operational threads within their environmental activities to mitigate them or somehow eliminate them. Simba (2017) bonds the Supply Chain resilience with Risk management. The resilience in the Risk

Management analysis is the response to disruption. Somehow the capability to enable the prepared risk assessment and return to stable operations as soon as possible. Resilience should not only be reactive but proactive. In other words, it is essential to be responsive to any risk that could be materialized and reductant, limiting the risks and its consequences. Going forward with setting a strategy to strengthen the Supply Chain, Ponomarov (2009) supports the idea that firms which count with an SCRM responds and recovers themselves from any unpredicted supply chain event.

Supply chain resilience has been growing because a hit to a company can cause disruptions and an adverse effect on supply chain productivity, profitability, and competitiveness. Simba (2017), divides the supply chain resilience in two main capabilities, the redundancy, and flexibility. The redundancy is the state where the companies should limit their risk and consequences in terms of working with reserves and anticipating any damage. Flexibility, on the other hand, is how quickly a company can respond to the risk that will materialize. When working with supply chain resilience, businesses should invest in mechanisms to improve redundancy and flexibility, Wieland (2013) suggest investing time and analysis in both capabilities, focusing on just one capability would increase the risk of occurrence cost. On that line, SCRM is a crucial process to set strategies that will support resilience Breuer (2013) states that SCRM is one of the most significant driver essentials to increase supply chain resilience.

15.2.5 Risk Mitigation Tools

The importance of the use of a tool for the Supply Chain Management, according to Simba (2017), is that the tool works on a risk-oriented action and can somehow intertwine the steps that compound the SCRM process; these processes are mentioned in Fig. 15.2.

Fig. 15.2 Supply Chain Risk Management Process, Simba (2017)

- Risk identification—Identification of the vulnerabilities due to some agent or potential risk.
- Risk Assessment—Determines the probability, frequency, and impact of disruptions.
- Risk Mitigation—How will the company deal with the risk.
- Risk Monitoring—To monitor the possibility of risk or disruption.

According to diverse researches, for the use of a correct tool to identify and asses risk within a company, varies depending on the type of manufacturer, however among several, it was possible to find the Failure Mode and Effect Analysis (FMEA), Supply Risk Matrix, feedbacks, brainstorming's, scorecards, workshops, and benchmarking, these are some of the tools that companies use for the Risk Management process, Wieland (2013).

For instance, to mention some of the tools, Sun (2018) used a Risk matrix to provide a mechanism to assign risks on a vessel traffic process, where he can establish the frequency and consequence of a potential risk, sorting the level of impact and the decision making for each risk. Blos (2009), as another example, worked on an empirical study, with feedbacks and brainstorming on an automotive and electronic industry to find vulnerabilities on a mapping process like financial, strategic, hazard, and operational.

15.3 Why is Resilience Remarkable in Supply Chain for the Tool Manufacturer Based in Mexico

According to the Economy Department of the Mexican Republic (2016), Puebla, the location where the Tool Manufacturer runs its business, has a total space of 34,306 km^2, with a total population of 6.1 M people living in the state. Its economic activity is one of the most important in all Mexico, the state has the participation of 3.4% GDP from the total of 32 states, being in the top ten of all the industrial and commercial activity of the country. The main activities that make the total economy of each country are usually divided into first, second, and third activities. The first one will be involved in land work and natural activities, the following activity is involved in the industry and manufacturing sector, while the third activity will be the commercial bond activity from the first and second economy activity. Puebla is an industrialized state; its primary economy workforce comes from the second and third activity, 35.2 and 60.8% accordingly, while the first activity only represents 4.1% of the economy, Fig. 15.3.

The tool manufacturer set its strategic location in the south-central part of Mexico, 120 km away from Mexico City, and 300 km away from the international port of Veracruz. The company exports to Europe, Asia, and the United States, the industry in the state of Puebla is stable in the automotive, auto parts, metal mechanic, chemicals, plastics, apparel, furniture, fresh and processed foods, mining and information technologies per the Department of Economy (2016). Therefore,

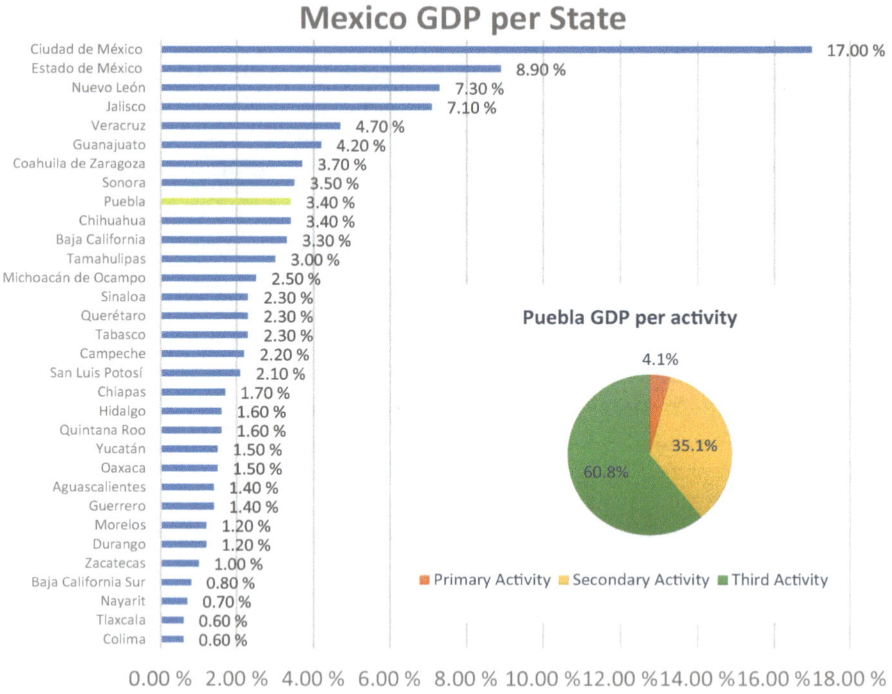

Fig. 15.3 Economy Department of the Mexican Republic (2016)

the company was established on an industrialized state within the country; Mexico is among the world's 20 largest economies and the second largest in Latin America, 80% of its exports are designated to the U.S. and should continue to be so in the future, World Economic Forum (2008). This paper is focused on Natural and Hazardous risks that can affect the company. However, there are other external risks for this company and others in Mexico that are always latent. For example, the reliance on the US economy, uncertainty on the US external commercial policy, political uncertainty for international companies setting up new businesses in Mexico, the high level of insecurity, being ranked as one of the five most dangerous countries for companies in LATAM is shutting down several companies and distribution centers such as Coca Cola, FEMSA or Lala El Universal (2018).

Natural disasters, according to Kreimer (1999), have had a powerful impact on the development in Mexico; the country has gone through social, physical, and capital destruction. Mexico is very vulnerable to a variety of natural disasters, such as earthquakes, hurricanes, and volcanos. The frequency in which these disasters strike is high. However, there have been significant efforts to a quick recovery relief and reconstruction. The National Center of Disaster Prevention (CENAPRED), for its acronym in Spanish, is an institution that has academics and government officials that alert residents in case of any natural disaster. The types of Risks are well explained by different international organizations such as the UNISDR or the United States Department of Homeland Security (USDHS), each organization focused on the different type of risks to support with data and relevant information. However, for the analysis of this paper, the data to be used will be from CENAPRED in coordination with the National Civil Protection System (SINAPROC) for its acronym in Spanish. The type of disaster risk classified by CENAPRED (2018), is called disturbing agents. A disturbing agent by the Mexican Government Department is a harmful action to a mass concentration of people. They are dividing these types of disturbing agents in Natural disturbing and Anthropogenic disturbing. The first one is produced by the mother nature and the second produced by the human action, these second agents will not be analyzed in this chapter, but among others, it is possible to find hazards such as terrorism, vandalism, mass people concentration, land, air accidents, and so forth. Table 15.1 shows all different types of disturbing agents classified by CENAPRED.

In Mexico, devastating natural disasters have affected vulnerable structures and had surpassed society the ability to respond. The precise time, location, and type of disaster will indeed occur. However, areas of vulnerability can indeed be identified, and losses in the society and infrastructure can somehow be prevented. According to Kreimer (1999) and some other facts such as Mexico is in the "Fire Belt," Mexico has gone through a long story with natural disasters, Earthquakes, and floods. Also, Mexico experienced from the 80s' till almost the end of the millennium around 79 disasters, among these disasters, disturbing agents such as earthquakes, volcanic eruptions, and landslides were presented, in other words, we can say that those are the most significant strikes hitting Mexico. It is true that much progress has been made, for example, the creation of the CENAPRED institution after the earthquake in 1985 and identifying vulnerable areas within cities to create strategies for reducing risk. However, there is not much on strategy for companies and commerce.

15.4 Brief Company Infrastructure

This tool manufacturer, located in Puebla City, Mexico, is part of a multinational firm, most significant in the manufacturing of tools and Storage headquarters based in the U.S.A. The branch based in the City of Puebla is strategically located in the center of the country, to import 98% of its total manufacturing to North America, Europe, and Asia, the other 2% stays in Mexico and goes to the center and south of America. Their primary production would be demolition tools; this is a forging and metal transformation company with three shifts working labor, with annual sales between USD 25 and 28 million, while corporate has been valued on USD 16 billion revenue. Although the company is part of a big corporation, they do not have the right processes and contingency response in case of any natural disaster; the company does not have any process to classify natural phenomena risks nor potentials failures within the value-added chain. There is indeed a general study in

Table 15.1 Type of disaster classification CENAPRED (2006–2018)

Disturbing Agents	Coordinator	Type of Impact	Disturbing Agents	Coordinator	Type of Impact
Hydro-meteorological	Ministry of Environmental, Natural Resources, and Fisheries	Freezing	Chemical-Technological	Ministry of Trade and Industrial Development	Fires
		Tornado			Explosions
		Extreme Rains			Toxic Leaks
		Extreme Sun Hit			Radiation
		Floods			Spills
		Tropical Cyclones	Sanitary-Ecological	Ministry of Environmental, Natural Resources	Air Pollution
		Droughts			Water Pollution
Geological	Ministry of Social Development	Earthquakes	Sanitary-Ecological	Ministry of Environmental, Natural Resources, and Fisheries	Soil Pollution
		Volcanic eruptions			Food Pollution
		Slope Instability			Epidemics
		Tsunamis			Pests
		Earth Cracking	Socio-Organizational	Ministry of the Interior	Interruption of Infrastructure Services
		Subsidence			Air, sea or land accidents
		Earth Sinking			Terrorism, sabotage, vandalism
		Flows			Mass population concentration
		Landslides			Non-conformity Social Demonstrations

Risk Management, but this one is focused on the operation response within the value-added chain. The City of Puebla is in the "Belt of Fire," and it can cause severe damage to the supply chain and it should be informed to the potential risk in order to set a strategy to mitigate these and create conscious of resilience within the company, not only for those in operation but also for externals that could end up in disruptions or total shutdowns.

15.5 Current Situation of the Company, Environment and Project Justification

So far, the company has not had significant disruption or shutdowns due to natural disasters, and for that, the firm has not gone more in-depth on the type of risk, some inside the corporation would say these risks will not be measured until something catastrophic happens. Nonetheless, risks are latent; this company several times has had intermittent breaks due to seismic alerts, electricity shutdowns because of hard rains, and low productivity in the heat treatment process for low temperatures, among others. One of the most significant threads for this company is the volcano named Popocatépetl; it is one of the most active volcanos within the Mexican Republic, and in case of any disaster could affect more than 9 million people in the states of Puebla, Mexico, and Morelos. Companies should prevent, for instance, risk management is only concentrated in operations in the case of this tool manufacturer, this could happen to be reasonable as only a few regions are threatened for natural disruption agents. In the case of Mexico, the Fund of Natural Disasters (FONDEN), for its acronym in Spanish, was created to support disaster relief and reconstruction. The government somehow has worked on this matter, social resilience, and it is time for companies to take this topic more seriously.

To justify the study of this paper, the FONDEN (2018), states that Mexico is exposed to a variety of geological and hydrogeological hazards; from the years of 1970–2009, a sum of 60 million people was affected by natural disasters. The types of risks found were earthquakes, volcanos, tsunamis, hurricanes, wildfires, floods, and landslides. Mexico is ranked as one of the 30 most exposed countries for three or more types of disturbing natural agents. On average, Mexico experiences more than 90 earthquakes per year with a magnitude of 4.0 or above on the Richter scale. Also, Mexico is located within the Trans-Mexican volcanic belt, which contains nine active volcanos along, and have tropical cyclones that affect the Pacifica and Atlantic coasts, plus massive rainfalls events occurring in all the territory. In Table 15.2, there is an analysis from CENAPRED (2018) regarding land and total population damage.

As seen in Table 15.2, almost all the country is somehow being affected for a hazard, which means that the risk of natural disasters in the country is high. Now, focusing in Puebla City, where the tool manufacturer is located. There is a study already made by CENAPRED along with SINAPROC, about the vulnerability in

Table 15.2 Land and Population exposure hazard, extracted from CENAPRED (2018)

Natural Hazard	Area		Population	
	Km^2	Territory Damage (%)	Millions	Population (%)
Storm, Hurricane, flood	815,353	41	31.3	27
Earthquake	540,067	27	31	27
Drought	573,300	29	21.2	19
Forest Fire	747,574	37	28.4	25

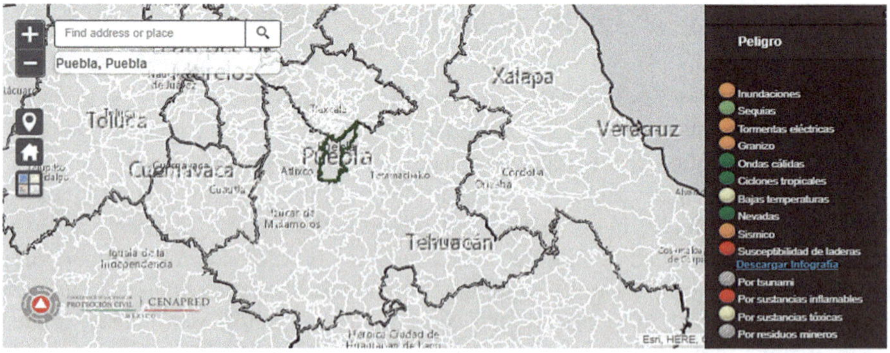

Fig. 15.4 Vulnerability in the City of Puebla, National Risk Atlas. Atlas_National_Risk. (2018)

the city in terms of natural disasters. Below, a general summary and general information to be used in our analysis. Figure 15.4 shows the danger of natural events.

In summary, we can say that:

There is a total of 1.5 million people living in the city.

- The city counts with all kind of facilities such as schools, hospitals, libraries, and others.
- The city of Puebla during the year 2018 has presented six different types of emergency cases, one disaster, and five contingencies.
- The city of Puebla is not vulnerable in social sectors, and its resilience level is categorized well.
- The city is vulnerable in flooding, electric storms, hail rain, low/high temperatures, and earthquakes.

In general, the state of Puebla is already well studied, the level of resilience as the society in the city is high. Nevertheless, this does not mean these disturbing agents did not affect at all; in fact, they were severe hits to the population. As an example, in the graphs below, it is possible to see an analysis extracted from data of the FONDEN (2018) in the last three years in the state of Puebla, 2015–2018 (Fig. 15.5).

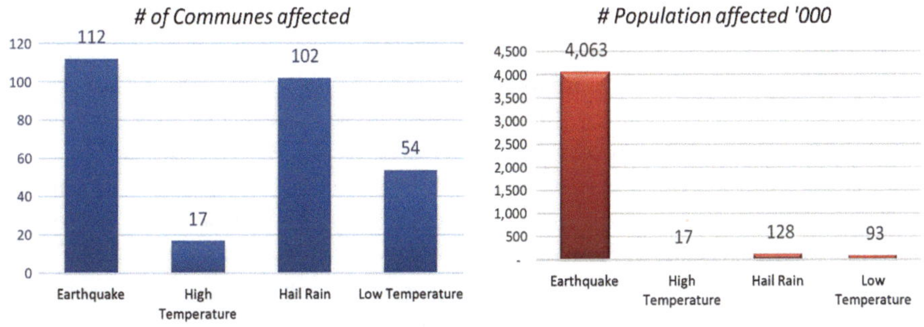

Fig. 15.5 Communes and Population affected FONDEN (2018)

There has been a total of 285 events and more than 4.3 million people affected, the risk is high, and companies should have a plan for when a natural disaster presents.

15.6 Methodology

Following the process to analyze potential risks according to Simba (2017), it is essential to identify risk, asses it, mitigate it, and monitor it, and to control it. This practice will be combined with the (Define, Measure, Analyze, Improve and Control) DMAIC methodology; this will support identifying significant risks due to natural disasters and find a solution to that potential risk. DMAIC is a useful framework for improvement applications in many companies. In the studied company, for instance, it is used as a methodology for continuous improvement and solving problems. The best approach, when working with such analysis is to share best practices with the team, experts within the day to day work know their processes well, and brainstorming is always useful for the generation of ideas, in the case of this analysis the input of the experts was precious. Companies will always want to avoid loss of productivity, cost increased, declining quality, and lost revenue. Recent disasters have highlighted the vulnerability to companies', disruptions in the developed world and emerging markets. The following steps were suggested on behalf of the Tool Manufacturer corporate when they do Risk Mitigation analysis and the information of Simba. Both methodologies follow a similar path, below how the methodology ended.

- Identify—this would happen to be the Define and Measure phase, in this part of the process, the main hazards and disturbing agents will be identified, along with the impact of the hazards. Companies would usually use afterward, an FMEA, or a Supply Matrix to asses and analyze the information. In the case of this study, both were used.

- Asses—this would be the Analyze phase, having now understood each of the potential risks, it is important to prioritize them, sort them, and work with them accordingly. In this stage, it always is essential to make the profit at risk analysis; in other words, Likelihood versus Impact. As it was said, an FMEA and then a Supply Matrix was used to assess.
- Mitigate—This is the improve phase, it is essential to identify the correct strategy per each of the risks found, along with the resources needed and create a formal project plan if it is needed, mainly for the top hazards that could be presented in the company. A complete plan will be established to use it on a Business Continuity Plan.
- Monitor—Finally, we have the Control phase, which is the same study continuously to update overall risks, status, and priority. It should be made by the company frequently.

This analysis tends to give a piece of general information to the decision-makers so that a Business Continuity Plan can be built with the analyzed information.

15.7 Problem Solving—Risk Mitigation Implementation

A Risk Management analysis was performed with the help of the experts within the company to increase resilience in the company due to natural disasters exposures. First, all the risks were identified and then assessed, then all the odds, like forecast or frequency of the risk were studied along with the severity of losses, and finally, with the output of the information, the company could be able to implement programs to control all the risks or mitigate them.

15.7.1 Identify

The risks to be considered in this project are the once researched from CENAPRED from 2015 to 2018. The city of Puebla is vulnerable in the following disasters that will be analyzed and customized according to the current Tool Manufacturer processes:

- Risks on flooding
- Risks on electrical storms
- Risk on hard hails
- Risks on earthquakes
- Risks on Volcanic eruption
- Risk of low temperatures
- Risk of toxic substances exposure
- Risk of drought effects.

These types of disasters are potential risks; each one has a probability of happening and of impacting directly to several processes and infrastructure in the company. A detailed study was performed to identify all the impacts, with the company employees as they know what type of exposures they might have in their operation to lose. These types of exposures can be of different kinds. Impacts can damage the whole property, the land, the buildings, structure, or tangible assets such as furniture, office computers, machinery, raw material inventory, and others.

The potential losses also have been studied according to the CENAPRED risks. We tried to ensure that the entire business does not ignore anything that can be affected by the exposure of the risks. Somehow, it was a very detailed process followed and supported by the employees of the company to identify all the hazards that can expose and generate losses in operations. The main idea of this activity was working with the experts in the company to discover all the hazards based on risks of the CENAPRED. As a general idea, each person was assigned to make a list of risks, trying to discover all the dangers and asses if there is a relationship to one on another, and this could become critical to the organization. First off in Fig. 15.6, is a risk map of the potential risks according to the company impact, e.g., the city of Puebla is very vulnerable to hillsides, but the company is not close to any hill that can cause damage to the infrastructure of the company.

Categorizing all kind of risks exposure and the relation between the frequency and danger is very important to understand that from all hazards within the company, Electrical Storms and Hard Hails should be mitigated right away. Nevertheless, it is essential to say that each risk classified will be mitigated accordingly. In summary, it is possible to see the resulting of the four quadrants. On them, we can visualize risks which in strict theory, could be avoided, transferred, mitigated, and ignored. It is essential to mention that the criteria that orient the impact and severity of each risk were divided into quadrants; in the first quadrant

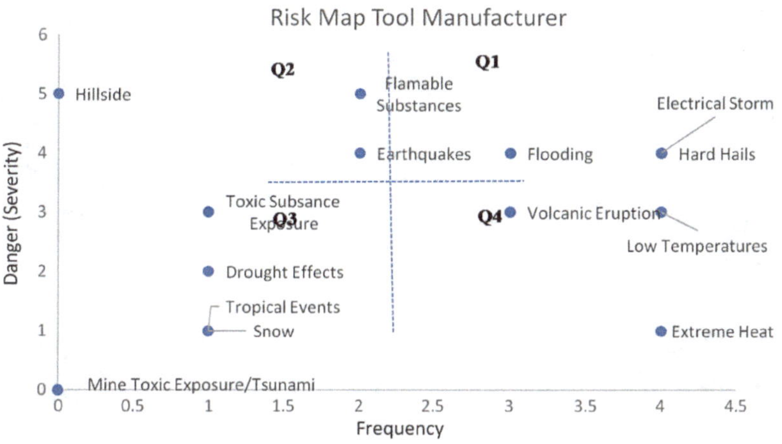

Fig. 15.6 Risk Map of the company

are the most critical risks, the once that have to be taken care and analyzed more in-depth; in the second quadrant are those risks that can be transferred, for instance, the company can hire some insurances that will manage that kind of risks; in the third quadrant are those in which a certain way can be ignored, in the case of this study will be analyzed, but considering them not on a high impact; and finally, in the last quadrant are those that have been mitigated immediately because of the high frequency, but not affecting the company.

Risks are mean to affect the company, and somehow those impacts are the ones to be studied and analyzed more in-depth. It is essential to understand that risks cannot be eliminated; for example, people cannot prevent hurricanes from happening or earthquakes to go beyond specific Richter scale. However, the company can develop strategies that can mitigate its impact. Following that line, Figs. 15.7 and 15.8 show the complete walk and mapping process of the production of the tool manufacturer, where impact risks were identified to later work with them on the Risk Matrix. Figure 15.7 shows the walking process of general tool production, the manufacturing process, as it can be seen it is not in line. It is a batch production on which material is carried from cell to another until the material is wholly manufactured, eventually, during the manufacturing process, the end of the process finishes to be under a one-piece flow production.

The production line is shown in the layout plant, the products manufactured in the company are those known as demolition tools; all the material is transformed from the metal bars until the complete hand tool product. Each cell was analyzed in order to identify impacts per the risks. The cells included in the production line are enumerated in the layout, they are steel cut, forging process A, forging process B, polish A, polish B, induction heat treatment, annealing heat treatment, sandblasting, liquid painting, bulk heat treatment A, powder coating, bulk heat treatment B, and packaging.

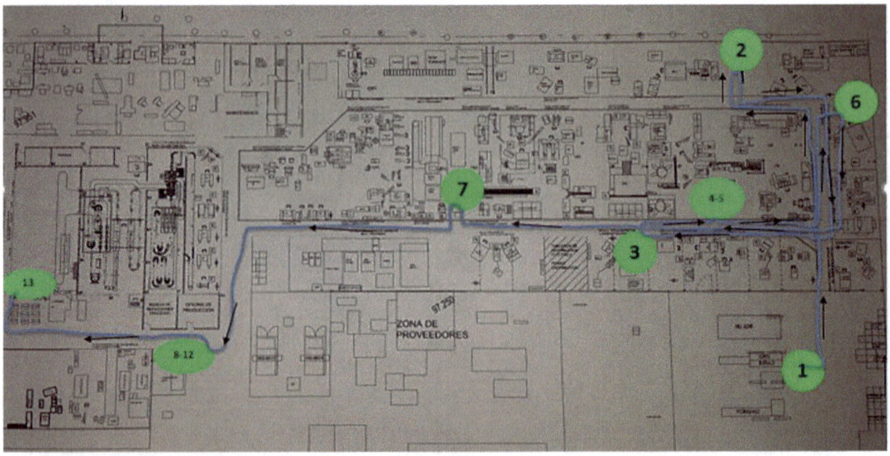

Fig. 15.7 Plant Layout, Tool Manufacturer (2018)

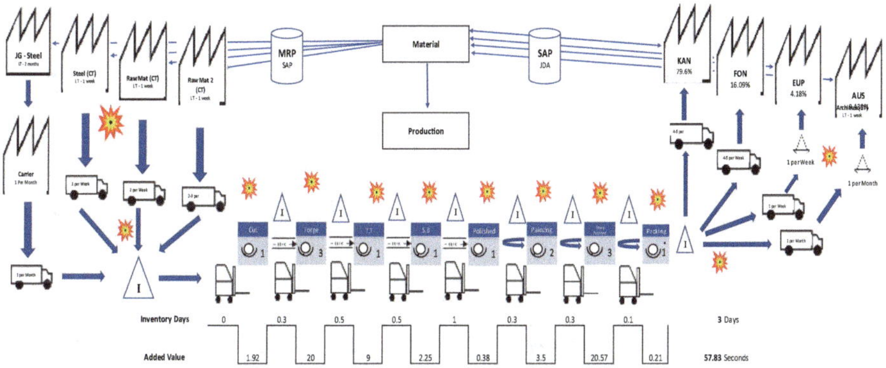

Fig. 15.8 VSM General Manufacturing process, self-production (2018)

The identification of each cell in the manufacturing process is critical to this project as these are key drivers that can lead to a disruption in the plant. In Fig. 15.8 is shown the complete mapping process of the tool manufacturing; in it, other vital drivers were also studied. It is essential to clarify that the process map was build according to the products that use all the cell manufacturers in the company so that the study can show all the production cells of importance. The Process map shows alerts in most of the processes, and this is normal since all the work made in the company is critical to the delivery of the product to the final customer, experts in the area decided to mark these alerts as critical.

In these alerts, experts considered the sourcing of the material, transportation, inventory in the warehouse, all the production lines, product inventory warehouse, and transportation to the end clients, these processes were analyzed deep. Table 15.3 shows all the risks and impacts per area, analyzed with the aid of experts.

Table 15.3 shows an extract of the potential impact risks that could be presented in the company due to natural disasters. As this is the first study approach, it is essential to continue working with this information and add potential risks found by experts for searching vulnerability in the future. The company had never done an analysis regarding natural disasters, all risks analyzed were just focused on the productivity of the company, for instance, operation controls, machinery, and labor that is linked on a day to day risk basis. In the case of this chapter, a complete mapping process of the company was studied with the CENAPRED risks the areas included were: sourcing area, raw material transportation, warehouse, data information (IT), all the production cells, warehouse for finish goods, and transportation delivery to the customer, the output was to get risk impacts in general.

Table 15.3 Extract of the Impact Risk Matrix

	RISK CENAPRED - PUEBLA CITY							
Work Area	**Flooding**	**Electric Storm**	**Hard Hails**	**Earthquake**	**Volcanic Eruption**	**Low Temp**	**Toxic Substances**	**Drought Effects**
Suppliers/Sourcing	-Damage Material/FG -Damage Machinery -Lack of Material/FG	-Short breaks Electricity -Damage Machinery	-Short breaks Electricity -Damage Infrastructure	-Damage Infrastructure -Damage Machinery -Damage Material/FG -Lack of Material/FG	-Damage Infrastructure -Damage Machinery -Damage Material/FG	- Labor Sickness - Absence -Lost of Efficiency	- Labor Sickness - Absence -Lost of Efficiency	- Labor Sickness - Absence -Lost of Efficiency
Raw Material Transportation	- Material undelivered -Accidents -Delays in deliveries	-	- Material undelivered -Accidents -Delays in deliveries	- Material undelivered -Accidents -Delays in deliveries	- Material undelivered -Accidents -Delays in deliveries	-	-	-
Warehouse Raw Material	-Damage Material/FG -Damage Forklifts - Human Injury	-Short breaks Electricity -Damage Computer	-Short breaks Electricity -Damage Infrastructure	-Damage Infrastructure -Damage Forklifts -Damage Material/FG	-Damage Infrastructure	- Labor Sickness - Absence -Lost of Efficiency - Environmental Legal Problems	- Labor Sickness - Absence -Lost of Efficiency - Environmental Legal Problems	- Labor Sickness - Absence -Lost of Efficiency - Environmental Legal Problems
IT/MRP/DATA	- Servers damaged	- Servers damaged	-	-Damage Infrastructure -Servers damaged		-	-	-
Steel Cut Cell	-Damage Machinery -Labor Sickness - Land Cracks - Infrastructure Damage	-Damage Machinery - Brakes in the process	-Noise excess -Ear problems -Breaks in the process	'-Machinery Damage -Land Cracks - Pipe Rupture	-Damage Infrastructure -Machinery Damage -Pollution Excess -Fires	- Energy over consumption	- Machinery substance leak - Emissions Excess	- Machinery Overheat

External

Internal

15.7.2 Asses

This part of the project is where the analysis of the information falls. Here it is essential to asses de information to later make decisions with each of the impacts of the risk. Natural disasters affect different than the day to day risks in the company, experts, before sending its information, were informed on how natural disasters impact to an unprepared firm. Natural disasters affect companies in four different ways. The first one is the infrastructure itself; whenever a natural disaster strikes, there is usually damages in physical assets such as buildings, equipment, machinery. Another way is damage to the raw material of the company. In this case, the cold weather can harm some adhesives that should not be stored under a specific temperature. The third way is the Supply Chain; floods can delay the delivery of raw material for several days. The last one is all the circumstances that do not allow the workers to perform its job. Because of the bad weather, some people would not be able to get to work even if they try. In this same line, all impact risks are aligned to the mentioned types of risks. Each risk impact was evaluated per the type of natural disaster, and Table 15.4 shows an extract of all the impact risks. There is a

Table 15.4 Impact List Risk, Self-production

Effect	Risk CENAPRED	External	Internal	Grand Total	Ext (%)	Int (%)	Total (%)
—Damage Machinery	Earthquake	1		1	2	0	0
	Electric Storm	1	11	12	2	4	4
	Flooding	1	11	12	2	4	4
—Damage Machinery Total		3	22	25	**7**	**8**	**8**
—Fires	Earthquake		5	5	0	2	2
	Flooding		5	5	0	2	2
	Volcanic Eruption		11	11	0	4	3
—Fires Total			21	21	**0**	**7**	**6**
—Damage Infrastructure	Earthquake	1	3	4	2	1	1
	Hard Hails	1	2	3	2	1	1
	Volcanic Eruption		13	13	0	5	4
—Damage Infrastructure Total		2	18	20	**4**	**6**	**6**
—Land Cracks	Earthquake		11	11	0	4	3
—Land Cracks Total			11	11	**0**	**4**	**3**
—Emissions Excess	Toxic Substances		11	11	0	4	3
—Emissions Excess Total			11	11	**0**	**4**	**3**

total of 37 impact risks, affecting the company 328 times, in different production cells. The extract only shows 88 affectations of the 328.

Then, for each of the 37-impact risk found, these were analyzed and evaluated with all the experts of the company, the objective was to identify opportunities to mitigate the impact of the natural phenomena risk begin studied. The tools used for this purpose were the Risk Matrix in combination with the FMEA. The first tool evaluates the project in terms of probability of occurrence, and the FMEA summarizes the impact of each risk. The impact was taken from the Risk Priority Number (RPN) factor, which is a metric, used to asses risks, and identify critical failure mode, the factors used on this matrices are: occurrence, detection, and impact, each of the variables was standardized into probabilities in the Risk Matrix so that the information could be used to the FMEA variables, in this case, the probability of occurrence. The first variable studied was the impact; it was tropicalized on how this risk affects the entire company with the help of Table 15.5.

Table 15.5 FMA Impact variable analysis

Effect	Split (%)	Rank	Impact AMEF (%)
—Lack of Material/FG	0.3	1	20.0
—Damage Forklifts	1.2	2	40.0
—Absence	2.7	3	60.0
—Infrastructure Damage	3.4	4	80.0
—Damage Machinery	7.6	5	100.0

In this table, it is possible to see that three significant risks like earthquakes, electric storms, and flooding can produce the machinery damage. These natural disasters can damage the machinery either with suppliers and internally in a total of 25 different areas of the company, this variable summarizes how, in general risks impact the company. In Table 15.5, there is an extract of five impacts translated into percentages; the total number of effects is 37.

The second variable is the occurrence, and this variable was analyzed according to the frequency of the risks shown in the graph below, this information is related on how often the risks studied is presented in the Puebla City. In Fig. 15.9, it is possible to see the occurrence of each event.

Finally, the last variable studied to calculate the RPN factor for the FMEA is the detection. The detection is the ability to identify the risks in the company. It is essential to understand that the analysis is not referring if we can forecast low temperatures, for instance, but how quick can we detect if the low temperature is affecting the company in general. In Fig. 15.10 is possible to see the ability of detection on each event and the standardization in percentages.

Having detected all the variables to calculate the RPN factor, it is now possible to asses each risk and evaluate how to start the solving process for each of the risks. The assessment helps to understand prioritization; the company should somehow attend all risks detected. However, the results of the first valuation are always essential to check on where the effort should be placed in terms of timing. The results were close to what excepted, for example, we can summarize from the information analyzed that a volcanic event which is somehow prevalent in the city can damage machinery outdoor, and can create an environment of pollution, affecting directly to the human resources or damage the infrastructure of the plant due to the excess of ashes in the roof. An action plan should exist for those that event and should also be registered in the Business Continuity Plan.

15.7.3 Mitigate

In this process of the project comes the action plan with the help of the Risk Matrix tool, all the risks that got in the first assessment stage, must be canalized on a plan so that each response can follow up and monitor each potential risk. The Risk

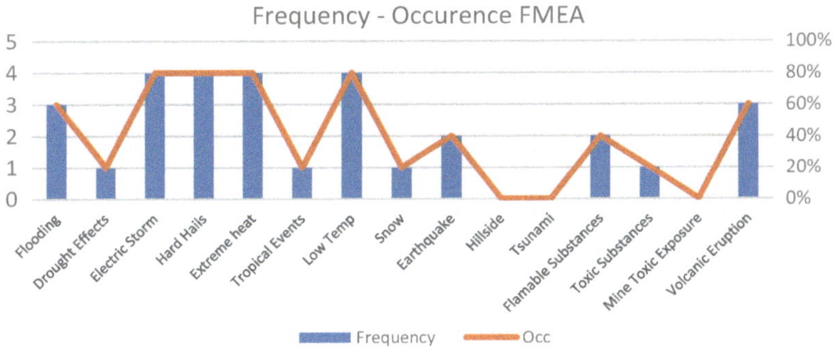

Fig. 15.9 FMEA Occurrence Variable Analysis

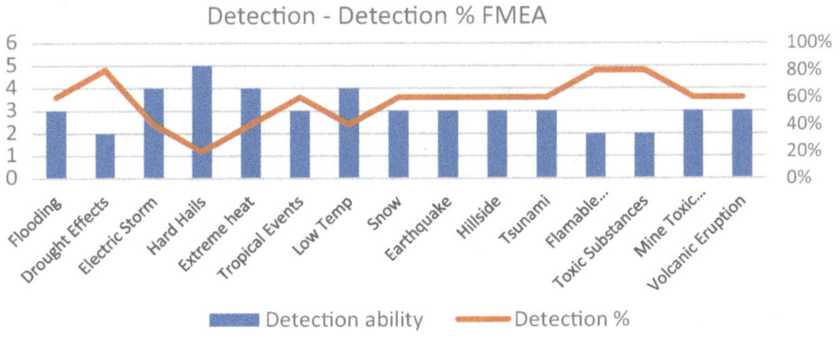

Fig. 15.10 FMEA Detection Variable Analysis

Matrix was built in three segments; the first one is the project information, start time, end time, risk category definition, and the financial project summary. The output on this first stage of the project was categorizing risks as Infrastructure, Resources, Supplier, Operation, Quality, and Business, and it is planned to start on Q3 2019 and have it implemented by Q3 2020, there is an estimated investment of USD 44,000 with a predicted risk impact of 1.2 M USD. The implementation of the project tends to reduce the impact cost to USD 500,000.

The second segment of the Risk Matrix measure the cost of risk with the probability of occurrence, this probability of occurrence was taken from the RPN factor, which was standardized in percentages, so at the end, the predicted risk value happens to be the probability of occurrence times the total risk value. Each of the risk impacts has a risk cost. All of them vary in terms of jeopardizing the cost of production per day. However, it was decided to take the most prominent impact production cost on each risk. The cost of transforming the raw material into finish good per day is USD 120,000.

The action plan composes the third segment of this Matrix, and the new risk cost value, as mentioned earlier, it has been forecasted to reduce the cost of risk from USD 1.2 M to 500 k.

Follow Appendix 15.1, it shows an extract of the Risk Matrix project worked in the company in this project, there is already an action plan in place, but this one is yet to be developed. The company can work on a global contingency plan with the information presented.

This Risk Matrix shows in general, risks, impact, categories, owner to follow up for each impact, the RPN assessment, the predicted risk cost value, the action plan, and the target to reduce the cost of risk, etc. The company can now implement this general plan on a Business Continuity Plan.

15.7.4 Monitor

This phase of the project is focused on trying to implement the complete analysis, along with a plan. It could either be a contingency or our suggestions, a Business Continuity Plan. This project is not finishing the complete Business Continuity Plan, but it will give the first roots to continue working with it. The continuity plan is a link to catastrophic events streaming from natural disasters, so it was suggested below points to the company frequently.

- First off, it is essential to have the analysis of potential impacts or disasters that could affect the company directly. This something it has already be done along with the scenarios of impact and the assessment on each risk.
- Secondly, it is critical to set in the continuity plan that can mitigate each of the risks that have been detected. It is essential to clarify that there is already a mitigation plan in place. However, there is not a complete preventive and corrective action control for each scenario, avoiding minimizing the business interruption. The goal in this step is to avoid any failures. All areas should be included or aware of the process, including the system and networks of information.
- Third, as we already have information of likelihood of occurrence and likelihood of loss, along with the total cost and potential risk cost reduction, each of the contingencies should be enlisted per the assessment studied in the paper, under this approach, the total risk analysis can be calculated per each of the hazards per event.

The business continuity plan in its mission must be focused on minimizing the risk to the employees, the service disruptions, and preserve all the company assessments along and customer information and needs. The business continuity plan is something that must be worked as soon as possible, and the plan should be added to the control plan of the company and its procedures.

15.8 Conclusion

As it has been reviewed in this study, the tool manufacturer is in the belt of fire, and so far, there is not a disaster recovery program established, nor actions or procedures that could support the operation of the company in terms of recovery on the areas of infrastructure, systems processes or IT. The objective of this paper was to reduce the impact that a catastrophic disruptive occurrence can have in this tool manufacturing. It let the company on time and coordinated response to act to any of these events, using the Risk Matrix, and work shortly on a Business Continuity Plan to reduce the cost of impact per each event.

Somehow, the objective of studying all the risks along with its impact risks is to increase the company recovery in case of any natural disaster, internally in production or process, identifying the critical systems and technical infrastructure required to support the business functions. In the end, the first information insight shows a risk cost mitigation improvement of around 58%.

Future works should focus on a different type of risk in the entire value-added, as the current Risk Mitigation plan of the company is just concentrated in the production area.

Appendix 15.1—Complete Risk Matrix Development, Self-production (2018)

Project Information	
Plant	Confidential
Project	Risk Mitigation
Project Manager	TBD
Project Start Date	Q1 2019
Planned Completion Date	Q4 2019
Estimated Completion Date	Q4 2019
Reporting Date	Weekly
Project Status	Open

Risk Category Definition						
I						Infraestructure
	R					Resources
		S				Supplier
			O			Operational
				Q		Quality
					B	Business

No	Impact of Risk	Risk Description	Category						Risk Owner	%			RPN (Prob) (%)
			I	R	S	O	Q	B		Occ (%)	Det (%)	Imp (%)	
1	—Absense	Drought Effects		X					HR, Prod	20	80	60	10
2		Low Temp		X					HR, Prod	80	40	60	19
3		Toxic Substances		X					HR, Prod	20	80	60	10
4	—Brakes in the process	Electric Storm	X			X	X	X	HR, Prod	80	40	80	26
5	—Emmisions Excess	Toxic Substances		X		X			HR, Prod	20	80	60	10
6	—Energy over consumption	Low Temp		X		X			HR, Prod	80	40	60	19
7	—Enviromental Legal Problems	Drought Effects						X	HR, Prod	20	80	40	6
8		Low Temp						X	HR, Prod	80	40	40	13
9		Toxic Substances						X	HR, Prod	20	80	40	6
10	—Human Injury	Flooding		X		X			HR, Prod	60	60	20	7
11	—Labor Sickness	Drought Effects		X		X			HR, Prod	20	80	60	10
12		Low Temp		X		X			HR, Prod	80	40	60	19
13		Toxic Substances		X		X			HR, Prod	20	80	60	10
14	—Land Cracks	Flooding				X		X	HR, Prod	60	60	80	29
15	—Machinery Overheat	Drought Effects	X			X		X	HR, Prod	20	80	80	13
16	—Machinery substance leak	Toxic Substances	X			X		X	HR, Prod	20	80	100	16
17	—Material undelivered	Earthquake						X	HR, Prod	40	60	60	14

(continued)

(continued)

No	Impact of Risk	Risk Description	Category						Risk Owner	%			RPN (Prob) (%)
			I	R	S	O	Q	B		Occ (%)	Det (%)	Imp (%)	
18		Flooding						X	HR, Prod	60	60	60	22
19		Hard Hails						X	HR, Prod	80	20	60	10
20		Volcanic Eruption						X	HR, Prod	60	60	60	22
21	—Pipe Ropture	Earthquake	X			X		X	HR, Prod	40	60	100	24

Project Financial Summary:	
Estimated Investment:	$44,000
Current Investment:	$0
Initial Predicted Risk Value:	$1,269,120
New Predicted Risk Value:	$495,360

Total Risk Value	Initial Risk Values		Action plan mitigation	Investment Req	%			RPN	Future Risk Values	
	Probability of Occurrence (%)	Predicted Risk Value			Occ (%)	Det (%)	Imp (%)	(Prob) (%)	Probability of Occurrence (%)	Predicted Risk Value
$120,000	10	$11,520	Water/Gatorade dispensers in the worksations	$1100	20	80	20	3	3	$3840
$120,000	19	$23,040	Special cold extractors in strategic manufacturing zones	800	80	40	20	6	6	$7680
$120,000	10	$11,520	Add in the OSHAS risks procidure a Toxic substance control plan for each cell	$0	20	80	20	3	3	$3840
$120,000	26	$30,720	Invest on an energy plant where needed to avoid disaster	$700	80	40	40	13	13	$15,360
$120,000	10	$11,520	Add in the OSHAS risks procidure a Toxic substance control plan for each cell	$0	20	80	20	3	3	$3840
$120,000	19	$23,040	Special cold extractors in strategic manufacturing zones	800	80	40	20	6	6	$7680
$120,000	6	$7680	Preventive action machinery check up	$0	20	80	0	0	0	$0
$120,000	13	$15,360	Special cold extractors in strategic manufacturing zones	800	80	40	0	0	0	$0
$120,000	6	$7680	Add in the OSHAS risks procidure a Toxic substance control plan for each cell	$0	20	80	0	0	0	$0

(continued)

(continued)

Total Risk Value	Initial Risk Values		Action plan mitigation	Investment Req	%			RPN (Prob) (%)	Future Risk Values	
	Probability of Occurrence (%)	Predicted Risk Value			Occ (%)	Det (%)	Imp (%)		Probability of Occurrence (%)	Predicted Risk Value
$120,000	7	$8640	Set a plan for exit emergency and alternative line manufacturer somewhere else by coorporate	1800	60	60	0	0	0	$0
$120,000	10	$11,520	Water/Gatorade dispensers in the worksations	$800	20	80	20	3	3	$3840
$120,000	19	$23,040	Special cold extractors in strategic manufacturing zones	800	80	40	20	6	6	$7680
$120,000	10	$11,520	Add in the OSHAS risks procidure a Toxic substance control plan for each cell	$0	20	80	20	3	3	$3840
$120,000	29	$34,560	Set a plan for exit emergency and alternative line manufacturer somewhere else by coorporate	1800	60	60	40	14	14	$17,280
$120,000	13	$15,360	Machine cooler's	$800	20	80	40	6	6	$7680
$120,000	16	$19,200	Add in the OSHAS risks procidure a Toxic substance control plan for each cell	$0	20	80	60	10	10	$11,520
$120,000	14	$17,280	Set up a contingency plan for earthquakes	$450	40	60	20	5	5	$5760
$120,000	22	$25,920	Set a plan for exit emergency and alternative line manufacturer somewhere else by coorporate	1800	60	60	20	7	7	$8640

(continued)

(continued)

Total Risk Value	Initial Risk Values		Action plan mitigation	Investment Req	%			RPN (Prob) (%)	Future Risk Values	
	Probability of Occurrence (%)	Predicted Risk Value			Occ (%)	Det (%)	Imp (%)		Probability of Occurrence (%)	Predicted Risk Value
$120,000	10	$11,520	Wear special protectors for hears when noice surpasses 85 decibels, NOM 011	$100	80	20	20	3	3	$3840
$120,000	22	$25,920	Plan in place, improve procidure applying a Kizen	$0	60	60	20	7	7	$8640
$120,000	24	$28,800	Set up a contingency plan for earthquakes	$450	40	60	60	14	14	$17,280
$120,000	6	$7680	Invest on an energy plant where needed to avoid disaster	$700	80	40	0	0	0	$0
$120,000	7	$8640	Set a plan for exit emergency and alternative line manufacturer somewhere else by coorporate	1800	60	60	0	0	0	$0
$120,000	7	$8640	Set a plan for exit emergency and alternative line manufacturer somewhere else by coorporate	1800	60	60	0	0	0	$0
$120,000	6	$7680	Special cold extractors in strategic manufacturing zones	800	80	40	0	0	0	$0
$120,000	24	$28,800	Set up a contingency plan for earthquakes	$450	40	60	60	14	14	$17,280
$120,000	10	$11,520	Set up a contingency plan for earthquakes	$450	40	60	0	0	0	$0
$120,000	14	$17,280	Set a plan for exit emergency and alternative line manufacturer somewhere else by coorporate	1800	60	60	0	0	0	$0

(continued)

(continued)

Total Risk Value	Initial Risk Values		Action plan mitigation	Investment Req	%			RPN (Prob) (%)	Future Risk Values	
	Probability of Occurrence (%)	Predicted Risk Value			Occ (%)	Det (%)	Imp (%)		Probability of Occurrence (%)	Predicted Risk Value
$120,000	6	$7680	Wear special protectors for hears when noice surpasses 85 decibels, NOM 011	$100	80	20	0	0	0	$0
$120,000	14	$17,280	Plan in place, improve procidure applying a Kizen	$0	60	60	0	0	0	$0
$120,000	16	$19,200	Wear special protectors for hears when noice surpasses 85 decibels, NOM 011	$100	80	20	60	10	10	$11,520
$120,000	6	$7680	Invest on an energy plant where needed to avoid disaster	$700	80	40	0	0	0	$0
$120,000	10	$11,520	Set up a contingency plan for earthquakes	$450	40	60	0	0	0	$0
$120,000	14	$17,280	Set a plan for exit emergency and alternative line manufacturer somewhere else by coorporate	1800	60	60	0	0	0	$0
$120,000	24	$28,800	Set up a contingency plan for earthquakes	$450	40	60	60	14	14	$17,280
$120,000	16	$19,200	Wear special protectors for hears when noice surpasses 85 decibels. NOM 011	$100	80	20	60	10	10	$11,520
$120,000	36	$43,200	Plan in place, improve procidure applying a Kizen	$0	60	60	60	22	22	$25,920
$120,000	24	$28,800	Set up a contingency plan for earthquakes	$450	40	60	60	14	14	$17,280
$120,000	32	$38,400	Invest on an energy plant where needed to avoid disaster	$700	80	40	60	19	19	$23,040

(continued)

(continued)

Total Risk Value	Initial Risk Values		Action plan mitigation	Investment Req	%			RPN (Prob) (%)	Future Risk Values	
	Probability of Occurrence (%)	Predicted Risk Value			Occ (%)	Det (%)	Imp (%)		Probability of Occurrence (%)	Predicted Risk Value
$120,000	36	$43,200	Set a plan for exit emergency and alternative line manufacturer somewhere else by coorporate	1800	60	60	60	22	22	$25,920
$120,000	10	$11,520	Set up a contingency plan for earthquakes	$450	40	60	0	0	0	$0
$120,000	14	$17,280	Set a plan for exit emergency and alternative line manufacturer somewhere else by coorporate	1800	60	60	0	0	0	$0
$120,000	10	$11,520	Set up a contingency plan for earthquakes	$450	40	60	0	0	0	$0
$120,000	14	$17,280	Set a plan for exit emergency and alternative line manufacturer somewhere else by coorporate	1800	60	60	0	0	0	$0
$120,000	6	$7680	Wear special protectors for hears when noice surpasses 85 decibels, NOM 011	$100	80	20	0	0	0	$0
$120,000	14	$17,280	Plan in place, improve procidure applying a Kizen	$0	60	60	0	0	0	$0
$120,000	13	$15,360	Wear special protectors for hears when noice surpasses 85 decibels, NOM 011	$100	80	20	40	6	6	$7680
$120,000	26	$30,720	Invest on an energy plant where needed to avoid disaster	$700	80	40	40	13	13	$15,360
$120,000	24	$28,800	Set up a contingency plan for earthquakes	$450	40	60	60	14	14	$17,280

(continued)

(continued)

Total Risk Value	Initial Risk Values		Action plan mitigation	Investment Req	%			RPN (Prob) (%)	Future Risk Values	
	Probability of Occurrence (%)	Predicted Risk Value			Occ (%)	Det (%)	Imp (%)		Probability of Occurrence (%)	Predicted Risk Value
$120,000	36	$43,200	Set a plan for exit emergency and alternative line manufacturer somewhere else by coorporate	1800	60	60	60	22	22	$25,920
$120,000	36	$43,200	Plan in place, improve procidure applying a Kizen	$0	60	60	60	22	22	$25,920
$120,000	29	$34,560	Set a plan for exit emergency and alternative line manufacturer somewhere else by coorporate	1800	60	60	40	14	14	$17,280
$120,000	29	$34,560	Set a plan for exit emergency and alternative line manufacturer somewhere else by coorporate	1800	60	60	40	14	14	$17,280
$120,000	7	$8640	Set a plan for exit emergency and alternative line manufacturer somewhere else by coorporate	1800	60	60	0	0	0	$0
$120,000	5	$5760	Set up a contingency plan for earthquakes	$450	40	60	0	0	0	$0
$120,000	14	$17,280	Set up a contingency plan for earthquakes	$450	40	60	20	5	5	$5760
$120,000	10	$11,520	Fans on workstations	$450	20	80	20	3	3	$3840
$120,000	19	$23,040	Special cold extractors in strategic manufacturing zones	800	80	40	20	6	6	$7680

(continued)

(continued)

Total Risk Value	Initial Risk Values		Action plan mitigation	Investment Req	%			RPN (Prob) (%)	Future Risk Values	
	Probability of Occurrence (%)	Predicted Risk Value			Occ (%)	Det (%)	Imp (%)		Probability of Occurrence (%)	Predicted Risk Value
$120,000	10	$11,520	Add in the OSHAS risks procidure a Toxic substance control plan for each cell	$0	20	80	20	3	3	$3840
$120,000	36	$43,200	Plan in place, improve procidure applying a Kizen	$0	60	60	60	22	22	$25,920
$120,000	13	$15,360	Wear special protectors for hears when noice surpasses 85 decibels, NOM 011	$100	80	20	40	6	6	$7680
$120,000	36	$43,200	Plan in place, improve procidure applying a Kizen	$0	60	60	60	22	22	$25,920
$120,000	5	$5760	Set up a contingency plan for earthquakes	$450	40	60	0	0	0	$0
$120,000	13	$15,360	Invest on an energy plant where needed to avoid disaster	$700	80	40	0	0	0	$0
$120,000	6	$7680	Wear special protectors for hears when noice surpasses 85 decibels, NOM 011	$100	80	20	0	0	0	$0

References

Atlas_National_Risk. (2018). Retrieved from www.atlasnacionalderiesgos.gob.mx.

Biggs, D. (2011). Understanding resilience in a vulnerable industry: The case of reef tourism in Australia. *1ARC Centre of Excellence for Coral Reef Studies, 18*. https://doi.org/10.5751/es-03948-160130.

Blos, M. F. (2009). Supply chain risk management (SCRM): A case study on the automotive and electronic a case study on the automotive and electronic. *Supply Chain Management: An International Journal*, 247–252. https://doi.org/10.1108/13598540910970072.

Breuer, C. S.-D. (2013). Collaborative risk management in sensitive logistics nodes. *Performance Management: An International Journal*, 331–351. https://doi.org/10.1108/TPM-11-2012-0036.

CENAPRED. (2018). Declaratorias sobre emergencia, desastre y contingencia climatológica a nivel municipal entre 2000 y 2017. National Center for Disaster. Retrieved from Prevention. Retrieved March 10, 2018, from https://datos.gob.mx/busca/dataset/declaratorias-sobre-emergencia-desastre-y-contingencia-climatologica/resource/41444ebe-6a35-4631-8f91-9237d5114488.

Chiang, W. L. (2003). The establishment of earthquake, typhoon and flood insurance. Research Reports for the Department of Insurance, Ministry.

Chung-Hung, T. C.-W. (2010). An earthquake disaster management mechanism based on risk assessment information for the tourism industry-a case study from the island of Taiwan. *Tourism Management*, 470–481. https://doi.org/10.1016/j.tourman.2009.05.008.

Cruz, A. M. (2015). Emerging issues for natech disaster risk management in Europe. *Journal of Risk Research*, 37–41. https://doi.org/10.1080/13669870600717657.

Economy_Department_of_the_Mexican_Republic. (2016, August). Cenapred. Retrieved from Coordinación Nacional de Protección Civil http://www.cenapred.gob.mx/es/documentosWeb/Enaproc/FenomenoAntrop.pdf.

El_ Universal, I. (2018). Businessmen urge the government to fight insecurity in Mexico. El Universal (p. 1).

FONDEN. (2018). 2015–2018. Recursos autorizados por declaratoria de desastre. National System of Civil Protection. Retrieved from http://www.proteccioncivil.gob.mx/es/ProteccionCivil/Recursos_Autorizados_por_Declaratoria_de_Desastre.

Hoa, W. (2015). Supply chain risk management: A literature review. *International Journal of Production Research*, 5031–5069. https://doi.org/10.1080/00207543.2015.1030467.

Kreimer, A. (1999). *Managing disaster risk in Mexico*. Washington DC: The International Bank for Reconstruction.

Leat, P., & Revoredo-Giha, C. (2013). Risk and resilience in agri-food supply chains: The supply chain management, 219–231. https://doi.org/10.1002/9781118937495.ch10.

Messner, F., and Green, C. (2007). Fundamental issues in the economic evaluation of flood damage. In Evaluating flood damages: Guidance and recommendations on principles and methods. Messner, F ., Penning-Rowsell, E., Green, C.H., Meyer, V., Tunstall, S. and van der Veen, A. (Eds) FLOODsite Report, T09-06-01, Wallington, 95–105.

Merza, M. (2013). A composite indicator model to assess natural disaster risks in industry on a spatial level. *Journal of Risk Research, 16*(9), 1077–1099. https://doi.org/10.1080/13669877.2012.737820.

Norrman, A. (2004). Ericsson's proactive supply chain risk management approach after a serious sub-supplier accident. *International Journal of Physical Distribution & Logistics Management*, 434–456. https://doi.org/10.1108/09600030410545463.

Pascaline, W. (2018). Economic losses poverty and disasters. United Nations Office for Disaster Risk Reduction UNISDR, 31. https://doi.org/10.13140/RG.2.2.35610.08643.

Ponomarov, S. Y., & Holcomb, M. C. (2009). Understanding the concept of supply chain resilience. *The International Journal of Logistics Management*, 124–143. https://doi.org/10.1108/09574090910954873.

Sertyesilisik, B. (2017). Building information modeling as a tool for enhancing disaster resilience of the construction industry. *Safety Engineering*, 9–18. https://doi.org/10.1515/tvsbses-2017-0002.

Simba, S. W. (2017). Supply chain risk management processes for resilience: A study of South African grocery manufacturers. *Journal of Transport and Supply Chain Management*, 1–13. https://doi.org/10.4102/jtscm.v11i0.325.

Sun, M.-Z. (2018). Vessel traffic risk assessment based on uncertainty analysis in the risk matrix. *Algorithms MDPI*, 1–12. https://doi.org/10.3390/a11050060.

UK-DFIF. (2011 November). GOV.UK. Retrieved from Department for International Development https://www.gov.uk/government/organisations/department-for-international-development#content.

Van-der-Veen, G. (2017). Indirect economic damage: Concepts and guideline in evaluating flood damages. Flood site report, 95–105.

Wieland, T. J. (2013). The influence of relational competencies on supply chain resilience: A relational view. *International Journal of Physical Distribution & Logistics Management*, 300–320. https://doi.org/10.1108/IJPDLM-08-2012-0243.

World_Economic_Forum. (2008, January). Building resilience to natural disasters: A framework for private sector engagement. WEF (pp. 1–20).

Chapter 16
Inclusive Short Chains as Strategy for Creating Resilience in Agricultural Economic Activity

Horacio Bautista-Santos, Fabiola Sánchez-Galván, Diana Sánchez-Partida, José-Luis Martínez-Flores, and Arely Del Rocio Ireta-Paredes

Abstract This case study was carried out in the Chontla municipality located in Veracruz, Mexico, belonging to the Huasteca Baja Region. The problem addressed in this research was the micro-regions design for a municipality that has 138 localities. The aim is to articulate the agricultural backyard producers into short commercialization circuits to minimize the distances traveled for their product collection and distribution. As a basis for this study, the p-median model used and a balanced constraint was included. It can hypothesize that this study will contribute to establishing inclusive agro-food shorts chains that can create resilience in agricultural economic activities. Computational results indicate that the Chontla municipality can segment into seven micro-regions.

Keywords Backyard-agricultural production system · p-median model · Balanced constraint · Short-commercialization circuits · Facility location

---

Electronic supplementary material The online version of this chapter (https://doi.org/10.1007/978-3-030-67295-9_16) contains supplementary material, which is available to authorized users.

H. Bautista-Santos · F. Sánchez-Galván (✉)
Research and Postgraduate Department, Tecnológico Nacional de México/ITS de Tantoyuca, Desviación Lindero Tametate S/N. Col. La Morita, C.P. 92100 Tantoyuca, Veracruz, Mexico
e-mail: fabiola.sanchez@itsta.edu.mx; fsgalvan01@gmail.com

H. Bautista-Santos
Tecnológico Nacional de México/ITS de Chicontepec, Calle Barrio Dos Caminos No. 22. Col. Barrio Dos Caminos, C.P. 72709 Chicontepec, Veracruz, Mexico

F. Sánchez-Galván · D. Sánchez-Partida · J.-L. Martínez-Flores
Department of Logistics and Supply Chain Management, Universidad Popular Autónoma del Estado de Puebla (Upaep University), 17 Sur 901, Barrio de Santiago, C.P. 72410 Puebla, Puebla, Mexico

A. D. R. Ireta-Paredes
Universidad Interserrana del Estado de Puebla—Ahuacatlán, Los Llanos Km 1, San Andrés Tlayehualancingo, C.P. 73330 Ahuacatlán, Puebla, Mexico

D. Sánchez-Partida et al. (eds.), *Disaster Risk Reduction in Mexico*,
https://doi.org/10.1007/978-3-030-67295-9_16

16.1 Introduction

Resilience-based on the community provides a bridge between traditional agriculture and natural resource management through relationship building and inclusiveness (King, 2008). Climatic conditions, economic processes, production diversification, and geography, characterize the agricultural production system (Wajszczuk, 2016).

It is necessary to encourage backyard agricultural producers to work collectively and organized to promote rural development (De los Ríos-Carmenado et al., 2011), generating changes in the production and distribution of food produced locally (WEF, 2017).

The Localized Agro-Food System (LAFS) approach promotes the local resources revaluation from the perspective of sustainable development (Fournier & Muchnik, 2012), promotes that small producers get involved in development dynamics under solidarity, coordination, and territorialized cooperation forms (Boucher & Poméon, 2010). It is required that within a context studied that the small producers are organized at least in a cooperative way to achieve it (Giacomini & Mancini, 2015).

Geographical proximity is not enough to generate cooperation among local producers, but it is essential to organize them in cooperative networks (Filippi & Torre, 2003). Short-commercialization circuits arise from the need to link local producers with final consumers in social proximity terms and physical closeness (Rodríguez & Riveros, 2016). These allow enhancing the family farming capacities by integrating them into inclusive productive chains (CEPAL-FAO-IICA, 2014), where the geographical area of production, marketing, and consumption occurs in a radius of between 20 and 100 km (JRC, 2013).

Sustainable agricultural development strategies need to focus on the heterogeneous and variability of resources that exist in a territory (Ruben & Pender, 2004). The local product diversity can stand out, generating a link between production and consumption through inclusive strategies into an agricultural food system (FAO, 2017).

Paasi (2004) indicates that there are several reasons that geographers interpret territorial development dynamics and this depending on the research context: economy structural relations analysis, networks, and governance; also, assures that the links between local economy, region/place, and boundaries are crucial for understanding the spaces of identity gendered, inclusion and exclusion, and multi-level citizenship.

Soto et al. (2007) defines the importance that territory could be studied through collecting information without any particular conceptual role, taking only as an example the notion of the region, and after that may arise a set of historically contingent social practices and discourses.

The potential of the backyard agricultural production system can be highlighted with supply chain modeling tools (Aghazadeh, 2004). To create sustainable and efficient logistics networks (Soysal et al., 2012) based on the producers' geographical concentration and their integration into a network of local food producers that improves potential markets, logistics efficiency and environmental impact (Bosona & Gebresenbet, 2011; Erickson et al., 2013).

In the United States of America, local food producers are part of an alternative network that interacts through retail outlets, food cooperatives, and community-supported agriculture programs (Evans, 2010). The marketing of local products is done in two primary forms: direct-to-consumer (farmer to the customer) and direct-to-retail (farmer to restaurants, hospitals, schools, and organizations) (Brain, 2012).

Due to the perishable nature of fresh food, strategic decisions such as the location of collection and distribution centers are essential during the agro-food chain design (Ahumada & Villalobos, 2011; Sanabria et al. 2016; Zhen et al., 2016). The logistics network design for fresh agricultural products significantly influences its network performance and product quality (Keizer et al., 2017).

The facilities' location and customer assignment are central themes of a logistics network design; models and their solution algorithms vary widely in mathematical complexity terms and computational performance (Allaoui et al., 2018; Klose & Drexl, 2005).

Tong et al. (2012) developed a model for locating a fixed farmers' market with constraints of time windows in travels based on a customer's daily activity (home and workplaces) and associated service schedules (weekday morning-noon, weekday afternoon-evening and weekends), resulting in a significant reduction in the overall travel cost.

Orjuela-Castro et al. (2017) argue that transporting perishable products can be exposed to changes in temperature and humidity, which affects product quality. They propose a mixed linear programming model for the localization of collection centers and companies processing perishable foods, based on a multi-product and multi-echelon transport system. This model minimizes the total cost of fresh product loss among the producers, collection centers, processing centers, wholesalers, warehouses, and markets.

The facility location problem in perishable commodities emergency system is a class of practical application problems during the process of decision-making, and it is studied under specific time constraints with the objective of logistics cost minimization (Zhang & Yang, 2007).

The p-median is a mathematical model that allow to solve facility location problem. A variant of the p-median model is the Capacitated p-median problem (CPMP), where facilities are economically selected in order to serve the total demand placed on each of the candidate medians and does not exceed their capacity (Fleszar & Hindi, 2008). Another variant is a bi-level capacitated p-median facility location problem with an approach to estimate the desirability and service probabilities (Abareshi & Zaferanieh, 2019).

Another optimization method used is a Tabu-search, which is a metaheuristic that offers a solution alternative to solve the CPMP (Romero et al., 2018). Moreover, a Multi-Objective Artificial Bee Colony (MOABC) was used to solve evacuation problems as an essential activity for reducing the number of casualties and the amount of damage to disaster management; two objective functions were considered to minimize the total traveling distance from an affected area to shelters and to minimize the overload capacity of shelters (Niyomubyeyi et al., 2019).

This chapter presents the case study of the backyard agricultural production system of Chontla municipality located in the North of Veracruz, Mexico, which presents high marginalization and indigenous degree (SEGOB-SNIM, 2010). Moreover, it is a municipality highly susceptible to the negative impacts of natural phenomena (flood, hurricane, and forest fires) (FONDEN, consulted May 2019).

The aim is to segment a municipality into rural production micro-regions, for the inclusion of backyard agricultural producers into short-commercialization circuits as a strategy for strengthening the local economy (Sánchez-Galván et al., 2019) and as a learning mechanism that allows identifying the elements that determine territorial development. It can hypothesize that this study will contribute to establishing inclusive agro-food shorts chains for creating resilience in agricultural economic activities.

As a basis for this study, a mathematical model named *p-median* was used, which minimizes the distance traveled from a customer i to the nearest facility location j, it also assigned customers to facilities. In this work, the *p-median* model is used to locate collection centers of backyard agricultural products, where rural localities represent customers and collection centers represent facilities, so that a set of n rural localities will be assigned into p collection centers to integrate micro-regions of rural production, such that the total distance within each micro-region is minimized, this represents a short-commercialization circuit. The dissimilarity of a micro-region is the sum of the distances between each locality who belongs to the short-commercialization circuit and the average distance associated with the micro-region. As a variant of the model, a balanced constraint designed, which assigns a similar number of localities to collection and distribution centers.

16.2 Case Study

Chontla municipality is located in Veracruz, México in the Huasteca Baja Region (97°55′26″ W, 21°17′40″ N), it has a population of 17,429, with 138 rural localities and a total area of 390 km^2 which represents 0.54% respect to the state (INEGI, 2010). Chontla is characterized by a high degree of marginalization; its scope is rural with an indigenous presence like Teenek or Huasteco (predominant), Nahuatl, Totonaca, and Mazateco (SEGOB-SNIM, 2010).

Of the total population of Chontla municipality, 21% live in extreme poverty, 50% in moderate poverty, 71.2% have an income below the welfare line, and 32.1% have educational lagging. Chontla lacks access to social security (84.3%), essential housing services (81.3%), and access to food (36.8%) (CONEVAL, 2015).

Chontla receives support from the government program named Zones Priority Development, which aims to contribute to providing an adequate environment for the territories that presents the greatest marginalization and social backwardness in the country, through to improve access to essential housing services and the community social infrastructure (SEDESOL, 2017).

Moreover, the Huasteca Baja region is highly susceptible to the negative impacts of natural phenomena (flood, hurricane, and forest fire). Only in four years of the period from 2010 to 2016 have added 76 emergency declarations according to the reports generated by the Natural Disasters Fund (FONDEN), that is, these regions represent 34% of the affectations of the 100% occurred throughout the State of Veracruz (FONDEN, consulted May 2019).

According to The National Agricultural Survey of Mexico, the production indicators include small, medium, and large producers (INEGI, 2017). The backyard agricultural producers are considered as an informal sector, and there are no production records, given the types of products it generates—the backyard producers market local products for regional or local consumption.

This study promotes the aggregation of backyard agricultural production from a facility location that serves as a backyard agricultural product collection and distribution center. Hence, the importance of designing inclusive agro-food short chains to incorporate backyard agricultural producers into a short-commercialization circuit.

For the collected data, a production survey was designed that was structured into two sections. The first section is intended to know the producer information such as education level, people living in their house, the economic activities carried out, his main income source, the reasons why he/she performs its activity, and the way to move to the nearest local market.

The second section is intended to identify the products and the destination of their production (each producer has one or more products in his/her backyard). For each product interviewer asked for its measure unit, its equivalent weight in kilograms, the approximate amount destined for sale, self-consumption, and the reasons by do not market; also, how many times do they harvest the product per year.

From January 2017 to June 2018, in Chontla municipality were collected 370 surveys. The sample size calculated with a maximum variance (95% confidence, 5% error, and 50% heterogeneity). The target population of the survey was defined as the number of dwellings number reported from the INEGI 2010 census (4227 homes in Chontla).

The localities such as Arranca Estacas, Las Cruces, San Francisco, San Juan Otontepec, Tezitlal, and Xochitlán belong to Chontla municipality where the surveys were applied. Each one of them is a locality with a high marginalization degree and rural scope, and they belong to a government program for priority areas development due to their poverty situation by at least one social deprivation and without enough income to satisfy their basic needs (SEGOB-SNIM, 2010).

According to surveys made, 75% of backyard producers surveyed are dedicated to agriculture only, 15% to agricultural and livestock activities, 7% to agriculture and handicrafts, 1% to livestock, and 2% to three activities (agriculture, livestock and artisanal). Additional income sources (39%), self-consumption (14%), job alternative (16%), family inheritance (14%), and others (7%) are primary reasons for developing these activities.

The products identified were nopal *(Opuntia ficus-indica)*, creole pumpkin *(Cucurbita maxima)*, coriander *(Coriandrum sativum)*, plum (*Spondias purpurea L.*), passion fruit *(Passiflora edulis S.)*, jobo *(Spondias mombin)*, creoli chili *(Capsicum spp.)*, litchi *(Nephelium litchi camb.)*, and tamarind (*Tamarindus indica*).

Table 16.1 Annualized estimated production of backyard agricultural products of Chontla municipality

Product	Annualized estimated production (Ton)						Mexican pesos thousands ($)[1]
	Arranca Estacas	Las Cruces	San Francisco	San Juan Otontep.	Tezitlal	Xochitlan	
Nopal	0.066	–	–	2.560	–	0.040	213.27
Creole pumpkin	0.270	0.615	0.120	4.746	–	0.045	150.70
Coriander	0.010	0.017	0.002	0.420	0.040	0.001	97.82
Plum	0.100	0.350	0.200	0.190	–	0.200	31.20
Passion Fruit	0.120	0.240	–	0.240	–	–	18.00
Jobo	–	–	–	0.650	–	0.130	15.60
Creole chili	0.378	0.038	0.114	0.204	–	0.040	15.45
Litchi	–	–	–	0.500	–	–	10.00
Tamarind	–	–	0.080	–	–	0.160	7.20

Source Own elaboration with information collected from surveys from January 2017 to June 2018
[1]An amount calculated based on the average price in the local market from January to June 2018

In this study, 22.0% of total production is destined for sale, 31.5% for self-consumption, and 42.9% is non-commercialized by reason for price, transport, and because the product cannot be collected.

Rodríguez and Riveros (2016) argue it is feasible to establish a short-commercialization circuit around agricultural backyard production characteristics. It was necessary to conduct semi-structured interviews face-to-face with agricultural backyard producers, intermediaries, agents, and municipal authorities, retail traders and wholesalers, and market premises owners to identify these characteristics.

Hence, the backyard agricultural producer organizes itself individually and informally, the products are sold in local markets or different streets recognized as community or municipality markets. The backyard producer does not make a pre-sale contract, does not generate production records, and products do not have third-party certification. The social proximity is, in general, much closer; there are affinity and sensitivity between the producer and the final consumer.

Table 16.1 shows the representative products of the locations where the information collected; also, it shows its respective economic value in the local market.

Fig. 16.1 Methodology for backyard-agricultural production micro-regions design. *Source* Own elaboration

16.3 Methodology and Methods

This section presents the methodology used to the agricultural backyard production micro-regions design as a strategy for articulating backyard producers within short-commercialization circuits for contributing to designing inclusive agro-food short chains. The methodology is shown in Fig. 16.1.

- The *distances matrix* built with a computational application developed in Java SE 8u221 Language, which interacting with the Google Maps API platform. The distance between nodes (origin locality and destination locality) was determined from the geographic coordinates (latitude and altitude) obtained from the INEGI database (INEGI, 2015). Chontla municipality has 138 localities. Each locality is an origin and a destination, therefore the matrix size of 138 × 138 see Appendix 16.A.
- *Scenario 1.* This scenario is constructed through discussion forums with municipal authorities and experts in the logistics area; the objective is to define the instances number that will be executed in this scenario. The instance definition determines the number of micro-regions in which the municipality would be segmented. Each instance uses the *p-median* model to determine mathematically the locality where a facility should be installed. In this case, the facility is a rural locality that will work as a collection and distribution center of backyard-agricultural products; also will determine the localities that will be assigned to each micro-region.
- *Geographical adjustment.* The localities selected as collection centers by the *p-median* model in scenario 1 are not necessarily localities with the better access routes or with, the better local markets, this is because the model simply seeks to reduce the average distances from the origin locality to each destiny locality. Therefore, in a discussion forum with municipal authorities and experts in the logistic area, should select the two better instances from scenario one, and after that, it is necessary established a geographical adjustment that consists into move the locality assigned by the p-median model to the nearest locality that has the largest population, best access routes, and better local market. To visualize the locations selected by the p-median model and the geographic adjustment can be used as the Google maps tool.
- *Scenario 2.* This is constructed with the best instances of scenario one and its geographical adjustment. This scenario uses the p-median model with a balanced constraint. This constraint assigns a homogenous number of clients to

nodes, which means assigns a homogenous number of localities to each micro-region. The mathematical model programming was carried out in an optimization language named LINGO that used a branch-and-bound (B&B) algorithm for its solution. The best instance of scenario two will define the micro-region design.

- *Micro-regions design.* The best instance of scenario two will define the micro-regions design. Google maps tool allows visualizing the municipality segmentation in backyard-agricultural production micro-regions and the short-commercialization circuit.

16.3.1 Mathematical Model

This section describes the p-median model. The p-median model is one of the basic mathematical models in facility location theory, which is a field of Operational Research (Mladenovic et al., 2007). Hakimi (1964) first presented the *p-median* model to facilities location *p*, minimizing the total distance between the demand nodes.

A p-median model variant considers external conditions, such as traffic, on travel time, which accounts for fluctuations in travel cost distance at different time intervals; for this purpose, Google Traffic and Foursquare data to respectively retrieve traffic information and estimate demand in a region (Noorian et al., 2018).

Another p-median model variant is to enumerate all of the possible combinations of P facilities and, after that, adopt simulated annealing to allocate resources. This variant was applied in a healthcare center that considers both geographic accessibility and service quality as a capacity constraint and spatial compactness constraint (Zhao & Dou, 2011).

The p-median model assigns customers to facilities, without considering a minimum or maximum customers number per facility, so that the customers' assignment is imbalanced.

In this study, the authors propose a variant for the p-median model. This variant considers a balanced constraint to be homogenous the number of clients assigned to each node; that is, the number of localities assigned to each micro-region can be homogenized.

16.3.2 P-Median Model

The objective function (16.1) minimizes the distance traveled from a rural town (*i*) to the nearest collection center (*j*). Constraint (16.2) controls the collection centers number required; (16.3) ensures that all rural locations are assigned to a collection center; (16.4) controls that rural locations are only assigned to open collection centers; (16.5) and (16.6) define that the decision variables are binary (1 if assigned, 0 if not assigned) (Revelle & Eiselt, 2005).

Constraint (16.7) is the contribution of this study; it balances the localities number assigned, that is, assigned a homogenous number to each collection center, and working together with p-median model constraint both determine which rural localities will be assigned to collection centers to integrate micro-regions of rural production.

$$\text{Objective function} : Min \sum_{i \in I} \sum_{j \in J} d_{ij} y_{ij} \tag{16.1}$$

$$\text{Subject to} : \sum_{j \in J} x_j = p \tag{16.2}$$

$$\sum_{j \in J} y_{ij} = 1 \quad \forall i \in I \tag{16.3}$$

$$y_{ij} - x_j \leq 0 \quad \forall i \in I, \ j \in J \tag{16.4}$$

$$x_j \in \langle 0, 1 \rangle \quad \forall j \in J \tag{16.5}$$

$$y_{ij} \in \langle 0, 1 \rangle \quad \forall i \in I, \ \forall j \in J \tag{16.6}$$

$$\sum_{j \in J} y_{ij} \leq \frac{J}{p} + 1 \quad \forall j \in J \tag{16.7}$$

where:

I = the rural localities set i.
J = the candidate collection centers set j.
d_{ij} = the distance between rural town i and the collection center located at site j.
p = the number of proposed collection centers.
x_j = a binary decision variable that will take the value 1 if the collection center j is opened or 0 in another case.
y_i = a binary decision variable that will take the value of 1 if the rural location i is assigned to the collection center j or 0 otherwise.

16.4 Results

The results presented in this section detail the process of the backyard-agricultural production micro-region design for the Chontla municipality.

16.4.1 Results of Scenario 1

A series of discussion forums were organized to determine in how many micro-regions the municipality should be segmented. The experts in the logistics area argued that the higher the micro-regions number, the lower the distance traveled from origin locality to destiny locality. Municipal authorities argued that the locality that should be work as a collection and distribution center should be the locality with the largest population. As a final decision, the participants of the forum decided to evaluate four instances for segmenting the municipality in 3, 5, 7, and 10 micro-regions with the p-median model to determine the localities that will work as a collection and distribution center.

Table 16.2 shows the results of the instances of scenario 1.

Table 16.2 P-median model results for instances of scenario 1. Chontla municipality

Instance 1: ***3 micro-regions***		*Instance 2:* ***5 micro-regions***	
Instance cost (km): **888.64**		*Instance cost (km):* **660.80**	
CDCBAP[1] location	Number of localities assigned	CDCBAP[1] location	Number of localities assigned
Las Canoas	41	Santa Rita	31
Villahermosa (El Crucero)	52	Las Cruces	30
El Nopal	45	La Laja	23
		Huayacocotla	17
		Zocohuite	37
Instance 3: ***7 micro-regions***		*Instance 4:* ***10 micro-regions***	
Instance cost (km): **527.83**		*Instance cost (km):* **414.36**	
CDCBAP[1] location	Number of localities assigned	CDCBAP[1] location	Number of localities assigned
Tlatemalco	22	San Francisco	12
La Laja	24	El Órgano	6
Paciencia	22	Santa Rita	16
Palo Alto	22	La Laja	17
El Arroyo	16	Chalahuite	8
El Deseo	18	Huayacocotla	13
La Peñita	14	Paciencia	19
		Palo Alto	16
		El Arroyo	13
		El Deseo	18

Source Own elaboration with information collected from surveys from January 2017 to June 2018
[1]Collection and Distribution Center of Backyard Agricultural Products (CDCBAP)

Under *Scenario 1*, instances show that to a higher quantity of facilities located, the lower is the instance cost. Further, the localities number assigned to each commercialization circuit is variable.

The results of this scenario do not include the municipal head (Chontla locality) as a collection center because the model simply seeks to reduce the average distances from the consolidation center to each destiny locality. However, the municipal head is the primary municipality market, and it is a locality with a larger population.

16.4.2 Geographical Adjustment

Municipal authorities and experts in the logistics area decided to select as the best instances of scenario 1, the instances of seven and ten micro-regions. The geographical location of each locality of each instance selected showed with the Google Maps tool. The geographical adjustment was defined as the locality nearest to the locality proposed by the p-median model. The criteria for making a geographical adjustment were that the locality should have 500 population and that the locality should have at least one rural road that communicates with the local market.

Figures 16.2 and 16.3 show the collection centers assigned by the *p-median* model and the collection centers proposed by geographical adjustment for seven and ten facilities that will work as a collection and distribution centers.

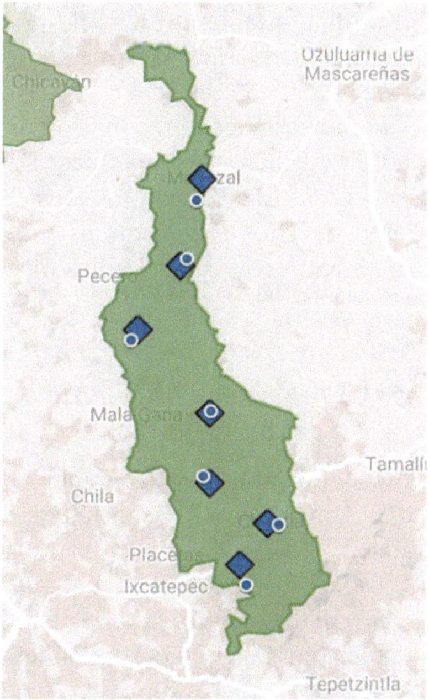

Fig. 16.2 Seven collection and distribution centers located in Chontla municipality, (•) *p-median* model results, and (♦) geographical adjustment proposed. *Source* Own elaboration based on the results of scenario 1

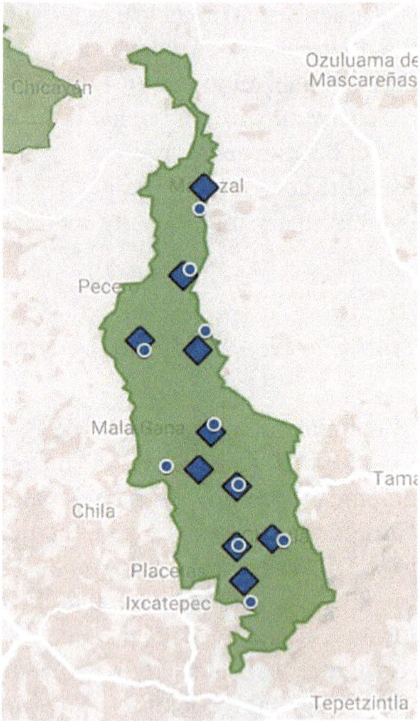

Fig. 16.3 Ten collections and distribution centers located in Chontla municipality, (•) *p-median* model results, and (♦) geographical adjustment proposed. *Source* Own elaboration based on the results of scenario 1

For an instance of seven micro-regions, the localities selected in a discussion forum that established a geographical adjustment are Chontla, Magozal, La Laja, Las Cruces, El Ebanal, Cruz Manantial, and La Floreña. Only one location coincided with being selected by both the p-median model and geographical adjustment.

For an instance of ten micro-regions, the localities selected in a discussion forum that established a geographical adjustment are Chontla, Magozal, San Francisco, Santa Rita, Las Cruces, El Higo, El Ebanal, Mata de Otate, La Floreña, and La Campana. Only two locations coincided with being selected by both the p-median model and geographical adjustment.

16.4.3 Results of Scenario 2

After visualizing the geographical adjustment to identify the localities where the collection center will be installed, the *p-median* model with a balanced constraint is runned.

Table 16.3 shows the results of the instances of scenario 2.

In a discussion forum, the results were analyzed. For example, the instance costs of scenario 2 are more considerable than scenario 1, but scenario 2 shows a homogenous allocation of the locations that will belong to the micro-region. The forum participants argued that the economic cost of installing seven centers is less

Table 16.3 P-median model with a balanced constraint results for instances of scenario 2 for Chontla municipality

*Instance: **7 micro-regions***		*Instance: **10 micro-regions***	
Instance cost (km): **578.72**		*Instance cost (km):* **576.35**	
CDCBAP[1] location	Amount localities assigned	CDCBAP[1] location	Amount localities assigned
Chontla	20	Chontla	14
Magozal	18	Magozal	14
La Laja	20	San Francisco	14
Las Cruces	20	Santa Rita	14
El Ebanal	20	Las Cruces	14
Cruz Manantial	20	El Higo	14
La Floreña	20	El Ebanal	14
		Mata de Otate	12
		La Floreña	14
		La Campana	14

Source Own elaboration based on the results of geographical adjustment
[1]Collection and Distribution Center of Backyard Agricultural Products (CDCBAP)

than that of installing ten centers. Therefore, the forum participants decided to select seven micro-regions due to the instance cost is relatively lower. The localities selected will work as collection and distribution centers of backyard agricultural products.

16.4.4 Micro-Regions Design

A short-commercialization circuit defines all activities of the agricultural production system (from agricultural food production to consumption) that are located within the same geographic region (Rodríguez & Riveros, 2016). Local consumption helps to traditions, culture, and even the community professions have not been lost (Soto, 2006).

Local product production contributing to nature, respect, and care for the environment through energy reduction, fuel, materials, and logistical or transportation efforts (JRC, 2013).

Table 16.4 shows the maximum distance that is traveled within the short-commercialization circuit, as well as the minimum and average distance. This distance represents the kilometers traveled from a locality origin to the locality that will serve as a collection center.

Figures 16.4 and 16.5 show the Chontla micro-regions designed and short-commercialization circuit design, respectively. These figures designed with Google Maps display tool.

Table 16.4 Distances from origin locality to destiny locality within micro-regions designed for Chontla municipality

CDCBAP[1] location	Maximum Distance (km)	Minimum Distance (km)	Average
Chontla	13.8	1.3	5.0
Magozal	13.5	0.7	0.7
La Laja	7.2	1.0	3.5
Las Cruces	2.7	1.4	5.2
El Ebanal	13.0	1.3	5.6
Cruz Manantial	10.4	1.0	4.9
La Floreña	6.0	0.9	2.4

Source Own elaboration based on the results of scenario 2
[1]Collection and Distribution Center of Backyard Agricultural Products (CDCBAP)

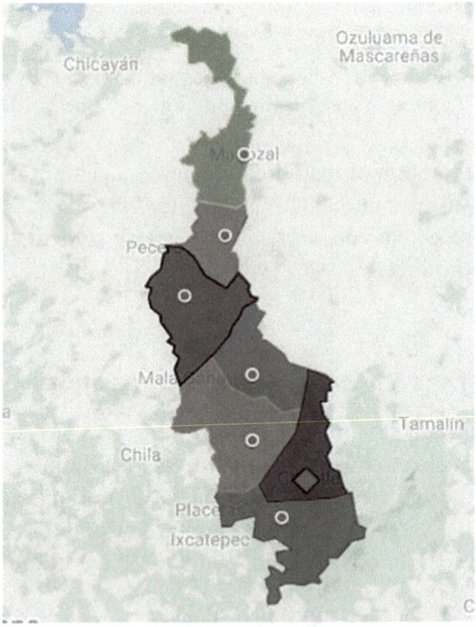

Fig. 16.4 Micro-regions design for Chontla municipality. (•) Facility location. (♦) Municipal Head. *Source* Own elaboration based on the results of scenario 2

According to work done in this chapter, seven short-commercialization circuits named Chontla, Magozal, La Laja, Las Cruces, El Ebanal, Cruz Manantial, and La Floreña are defined into seven micro-regions designed for Chontla municipality; each of them does not have more than twenty localities, with a maximum affluence radius of 14 km.

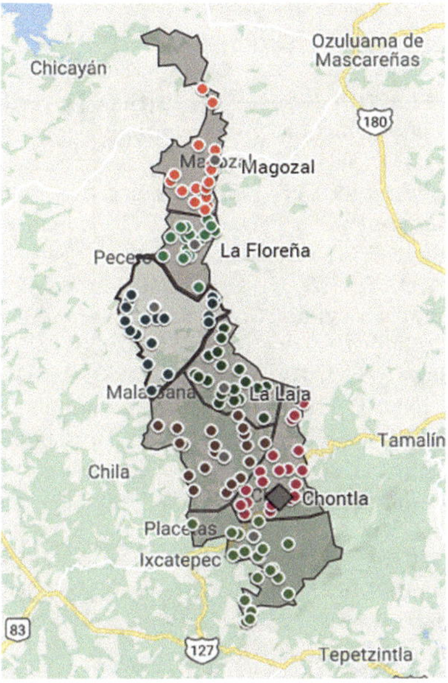

Fig. 16.5 Short-commercialization circuit design for Chontla municipality. *Source* Own elaboration based on the results of scenario 2

Minimizing the distances for food distribution and consumption, the use of natural resources, emissions of waste, and contaminants are minimized too throughout the product life cycle in accordance with the National Institute of Ecology and Climate Change (INECC, 2019).

16.5 Concluding Remarks

The territories analysis is used from measurable variables within a production system. In Chontla, Veracruz-Mexico, there are no production records of backyard agricultural products because it is considered an open production system, which characterized by seasonal and perishable products. From this perspective, the territory segmentation into rural production micro-regions will help each micro-region be visualized as a research topic in which agricultural family census can be carried out to identify the products that exist in its backyard as learning mechanism that allows leading to territorial development through the backyard agricultural producer inclusion into productive short chains.

Due to the perishable nature of backyard agricultural products, strategic decisions such as the location of collection and distribution centers are essential during the design of short marketing circuits. Hence, the importance of articulating backyard agricultural producers in inclusive agro-food chains as a strategy to create resilience in economic activities.

References

Abareshi, M., & Zaferanieh, M. (2019). A bi-level capacitated p-median facility location problem with the most likely allocation solution. *Transportation Research Part B, 123,* 1–20. https://doi.org/10.1016/j.trb.2019.03.013.

Aghazadeh, S.-M. (2004). Viewpoint improving logistics operations across the food industry supply chain. *International Journal of Contemporary Hospitality Management, 16,* 263–268. https://doi.org/10.1108/09596110410537423.

Ahumada, O., & Villalobos, J. R. (2011). Operational model for planning the harvest and distribution of perishable agricultural products. *International Journal of Production Economics, 133,* 677–687. https://doi.org/10.1016/j.ijpe.2011.05.015.

Allaoui, H., Guo, Y., Choudhary, A., & Bloemhof, J. (2018). Sustainable agro-food supply chain design using two-stage hybrid multi-objective decision-making approach. *Computers & Operations Research, 89,* 369–384. https://doi.org/10.1016/j.cor.2016.10.012.

Bosona, T. G., & Gebresenbet, G. (2011). Cluster building and logistics network integration of local food supply chain. *Biosystems Engineering, 108,* 293–302. https://doi.org/10.1016/j.biosystemseng.2011.01.001.

Boucher, F., & Poméon, T. (2010). Reflexiones en torno al enfoque SIAL: Evolución y avances desde la Agroindustria Rural (AIR) hasta los Sistemas Agroalimentarios Localizados (SIAL). In *International EAAE-SYAL seminar "spatial dynamics in agri-food systems: Implications for sustainability and consumer welfare."* Parma, Italy (pp. 1–13).

Brain, R. (2012). The local food movement: Definitions, benefits & resources. Utah State University Extension Sustainability, 1–4.

CEPAL-FAO-IICA. (2014). Fomento de circuitos cortos como alternativa para la promoción de la agricultura familiar. Retrieved June 3, 2019, from https://bit.ly/2lx4DWV. [In Spanish].

CONEVAL. (2015). Consejo Nacional de Evaluación de la Política de Desarrollo Social. Medición de la pobreza, Estados Unidos Mexicanos. Retrieved June 15, 2018, from https://www.coneval.org.mx/Paginas/principal.aspx. [In Spanish].

De los Ríos-Carmenado, I., Cadena-Iñiguez, J., & Díaz-Puente, J. M. (2011). Creación de grupos de acción local para el desarrollo rural en México: Enfoque metodológico y lecciones de experiencia. *Agrociencia, 45,* 815–829. [In Spanish].

Erickson, D. L., Taylor Lovell, S., & Méndez, V. E. (2013). Landscape and urban planning identifying, quantifying and classifying agricultural opportunities for land use planning. *Landscape and Urban Planning, 118,* 29–39. https://doi.org/10.1016/j.landurbplan.2013.05.004.

Evans, T. L. (2010). Assessing the pocket market model for growing the local food movement: A case study of Metropolitan Vancouver. *Journal of Agriculture, Food Systems, and Community Development, 1,* 129–144. https://doi.org/10.5304/jafscd.2010.012.011.

FAO. (2017). Strategic work of FAO for inclusive and efficient food system. The importance of enabling inclusive and efficient agricultural and food system. Retrieved June 5, 2018, from http://www.fao.org/3/a-i6627e.pdf.

FONDEN. (2019). Fondo de Desastres Naturales. Retrieved May 1, 2019, from http://www.proteccioncivil.gob.mx/en/ProteccionCivil/Fonden. [In Spanish].

Filippi, M., & Torre, A. (2003). Local organisations and institutions. How can geographical proximity be activated by collective projects? *International Journal of Technology Management, 26,* 386–400.

Fleszar, K., & Hindi, K. S. (2008). An effective VNS for the capacitated p-median problem. *European Journal of Operational Research, 191,* 612–622. https://doi.org/10.1016/j.ejor.2006.12.055.

Fournier, S., & Muchnik, J. (2012). El enfoque SIAL (Sistemas agroalimentarios localizados) y la activación de recursos territoriales. *Agroalimentaria, 18,* 133–144. [In Spanish].

Giacomini, C., & Mancini, C. (2015). Organisation as a key factor in localised agri-food systems. *Bio-based* and *Applied Economics, 4,* 17–32. https://doi.org/10.13128/BAE-15088.

Hakimi, S. L. (1964). Optimum locations of switching centers and the absolute centers and medians of a graph. *The Institute for Operations Research and the Management Sciences, 12,* 450–459.

INECC. (2019). Instituto Nacional de Ecología y Cambio Climático. Retrieved May 20, 2019, from https://www.gob.mx/inecc. [In Spanish].

INEGI. (2017). Instituto Nacional de Estadística y Geografía. Encuesta nacional agropecuaria. Retrieved November 4, 2018, from https://www.inegi.org.mx/rnm/index.php/catalog/498. [In Spanish].

INEGI. (2015). Instituto Nacional de Estadística y Geografía. Centro distribuidor de metadatos (Clearinghouse). Retrieved September 20, 2018, from https://www.inegi.org.mx/app/geo2/ntm/. [In Spanish].

INEGI. (2010). Instituto Nacional de Estadística y Geografía. Censo de Población y Vivienda. Retrieved April 9, 2018, from https://www.inegi.org.mx/programas/ccpv/2010/. [In Spanish].

JRC. (2013). Short food supply chains and local food systems in the EU. A state of play of their socio-economic characteristics. European commission. Joint research centre. https://doi.org/10.2791/88784.

Keizer, M., Akkerman, R., Grunow, M., & Bloemhof, J. (2017). Logistics network design for perishable products with heterogeneous quality decay. *European Journal of Operational Research*, 1–39. https://doi.org/10.1016/j.ejor.2017.03.049.

King, C. A. (2008). Community resilience and contemporary agri-ecological systems : Reconnecting people and food, and people with people, *124*, 111–124. https://doi.org/10.1002/sres.

Klose, A., & Drexl, A. (2005). Facility location models for distribution system design. *European Journal of Operational Research, 162,* 4–29. https://doi.org/10.1016/j.ejor.2003.10.031.

Mladenovic, N., Brimberg, J., Hansen, P., & Moreno-Pérez, J. A. (2007). The p-median problem: A survey of metaheuristic approaches. *European Journal of Operational Research, 179,* 927–939. https://doi.org/10.1016/j.ejor.2005.05.034.

Niyomubyeyi, O., Pilesjö, P., & Mansourian, A. (2019). Evacuation planning optimization based on a multi-objective artificial bee colony algorithm. *ISPRS International Journal of Geo-Information, 8,* 110.

Noorian, S. S., Psyllidis, A., & Bozzon, A. (2018). A time-varying p-median model for location-allocation analysis. *AGILE*, 1–5.

Orjuela-Castro, J. A., Sanabria-Coronado, L. A., & Peralta-Lozano, A. M. (2017). Coupling facility location models in the supply chain of perishable fruits. *Research in Transportation Business & Management*, 0–1. https://doi.org/10.1016/j.rtbm.2017.08.002.

Paasi, A. (2004). Human geography place and region: Looking through the prism of scale. *Progress in Human Geography, 28,* 536–546. https://doi.org/10.1191/0309132504ph502pr.

Revelle, C. S., & Eiselt, H. A. (2005). Location analysis: A synthesis and survey. *European Journal of Operational Research, 165,* 1–19. https://doi.org/10.1016/j.ejor.2003.11.032.

Rodríguez, D., & Riveros, H. (2016). Esquemas de comercialización que facilitan la articulación de productores agrícolas con los mercados. IICA.

Romero, M., González, R., Estrada, M., Martínez-Flores, J. L., & Bernábe-Loranca, M. B. (2018). Solution search for the capacitated P-median problem using tabu search. *International Journal of Combinatorial Optimization Problems and Informatics*, *10*(2), 17–25. Retrieved from https://ijcopi.org/index.php/ojs/article/view/118.

Ruben, R., & Pender, J. (2004). Rural diversity and heterogeneity in less-favoured areas: The quest for policy targeting. *Food Policty, 29,* 303–320. https://doi.org/10.1016/j.foodpol.2004.07.004.

Sanabria, L. A., Peralta, A. M., & Orjuela, J. A. (2016). Modelos de Localización para Cadenas Agroalimentarias Perecederas: Una Revisión al Estado del Arte. *Rev. Ing., 22,* 23–45.

Sánchez-Galván, F., Bautista-Santos, H., Martínez-Flores, J. L., Sánchez-Partida, D., Ireta-Paredes, A. D. R., & Fernández-Lambert, G. (2019). Backyard agricultural production as a strategy for strengthening local economy: The case of chontla and tempoal, Mexico. *Sustainability, 11*(19), 5400.

SEDESOL. (2017). Secretaría de Desarrollo Social. Retrieved June 9, 2018, from bit.ly/2ZeS5W4. [In Spanish].

SEGOB-SNIM. (2010). Instituto Nacional para el Federalismo y el Desarrollo Municipal. Retrieved May 25, 2019, from www.snim.rami.gob.mx. [In Spanish].

Soto, F., Beduschi, L. C., & Falconi, C. (2007). *Desarrollo territorial rural: Análisis de experiencias en Brasil, Chile y México*. Santiago, Chile: FAO.

Soto, D. (2006). La identidad cultural y el desarrollo territorial rural, una aproximación desde Colombia.

Soysal, M., Bloemhof-Ruwaard, J. M., Meuwissen, M. P. M., & Van der Vorst, J. G. A. J. (2012). A review on quantitative models for sustainable food logistics management. *International Journal on Food System Dynamics, 3,* 136–155.

Tong, D., Ren, F., & Mack, J. (2012). Locating farmers' markets with an incorporation of spatio-temporal variation. *Socio-Economy Planning Sciences, 46,* 149–156. https://doi.org/10.1016/j.seps.2011.07.002.

Wajszczuk, K. (2016). The role and importance of logistics in agri-food supply chains: An overview of empirical findings. *Logistics Transport, 2,* 47–56.

WEF. (2017). World Economic Forum. Una Nueva Visión para la Agricultura. Una iniciativa del Foro económico Mundial. Retrieved July 25, 2019, from https://bit.ly/2lCxrgo.

Zhang, M., & Yang, J. (2007). Optimization modeling and algorithm of facility location problem in perishable commodities emergency system. In *Third international conference on natural computation* (pp. 5–9).

Zhao, X., & Dou, J. (2011). A hybrid particle swarm optimization approach for design of agri-food supply chain network. *IEEE* (pp. 162–167).

Zhen, L., Wang, W., & Zhuge, D. (2016). Optimizing locations and scales of distribution centers under uncertainty. *IEEE Transactions on Systems, Man, and Cybernetics: Systems*, 1–12.